Introduction to Microelectronic Systems

The PIC 16F84 Microcontroller

Introduction to Microelectronic Systems
The PIC 16F84 Microcontroller

Martin Bates
Hastings College of Arts & Technology

A member of the Hodder Headline Group
LONDON

First published in Great Britain in 2000 by
Arnold, a member of the Hodder Headline Group,
338 Euston Road, London NWI 3BH

http://www.arnoldpublishers.com

British Library Cataloguing in Publication Data
A catalogue record for this book is available from the British Library

ISBN 0 340 75920 8

1 2 3 4 5 6 7 8 9 10

Publisher: Siân Jones
Production Editor: Julie Delf
Production Controller: Priya Gohil
Cover Design: Terry Griffiths

Typeset in 10/12 Times by Academic and Technical, Bristol
Printed and bound in Great Britain by J. W. Arrowsmith Ltd, Bristol

Contents

Preface

The Microchip^{TM} PIC 16F84 microcontroller is an unremarkable looking 18-pin chip; so why write a whole book about it? The answer is that it contains within its little plastic case much of the technology that we need to know about to understand microprocessor and computer systems. It also represents a significant new development in microelectronics and, more importantly, it offers an easier introduction to the world of digital processing and control than conventional microprocessors. The microcontroller is a self-contained, programmable device, and the student, hobbyist or engineer can put it to use without knowing in detail how it works. On the other hand, we can learn a lot from looking inside.

Studying the PIC chip will give the user valuable insight into the technology behind the explosion in microprocessor controlled applications that has occurred in recent years, which has been based on cheap, mass produced digital circuits. Examples of these are mobile phones, video-cameras, digital television, satellite broadcasting and microwave cookers – in fact there are not many current electronic products which do not contain some kind of microprocessor. Industrial control systems have seen similar developments, where complex computer control systems have steadily increased productivity, quality and reliability. The key, of course, is the increase in power of microprocessors and related technology, while the costs continue to fall.

The microcontroller is essentially a computer on one chip, which can be connected directly to the other components to carry out a complex programmed sequence of actions, with the minimum of additional components. As an example, in this book a motor control circuit will be described which allows the position, speed or acceleration of a small dc motor to be programmed and controlled by the PIC chip. The only additional major components required are power transistors to provide the current drive to the motor. In the past, the control and interface circuits for such an application would have required many more components, and been much more complicated and expensive to design and produce. The small microcontroller also makes it easier for a device such as a motor to be individually controlled as part of a larger system.

I became aware of the PIC 16C84 chip when a student first used it for an industrial project. It was immediately obvious that this would be an ideal device for learning microprocessor software techniques, especially for students with minimal prior knowledge. It is relatively cheap and, even better, it is reusable because the program memory is easily reprogrammable. In addition, the manufacturers, Arizona Microchip (now Microchip Technology), had the foresight to make the development system software for programming the chip freely available. It can run on any PC with any version of DOS or Windows, and is regularly updated. Professional users can buy more powerful development tools when required.

Both the DOS and Windows versions of the PIC development system have been used to prepare the sample applications in this book, and the programs downloaded using the PICSTART-16B programming unit. However, there are many designs for inexpensive programmers available in magazines and on the Internet, usually with their own software. The current Windows version of the program development package, MPLAB, can be downloaded free of charge from the Internet at www.microchip.com, along with data sheets and all the latest product developments and information.

Most of the data sheet for the PIC 16F84, the successor to the 16C84, has been included in Appendix A because it is an excellent document which contains the definitive information on the chip, presented in a clear and concise manner. My objective is that any beginner, student or engineer, having studied this book, will be able to start using this chip immediately for their own projects and designs. When I started using it in my teaching, I put together a teaching pack and was expecting a range of suitable reference books to appear quickly. Indeed, the chip soon started to feature in numerous electronics magazine projects and was clearly popular, but all the books that I saw seemed to assume quite a quite a lot of prior knowledge of microprocessors. I wanted to use the PIC with students who were new to the subject, and eventually I realized that if I wanted a suitable book, I would have to do it myself! I hope that the reader finds the result useful.

Martin Bates
2000

Part A
Microelectronic Systems

Chapter 1
Computer Systems

Let's admit one thing straight away – microprocessors *are* very complicated! However, they are now found in such a huge variety of domestic and commercial products that all students of engineering must know something about how they work.

In this book, we are going to look in detail at a microcontroller, the PIC 16F84, that has all the essential features of a computer, but built into a single chip, or IC (integrated circuit). Conventional microprocessor systems, such as the personal computer (PC), are built with separate processor, memory, input and output chips. The extra hardware and software required to make these chips work together makes the system much more complicated than our single chip microcontroller. As well as being easier to understand, programmable microcontrollers are very important in their own right. They are now used instead of 'hard-wired' electronics in more and more applications, because they reduce the number of components required, making the hardware easier to design and more adaptable. In general, software is increasingly replacing hardware in electronic designs which carry out similar circuit operations.

With the PIC microcontroller, we can quite quickly work out some simple, but useful, applications to run on the chip. These will illustrate the universal principles of microprocessor systems that apply to more complex computer and control systems. We do not have to worry too much, at first, exactly how the hardware works.

Nevertheless, we will begin our analysis of microprocessor systems by looking at how the more complex PC operates when running a wordprocessor, because most readers will have used such a system, and will know how the system works from the user's point of view. This will allow some basic microprocessor system concepts to be introduced by looking at the way that the wordprocessor application software uses the computer hardware to store and process documents. For example, it will allow the operation of different types of memory devices, such as ROM and RAM, to be introduced in a familiar context. In addition, it is useful to get some idea of how a PC works because it is used as the hardware platform for the PIC microcontroller development system itself. The programs for the

(a)

(b)

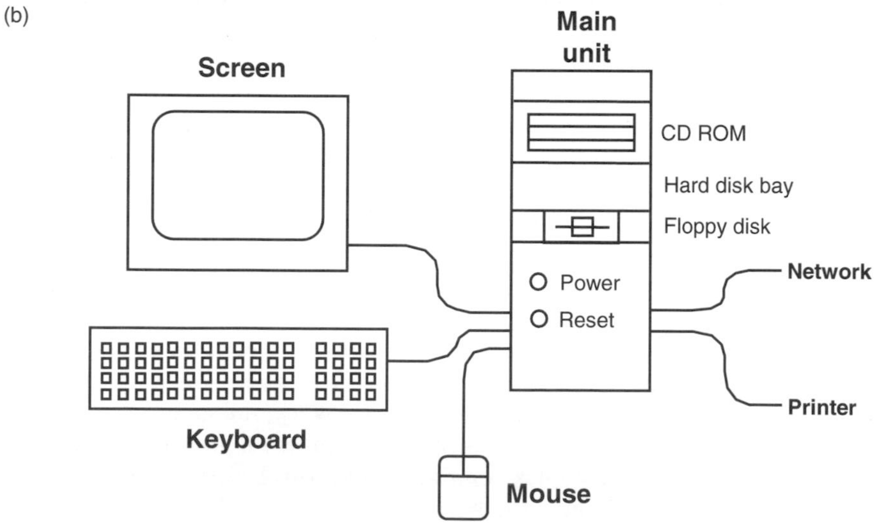

Figure 1.1 (a) PC system with PIC development kit; (b) PC system diagram.

PIC will be written using a text editor, and the assembly code program created and downloaded, using the PC. Hardware for programming and testing a PIC chip can be seen in Fig. 1.1(a).

At the end of this introductory chapter, we will look briefly at a microcontroller system, set up to operate as a simple equivalent of the wordprocessor. The microcontroller uses a keypad instead of a keyboard, and a seven-segment display instead of a screen. Its memory is much smaller than the PC, and it may not be as fast. However, the microcontroller costs only a few pounds or dollars, and yet can carry out all the same basic tasks as the more expensive Pentium processor in a PC, and it can be used in a vast array of different applications. The Pentium, on the other hand, is designed specifically to run standard PC systems, and is not often used for anything else.

It will be assumed here that the reader may have only a minimal existing knowledge of electronics. The basic building blocks of digital electronic circuits will be covered, but without going into too much detail at component level. The PIC data sheet, extracts of which are found in Appendix A, contains details of the 16F84 chip in a reasonably readable form, and is a great asset in our quest to unravel the mysteries of the microcontroller and microprocessor systems in general.

1.1 The PC System

Many of us now use a personal computer routinely in our daily work or studies, and most readers will have used a wordprocessor. It is probably the most familiar microprocessor based system, but is quite complex, and getting more so all the time! Nevertheless, it lends itself to the explanation of the common features of such systems – information input, storage, processing and output. In other domestic products, such as the video-recorder, the action of the microprocessor is not so obvious.

The standard PC hardware, originally introduced by IBM™ is based on the Intel™ series of microprocessors, usually using Microsoft™ operating systems software. This is sometimes known as the 'Wintel' (Windows & Intel) system. The PC was originally developed by IBM, but soon PC hardware started to be produced by numerous competing manufacturers. On the other hand, the software market continues to be dominated by Microsoft.

The standard PC hardware comprises a main unit, separate keyboard and mouse, VDU (visual display unit) and possibly a printer and a connection to a network (Fig. 1.1(b)). The circuit board (motherboard) in the main unit carries a group of chips, including the microprocessor, which work together to provide digital processing of information and control of input and output devices. A power supply for the motherboard and the peripheral devices is included in the main unit.

The processor must have operating system (Windows) software and an application program (in this case, a wordprocessor) available to allow useful work to be done by the system. The software is usually stored on, and retrieved from, a hard disk inside the main unit, which can hold large amounts of data, and which is retained when the power is off. The keyboard is used for data input, and the VDU screen displays the resulting document. The mouse provides an additional input device, allowing operations to be selected from menus or by clicking on icons and buttons. This is called the graphical user interface (GUI). The data files created (documents in the wordprocessor), are stored on the hard disk, or floppy disk if the files need to be taken away. There may be a network card installed in the PC to exchange information with other users, download data or applications, or share resources such as printers over a local area network (LAN). In addition, a modem can give direct access to a wide area network (WAN), most commonly, the Internet. A CD ROM drive allows large volumes of reference information on optical disk to be accessed, and is currently the most common medium from which to load application software and operating system upgrades.

(a)

(b)

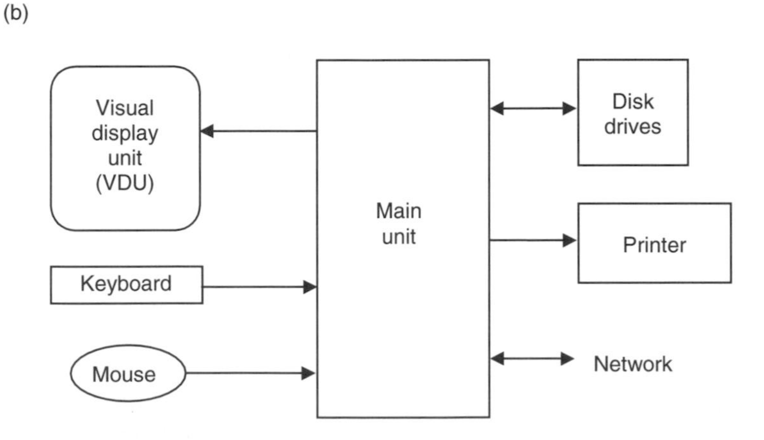

Figure 1.2 (a) Internal view of the PC main unit; (b) Block diagram of the PC system.

A photograph of the PC (Fig. 1.1(a)) cannot tell us very much about how it works, and it looks even more confusing (Fig. 1.2(a)) with the cover off! A block diagram (Fig. 1.2(b)) allows the basic subsystems to be more clearly identified. Block diagrams are useful for showing the main parts of a complex system, and the information flow between them, in a clearer, simplified form.

The arrows connecting the blocks show the direction of the information flow, in or out of the main unit. In the case of the disk block it is bidirectional (flowing in both directions), representing the process of saving data to, and retrieving data from, the hard disk or floppy disk. On the other hand, the data flow from the keyboard is in one direction only.

In the photograph of the main unit, the power supply is top left, with the hard disk drive below, and the motherboard is vertical at the back of the case. The disk and video interface expansion cards are visible at the bottom, with a modem in the middle in a dark case. The connections to the expansion boards are available at the rear of the machine to the left, with the floppy disk drive top right.

1.1.1 PC Hardware

Inside the PC main unit, the motherboard has slots for expansion boards and memory modules to be added to the system. The power supply and disk drives are installed separately into the main unit frame. The keyboard and mouse interfaces are usually on the motherboard. In older designs, the expansion boards carried interface circuits for the disk drives and external peripherals such as the display and printer, but these functions are now increasingly incorporated into the motherboard itself. Note that the functional block diagram does not show any difference between internally and externally installed peripherals, because it is not relevant to the overall system operation.

The PC is a modular system, which allows the hardware to be put together to meet the individual user's requirements, and allows subsystems, such as disk drives and keyboard, to be easily added or replaced if faulty. The modular design also allows upgrading (for instance, installing extra memory chips) and it is therefore well suited to industrial applications. In this case, the PC can be 'ruggedized' (put into a more robust casing) for factory floor usage. The modular architecture is one of the reasons for the success of the PC as a universal hardware platform.

1.1.2 PC Motherboard

The main features of a typical motherboard are shown in Fig. 1.3. The heart of the system is the microprocessor, a single chip, which is also called the central processing unit (CPU). This name refers back to the days when the CPU was built from discrete components and could be the size of a washing machine! In Fig. 1.3(a), the CPU is hidden under the cooling fan at the bottom right. The CPU controls all the other system components, but must have access to a suitable program in memory before it can do anything useful. The blocks of program required are provided by the operating system software (Windows) and the application software, which are both normally downloaded to memory from the hard disk when the computer is started up. In the example used here, it will be assumed to be a Windows version of Word, the Microsoft wordprocessor.

The Intel CPU has undergone continuous development since the introduction of the PC in 1981, with the Pentium processor being the current standard. The Intel processors are classified as CISC (complex instruction set computer) chips, which means they have many instructions which can be used in a number of different ways. This makes them powerful, but relatively slow compared with simpler designs which have fewer instructions. These are called RISC chips (reduced instruction set computer), of which the PIC microcontroller is an example.

The CPU cannot work on its own; it needs some memory and input/output devices for getting data in, storing it and sending it out again. The main memory block is made from RAM (read-and-write memory) chips, which are mounted in SIMMs (single in-line memory modules). These are seen at the top of the photograph in Fig. 1.3(a). Additional peripheral interfacing boards are installed in the expansion card slots to connect the

(a)

(b)

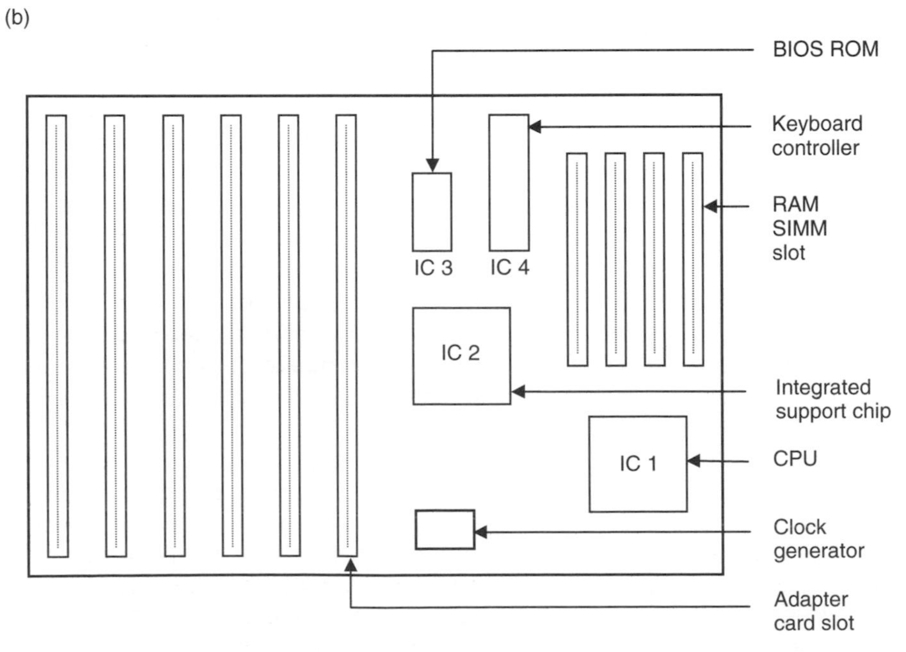

Figure 1.3 (a) PC motherboard within main unit; (b) General PC motherboard layout.

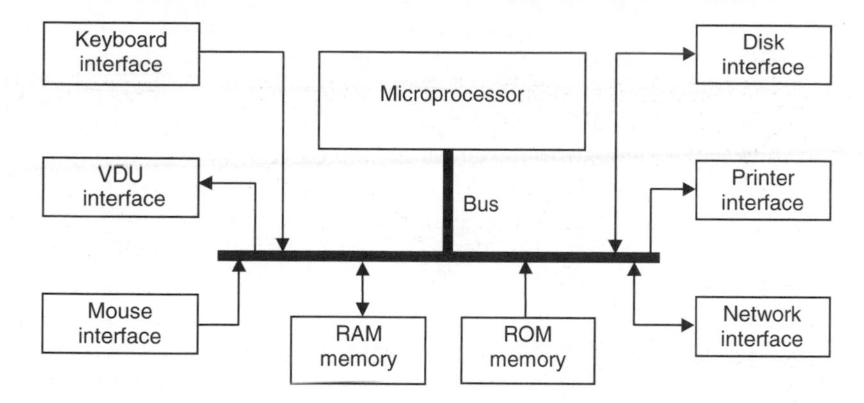

Figure 1.4 Block diagram of PC motherboard.

main board to the disk drives, VDU (visual display unit), printer and network. Spare slots allow additional peripheral interfaces and more memory to be added if required. Spare expansion slots can be seen at the lower left in Fig. 1.3(a). Each peripheral interface is a sub-circuit which is built around an input/output chip which handles the data transfer.

The integrated support device (ISD) is a chip specially designed to provide the necessary system control and memory management functions in one chip for a particular motherboard. The motherboard itself can be represented as a block diagram (Fig. 1.4) to show how the components are interconnected.

The block diagram shows that the CPU is connected to the peripheral interfaces by a set of bus lines. These are groups of connections on the motherboard which work together to transfer the data from an input such as the keyboard, to the processor, from the processor to memory, and then to an output peripheral, for example, the screen. Busses connect all the main chips in the system together, but, because they operate as shared connections, they can only pass data to or from one peripheral interface or memory location at a time. This arrangement is used because separate connections to all the main chips would require an impossible number of tracks on the motherboard. The disadvantage of bus type connections is that it slows down the program execution speed, because all data transfers use the same set of lines, and only one data word can be present on the bus at any one time. To help compensate for this, the bus connections are typically 16 or 32 bits wide, that is, there are 16 or 32 connections working together, each carrying one bit of a data word simultaneously. This parallel data connection is faster than a serial connection, such as the keyboard input or network connection which can only carry one bit at a time. In the microcontroller, these bus connections are internal, and therefore their operation is conveniently hidden from the user.

1.1.3 PC Memory

There are two types of memory in the PC system. The main memory block is RAM, where input data is stored before and after processing in the CPU. The operating system and application program are also copied to RAM from disk for execution, because access to data in RAM is faster. Unfortunately, RAM storage is 'volatile', which means that the current data, application and operating system software disappear when the PC is switched off, and these have to be reloaded each time the computer is switched back on. This means that some ROM (read-only memory), which is non-volatile, is needed to get the system started at switch on. The BIOS (basic input/output system) ROM chip, seen at the left of Fig. 1.3(a), contains enough code to check the system hardware and start loading the main operating system (OS) software from disk. It also contains some basic hardware control routines so that the keyboard and screen can be used before the main OS has been loaded. MSDOS™, at one time the most common operating system used in the PC, stands for MicroSoft Disk Operating System, because the main storage medium is hard and floppy disk. Windows does essentially the same job, but with a GUI.

The hard disk is a non-volatile, read/write storage device, consisting of a set of metal disks with magnetic recording surfaces, read/write heads, motors and control hardware. It provides a large volume of data storage for the operating system, application and user files. A number of applications can be stored on disk and then selected as required for loading into memory. Because the disk is a read and write device, user files can be stored, applications added and software updates easily installed.

1.2 Wordprocessor Operation

In order to illustrate the operation of the PC microprocessor system, we will look at how the wordprocessor application uses the hardware and software resources.

1.2.1 Starting the Computer

When the PC is switched on, the BIOS ROM program starts automatically. It checks that the system hardware is working properly, and displays messages to report the results. If there is a problem, the BIOS program attempts to diagnose the fault. If all is well, it loads the main operating system software (MSDOS/Windows) from disk into RAM. This all takes some time, which is an indication of the amount of processing and data transfer required, and the relatively slow access to the hard drive.

1.2.2 Starting the Application

Windows displays an initial screen with icons and menus which allows the application, the wordprocessor, to be selected using the mouse and on-screen pointer. Word is started by simply clicking on its icon, which is linked to its executable application file (WINWORD.EXE) stored on disk. The application is transferred from disk to RAM, or as much of it as will fit in the available memory. If necessary, application program blocks can be swapped into memory when needed. The wordprocessor screen is displayed and a new document file can be created or an existing one loaded by the user from disk for updating.

1.2.3 Data Input

The main input is from the keyboard, which consists of a grid of switches which are scanned by a chip within the keyboard unit. This chip detects when a key has been pressed, and sends a corresponding code to the CPU via a serial data line in the keyboard cable. The serial data is a sequence of high and low voltages on a single wire, which represent a binary code, and each key generates a different code. The keyboard interface converts this serial code to a parallel form for transfer to the CPU via the system data bus. It also signals separately to the CPU that a keycode is ready to be read into the CPU, by generating an 'interrupt' signal. A serial-to-parallel data conversion process is required in all the interfaces that use serial data transfer, namely, the keyboard, VDU, network and modem. Binary coding, interrupts and other such technicalities will be explained in more detail later.

In Windows, and other GUIs, the mouse can be used to select commands for managing the application and its data. It controls a pointer on the screen; when the mouse is moved, the ball turns two rollers, which have perforated wheels attached. The holes are detected using an opto-detector, which send pulses representing movement in two directions. These pulse sequences are passed to the CPU via the mouse interface and used to modify the position of the pointer on the screen. The buttons, used to select an action, are detected separately.

1.2.4 Data Storage

Each character of the text being typed into the wordprocessor is stored as an 8-bit binary code, which occupies one location in RAM. The parallel data is received by the CPU, then sent back via the same data bus lines from the CPU to the RAM. The RAM stores the data bytes at numbered locations. These are identified by the CPU using the system address bus. The data is transferred on the data bus to the address in RAM selected by the CPU and ISD, which provides the additional logic necessary to handle the data transfers.

1.2.5 Data Processing

In the past, programs running on DOS required less processing power, partly because the screen was simpler, being divided up into one space for each character. The video interface would convert the stored character code to the pattern for the character, and output it to the correct lines on the screen.

The Windows screen is more complicated, because the text is displayed in graphics (drawing) mode, at a higher resolution, so that the text size, style and layout is the same on screen as it will be when printed. Graphics, tables and special characters can all be embedded in the text. This means the CPU has far more work to do in displaying the page, and this is one reason why Windows needs more memory and a more powerful CPU than DOS wordprocessors. The processor must also manage the WIMP (windows, icons, mouse, pointer) interface that allows actions to be selected without using the keyboard. Word now has many more features than earlier wordprocessors, and there is now little difference between a wordprocessor and so called desk-top publishing (DTP) programs.

1.2.6 Data Output

The characters must be displayed on the screen as they are typed in, so the character codes stored in memory are also sent to the VDU via the system data bus and video interface. The display is made up of single coloured dots (pixels) organized in lines across the screen, which are output in sequence from a video amplifier (scanning). The shape of the character on screen must be generated from its code in memory, and sent out on the correct set of lines at the right time on the video signal. The display is therefore a two dimensional image made up from a serial data stream which sets the colour of each pixel on the

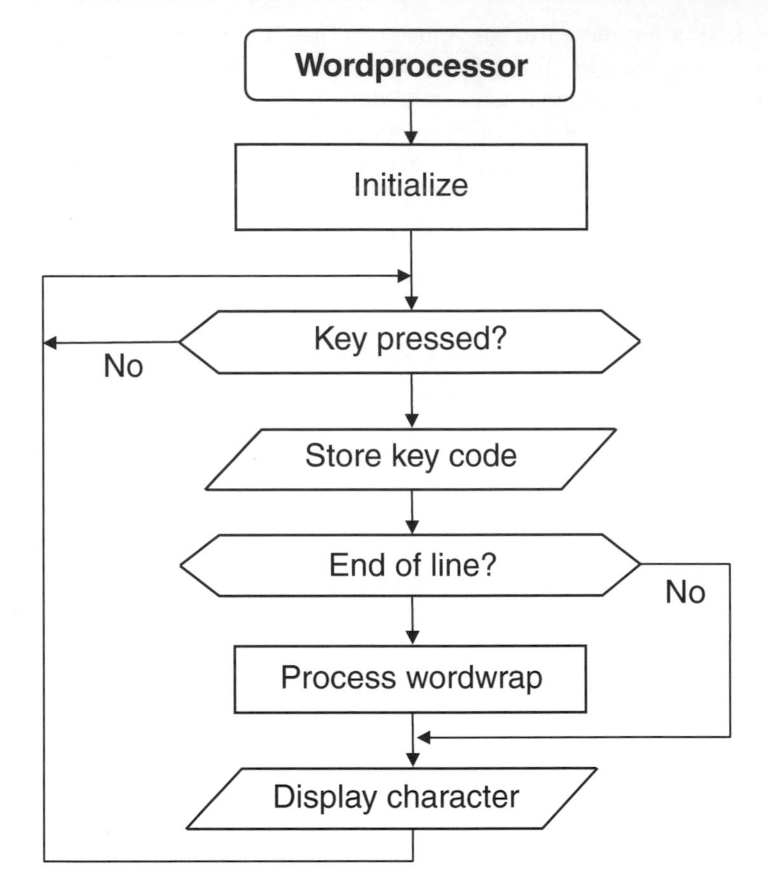

Figure 1.5 Wordprocessor flowchart.

screen in turn, line by line. If a file is transferred on a network, it must also be sent in serial form. A text file would normally be sent as ASCII codes, along with formatting information and network control codes. ASCII code represents one character (letter) as one byte (8 bits) of binary code, and is therefore quite compact.

The printer works in a similar way to the screen, except that the output is generated as lines of dots of ink on a page. If you watch an inkjet or dot matrix printer working, you can see the scanning operation take place. The characters are usually sent in 8-bit parallel form via the printer port.

The operation of the wordprocessor can be illustrated using a flowchart, which is a graphical method of describing a sequential process. Figure 1.5 shows only the basic process of text input and word wrapping at the end of each line. Flowcharts will be used later to represent microcontroller program operation.

1.3 The PC Microprocessor System

As we have seen, the PC working as a wordprocessor carries out the following functions: data input, data storage, data processing and data output. All microprocessor systems

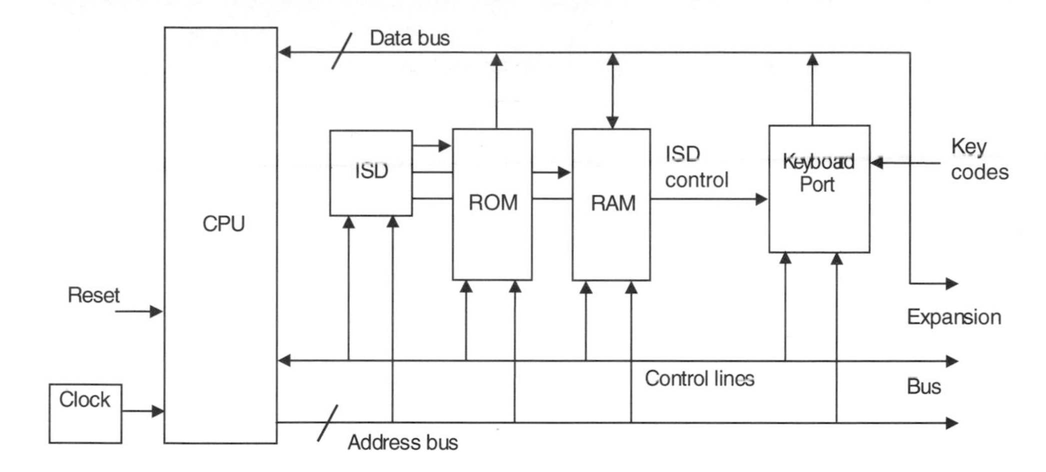

Figure 1.6 Block diagram of PC microprocessor system.

perform these same basic functions. The microprocessor system components include a CPU, RAM, ROM, I/O (input/output) ports, ISD, and XTAL (crystal) clock generator. The microprocessor system interconnections comprise an address bus, a data bus, and control lines.

1.3.1 System Components

The system components are connected as shown in Fig. 1.6. The address and data busses, control lines and support chip are required to handle the data transfer between the CPU, memory and ports. The clock input to the CPU is a precise fixed frequency signal from a crystal oscillator which drives the microprocessor by initiating operations on the clock edges as the signal rises and falls (clock period = 1/frequency). The CPU then generates all the main control signals based on this common timing reference.

The ISD is designed to generate control signals for the circuit arrangement on a particular motherboard. For simplicity, only the keyboard port is shown in the block diagram, as this was sometimes (in older designs) the only I/O device on the main board. The signal connections to the other peripheral interfaces would be the similar, that is, all must be connected to the system busses and the relevant control lines.

1.3.2 Program Execution

The ROM and RAM contain program information and data in numbered locations. The ISD, where it incorporates memory management, contains address decoding logic which allocates sections of memory to different program blocks in ROM and RAM. The I/O port registers, which are used to manage the data transfer in and out of the system, are also selected when required by the address decoding logic.

The wordprocessor program consists of a list of instructions in binary code stored in memory, with each instruction and any associated data (operands) being stored in sequential locations. The program instruction codes are fetched into the CPU and decoded (interpreted). The CPU then sets up the internal and external control lines as

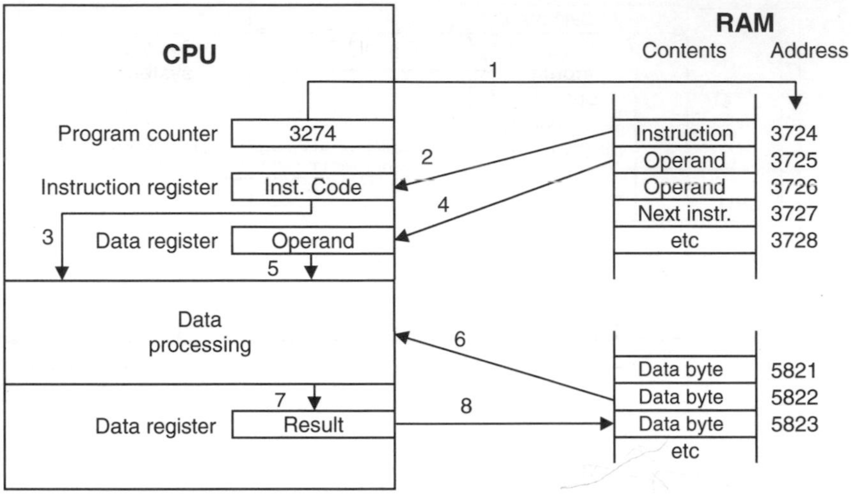

Figure 1.7 Program execution sequence.

necessary and carries out the operation specified in the program. The instructions are executed in order of their addresses, unless the instruction itself causes a jump to another point in the program, or an interrupt is received. When the system is started or reset, the execution point goes back to the beginning of the ROM boot program, and the application must be restarted. The principle of program execution is illustrated in Fig. 1.7.

1.3.3 Execution Cycle

Assuming that the application program code is in RAM, the program execution cycle proceeds as follows.

1. The CPU outputs (1) the address of the location (memory slot) containing the required instruction. The sample address is shown in decimal (3724) in Fig. 1.7, but it is output in binary on the address lines from the processor. The ISD uses the address to select the RAM chip which has been allocated to this address. The address bus also connects directly to the RAM chip to select the individual location.

2. The instruction code is returned to the CPU from the RAM chip via the data bus (2). The CPU reads the instruction from the data bus into an Instruction register. The CPU then decodes and executes the instruction (3). The operands are fetched (4) from the next locations in RAM via the data bus, in the same way as the instruction.

3. The instruction execution continues by feeding the operand(s) to the data processing logic (5). Additional data can be fetched from memory (6) (this would be the text data in our wordprocessor). The result of the operation is stored in a data register (7), and then, if necessary, in memory (8) for later use. The address of the next instruction is then output and the sequence repeats from step 2.

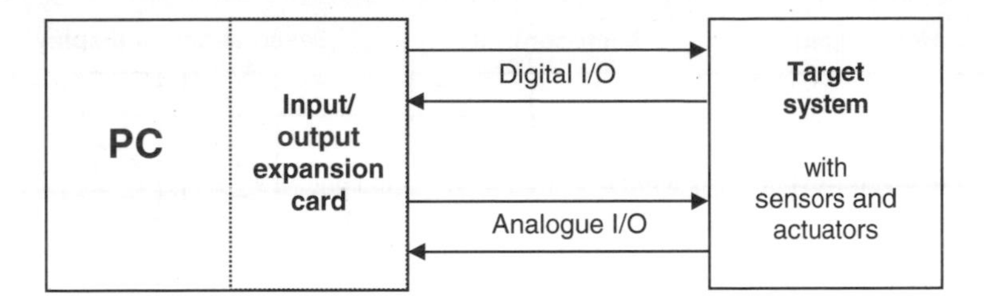

Figure 1.8 PC engineering application.

The operating system, the wordprocessor program and the text data are stored in different parts of RAM during program execution, and the wordprocessing application program calls up operating system routines as required to read in, process and store the text data. Current CISC processors such as the Pentium series have instructions which are more than 8 bits in size which are stored in multiple locations, and use complex memory management techniques, to speed up program execution.

1.4 PC Engineering Applications

The PC can be used as a standard hardware platform in a variety of engineering systems by fitting special interfacing hardware and programming the PC to control an external system through this I/O hardware (Fig. 1.8). The PC is increasingly used in manufacturing systems where it might control a machine tool, robot or assembly system, or be used to run an instrumentation or data logging application. The PC can provide network interfacing so that commands or design data can be sent to the PC and status information and other measurement data can be returned to a supervisory computer via a standard network, such as Ethernet.

The PC also has the advantage of using a standard operating system and programming languages that allow control programs to be written in high level languages such as 'C' or Visual Basic. Graphical programming tools are also available for designing control and instrumentation applications without any conventional programming at all.

1.5 The Microcontroller

We have now looked at some of the main ideas to be used later in explaining microcontroller operation; hardware, software, how they interact, and how the function of complex systems can be represented in a simplified form such as block diagrams and flowcharts. We can now compare the PC system with an equivalent microcontroller system.

The microcontroller can provide, in a simplified form, all the main elements of the conventional microprocessor system on one chip. As a result, less complex applications can be designed and built quickly and cheaply. A working system can consist of a microcontroller chip and just a few external components for feeding data in and out, and generating the clock.

1.5.1 A Microcontroller Application

A simple equivalent of the word-processing application described above could be built as shown in Fig. 1.9.

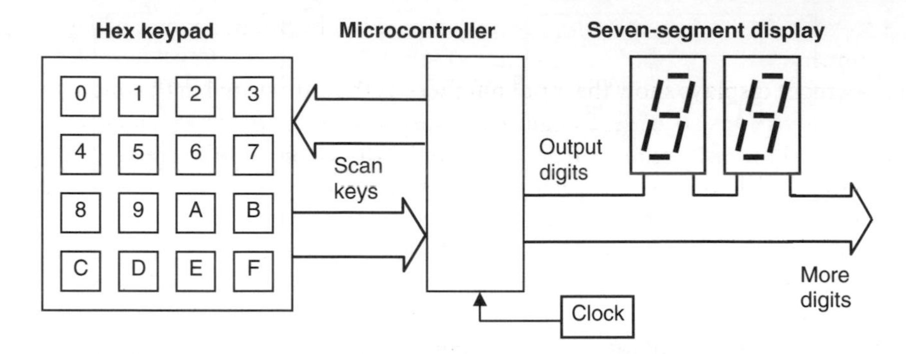

Figure 1.9 Microcontroller keypad display system.

The basic function of the system shown is to store and display numbers which are input on the keypad. The microcontroller chip can be programmed to scan the keyboard and identify any key which has been pressed. The keys are connected in a 4×4 grid of rows and columns, so that a row and column are connected together when the key is pressed. The microcontroller can identify the key by selecting a row and checking each column for a connection with the selected row. Thus, four output and four input lines are required.

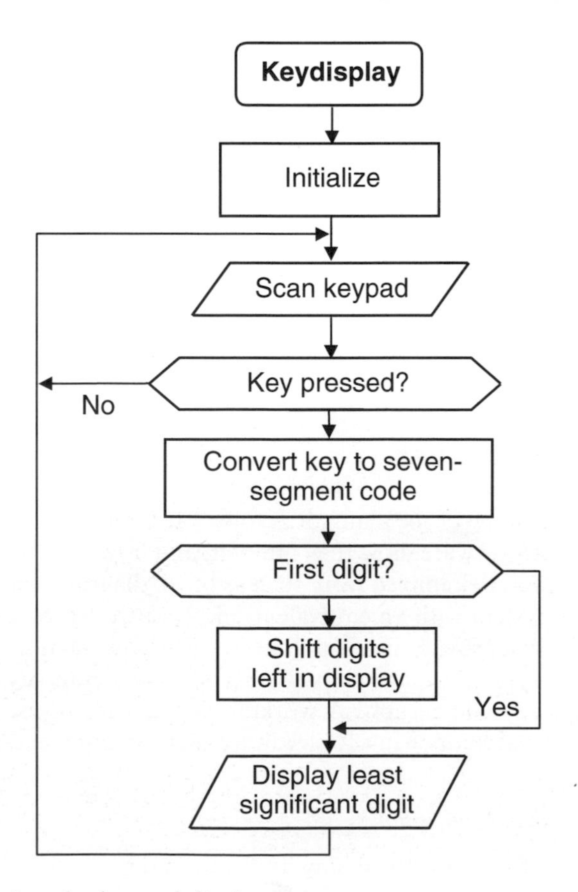

Figure 1.10 Flowchart for keypad display system.

In order to simplify the drawing, they are represented by the broad arrows, which indicate a parallel connection.

Seven-segment displays show the input numbers as they are stored in the microcontroller. Each display digit consists of seven light-emitting diodes (LEDs) which show as a segment of the digit when lit. Each number from 0 to 9 and letters from A to F can be displayed as a suitable pattern of illuminated segments (the need for letters A–F will be explained later). Digital watches use liquid crystal displays in the same format.

The basic display program could work as follows: when a key is pressed, the digit is displayed on the right (least significant) digit, and subsequent keystrokes will cause the previously entered digit to shift to the left, to allow decimal numbers up to 99 to be displayed. More display units could be added for larger numbers, if required, but the program would be the same in principle.

The starting point for writing the program for the microcontroller is to convert the general description given above into a description of the operations that can be programmed into the chip using the set of instructions that are available for that microcontroller. The instruction set is defined by the manufacturer of the device. The process

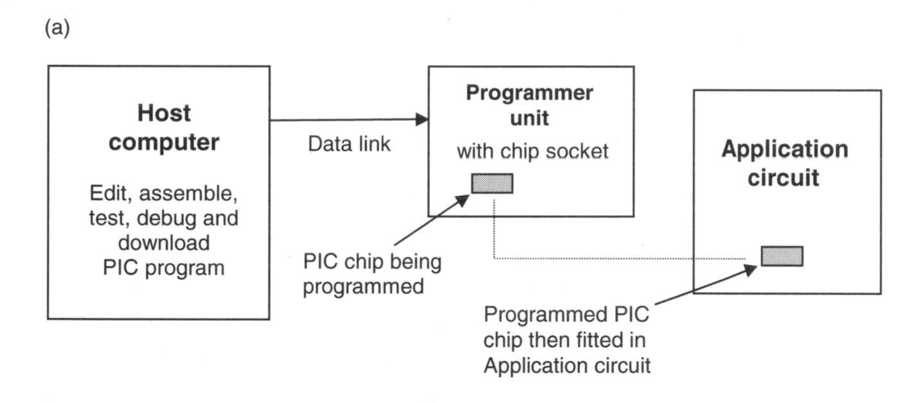

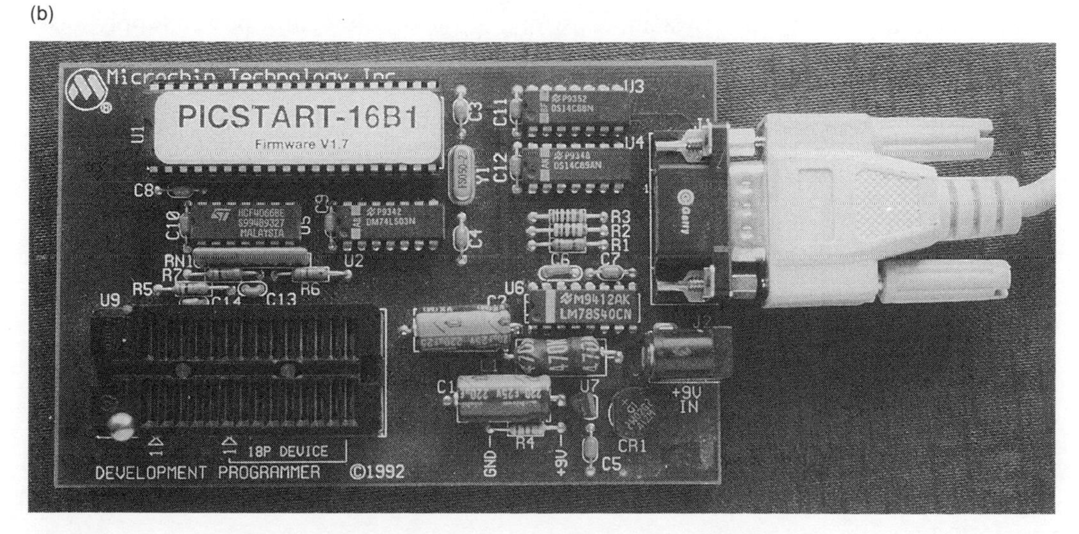

Figure 1.11 (a) PIC program downloading; (b) PIC programming unit.

whereby the required function is obtained is called the program algorithm, which can be graphically represented by a flowchart (Fig. 1.10).

With suitable development of the software and/or hardware, the system could be modified to work as a calculator, message display, electronic lock or similar application. Keyboard scanning and display driving are standard operations for microcontrollers, and we will return to these later. A range of demonstration programs for the PIC 16F84 microcontroller will be described in Part B.

1.5.2 Programming a Microcontroller

Microcontrollers normally have ROM program memory, into which the program must be loaded before the chip is placed in the application circuit. The PIC 16F84 is typical, in that it must be programmed by placing it in a programming unit (Fig. 1.11(b)), which is run from a host PC. Note the zero insertion force (ZIF) socket which will accept different size PIC chips for programming, and that the programming unit itself is controlled by a PIC chip. A power supply must be connected as well as the serial lead on the right-hand side.

The 16F84 is *not* typical in that it has flash ROM, which can be re-programmed if it does not work correctly at the first attempt. This is one of the reasons that we are going to use it! The program is written and converted to machine code in the host, and downloaded via a serial data link. The chip is then removed from the programming unit and inserted into the application hardware, in this case, the keypad and display unit.

Having introduced some basic ideas concerning microprocessors and microcontrollers, we will now review some principles of digital circuits and microprocessor systems. The process of creating microcontroller applications, such as the example above, can then be tackled.

Summary

- The PC consists of data input, storage, processing and output devices.
- The main unit is a modular system, consisting of the motherboard, power supply, disk drives and expansion cards containing interfacing circuits plugged into the motherboard.
- The motherboard carries the microprocessor (CPU) chip, RAM modules, a BIOS ROM, ISD and keyboard interface.
- The CPU communicates with the main system chips via a shared set of address and data bus lines. The address lines select the device and location for the data to be transferred on the data bus.
- The microcontroller provides, in simplified form, most of the features of a conventional microprocessor system on one chip.

Questions

1. Name at least two PC input devices, two output devices and two magnetic storage devices.
2. Why is the BIOS ROM needed when starting the PC?

3. Why are shared bus connections used in the typical microprocessor system, even though it slows down the program execution?

4. State two advantages of the modular PC hardware design.

5. Why does the PC take so long to start up?

6. Sort the following data paths into serial and parallel:
   (a) internal data bus,
   (b) keyboard input,
   (c) VDU output
   (d) printer output,
   (e) modem I/O.

7. State the function, in 10 words or less, of the following:
   (a) CPU,
   (b) ROM,
   (c) RAM,
   (d) ISD,
   (e) address bus,
   (f) data bus,
   (g) program counter,
   (h) instruction register.

8. Explain the difference between a typical microprocessor and microcontroller.

Activities

1. Study the messages which appear on the screen when a PC is switched on, and try to relate them to the system description provided.

2. Under supervision if necessary, and with reference to relevant manuals: disconnect the power supply and remove the cover of the main unit of a PC and identify the main hardware subsystems, i.e., power supply, motherboard and disk units. On the motherboard, identify the CPU, RAM modules, expansion slots, keyboard interface, VDU interface, disk interface, printer interface. Is there an internal modem or network card? Are there any other interfaces installed?

3. Run the wordprocessor and study the process of word-wrapping which occurs at the end of each line. Describe the algorithm that determines the word placement, and the significance of the space character in this process.

Chapter 2
Information Coding

This chapter introduces some methods for representing information within microprocessor systems. Binary and hexadecimal number systems will be outlined, so that data storage and program coding methods can be then explained.

Much of modern technology is based on the use of mathematics to represent information and processes in the real world. These mathematical models can be used in engineering to design new systems and products. For instance, the three-dimensional drawing of a car on a CAD (computer aided design) system screen is generated from a digital model of the car in the memory of the computer. The advantages of the computer model are fairly obvious – it can be stored in digital form, transferred electronically and changed more quickly than the equivalent information on paper. The design can also be mathematically analysed in the computer prior to construction. For example, the stresses and strains in load-bearing components in the body shell can be studied. Furthermore, when the design is finished, the design data can be sent directly to the next process, such as machine tools for cutting a three-dimensional solid model of the car, or manufacturing the actual body panels. We therefore need to know something about the data that is stored in the computer.

2.1 Number Systems

Mathematics is based on number systems that use a set of characters to represent numerical values. The characters used are simply symbols, just patterns on a page, but the number systems they are part of have been refined over thousands of years – because they are useful. In microprocessors, numerical processing is carried out using binary codes. This number system has only come into common use with the advent of electronic computers. In these machines, numbers are represented by electrical voltages, high and low. We therefore have to understand binary numbering in order to use a microcontroller. Another number system, hexadecimal, is also useful here because it provides a more compact way of representing binary code.

2.1.1 Decimal: Base 10

The name of each number system refers to the 'base' of the number system, which corresponds to the number of symbols used in representing values. In decimal, 10 symbols are used:

0 1 2 3 4 5 6 7 8 9

Why use a particular base? The reason for using 10 is simple – we humans have 10 fingers which can be used for counting, so the decimal system was developed as a way of writing this down and doing calculations on paper (or stone!) instead of on our fingers.

Assuming that we know how to count and write down numbers in decimal, let us analyse what a typical number means. Take the number 274; in words, it is two hundred and seventy four. This means: take two hundreds, seven tens and four units and add them together. The position of each digit in the number is literally significant; each column has a weighting which applies to the digit in that column. As you know, the least significant is conventionally placed at the right, and the most significant at the left. More digits are added at the left-hand end as the number size increases. In decimal, the columns have a weight 1, 10, 100, etc. Note that these correspond to a power series of 10, the number system base. Another example is detailed in Table 2.1.

A number system can be used with any base you like, but some are more useful than others. For instance, relics of the base 12 system are still in use – think of clocks, boxes of eggs and measurement of angles. Base 12 is useful because 12 is divisible by 2, 3, 4 and 6, giving lots of useful fractions – a half, a third, a quarter and one sixth. However, the decimal system is our standard system, so the analysis of other systems will still be based on decimal for comparison of number values.

2.1.2 Binary: Base 2

Binary is used in digital computer systems because it represents the way that values are actually stored and processed. The binary digits, 0 and 1, represent the two voltage levels used in digital circuits.

In binary, the base is 2, so the column weighting is a power series of 2, as shown in Table 2.2 (note that any number to the power zero has the value 1). With a base of 2, only the digits 0 and 1 are available, so the numbers tend to have lots of digits. For instance, a 32-bit computer uses 32 digit binary numbers. An example with 8 digits is given showing what the digits represent and how to convert the value back to decimal.

The decimal equivalent in all number systems can be calculated by multiplying the digit value by its weighting in decimal, and then adding the resulting column products. In binary,

Table 2.1 Structure of a decimal number

Column Weight	1000	100	10	1
Power of Base	10^3	10^2	10^1	10^0
Digits	**3**	**6**	**5**	**2**
Total value	(3 × 1000)	+(6 × 100)	+(5 × 10)	+(2 × 1)

Table 2.2 Structure of a binary number

Column Weight	2^{7*}	2^6	2^5	2^4	2^3	2^2	2^1	$2^{0\dagger}$
Decimal Weight	128	64	32	16	8	4	2	1
Example Number	1	0	1	0	0	0	1	1
Decimal Equivalent	128 + 0 + 32 + 0 + 0 + 0 + 2 + 1 = 163							

* Most significant bit (MSB).
† Least significant bit (LSB).

because the digit value is 1 or 0, the result can be obtained by simply adding the digit weight where the digit value is a '1', because any number multiplied by zero is zero. When decimal data is entered into a computer, the values are converted to binary. The program instructions which process input and output data are also stored as binary codes.

2.1.3 Hexadecimal: Base 16

Binary numbers have lots of digits, so they are not very convenient for data entry and display. However, conversion to decimal is not particularly easy, so hexadecimal is used as a way to represent binary numbers in a compact way, while allowing easy conversion back to binary.

Hexadecimal (base 16), or 'hex' for short, uses the same digits as the decimal system from 0 to 9, then uses letters A–F, as a single character representation for numbers 10–15. Thus, characters which are normally used to make text words, are used here as numbers. One advantage is that the symbols are already available on the computer keyboard. A binary number can be easily converted to hex by writing it down in groups of 4 bits, and then converting each group to its equivalent hex digit, as in Table 2.3.

Table 2.3 Hexadecimal digits

Decimal	*Binary*	*Hexadecimal*
0	0000	0
1	0001	1
2	0010	2
3	0011	3
4	0100	4
5	0101	5
6	0110	6
7	0111	7
8	1000	8
9	1001	9
10	1010	A
11	1011	B
12	1100	C
13	1101	D
14	1110	E
15	1111	F

Table 2.4 Examples of equivalent values

Decimal	Binary	Hexadecimal
16_{10}	10000_2	10_{16}
31_{10}	$1\ 1111_2$	$1F_{16}$
100_{10}	$110\ 0100_2$	64_{16}
169_{10}	$1010\ 1001_2$	$A9_{16}$
255_{10}	$1111\ 1111_2$	FF_{16}
1024_{10}	$100\ 0000\ 0000_2$	400_{16}

The base of the number shown can be shown as a subscript where necessary to avoid confusion. All number systems use the same set of characters, so if the base of the number given is not obvious from the context, it can be specified. For example, the number 100 (one, zero, zero) could have the decimal value 4 in binary, 100 (one hundred) in decimal or 256 in hexadecimal.

Some examples of equivalent values are given in Table 2.4. The numbers are printed in 'Courier' type, as used on old-fashioned typewriters, because it is not proportionally spaced, so each character occupies the same width space, and all the digits line up in columns.

2.1.4 Counting

A list of equivalent numbers, counting from zero, is given in Table 2.5, with some comments on important values. This table also refers to the representation of memory size in microprocessor systems; for instance, '1k' of memory corresponds to 1024_{10} locations.

Table 2.5 Significant equivalent numbers

Decimal (Base 10)	Binary (Base 2)	Hexadecimal (Base 16)	Comment
0	0	0	All the same.
1	1	1	All the same.
2	10	2	Use 2nd column in binary.
3	11	3	Maximum 2-bit count.
4	100	4	[2^2] Use 3rd column in binary.
5	101	5	
6	110	6	
7	111	7	Maximum 3-bit count.
8	1000	8	[2^3] Use 4th column in binary.
9	1001	9	Decimal and hex same until 9.
10	1010	A	Use letters in hex.
11	1011	B	
12	1100	C	
13	1101	D	
14	1110	E	

Continued . . .

Table 2.5 Significant equivalent numbers

Decimal (Base 10)	*Binary (Base 2)*	*Hexadecimal (Base 16)*	*Comment*
15	1111	F	Maximum 4-bit count.
..	....	..	
16	1 0000	10	[2^4] Use 2nd column in hex.
17	1 0001	11	Use space to clarify binary.
18	1 0010	12	
19	1 0011	13	
20	1 0100	14	
21	1 0101	15	
22	1 0110	16	
23	1 0111	17	
24	1 1000	18	
25	1 1001	19	
26	1 1010	1A	
27	1 1011	1B	
28	1 1100	1C	
29	1 1101	1D	
30	1 1110	1E	
31	1 1111	1F	Maximum 5-bit count.
32	10 0000	20	[2^5]
33	10 0001	21	
34	10 0010	22	
..	.. ....	..	
62	11 1110	3E	
63	11 1111	3F	Maximum 6-bit count.
64	100 0000	40	[2^6]
65	100 0001	41	
..	... .	..	
127	111 1111	7F	Maximum 7-bit count.
128	1000 0000	80	[2^7]
129	1000 0001	81	
..	....	..	
254	1111 1110	FE	
255	1111 1111	FF	Maximum 8-bit count.
256	1 0000 0000	100	[2^8]
..	. ...	..	
511	1 1111 1111	1FF	Maximum 9-bit count.
512	10 0000 0000	2FF	[2^9]
..	.. ..	...	
1023	11 1111 1111	3FF	Maximum 10-bit count.
1024	100 0000 0000	400	[2^{10}] = 1k
..	....	...	
2047	111 1111 1111	7FF	Maximum 11-bit count.
2048	1000 0000 0000	800	[2^{11}] = 2k
..	....	...	
4095	1111 1111 1111	FFF	Maximum 12-bit count.
4096	1 0000 0000 0000	1000	[2^{12}] = 4k
..	. ...	..	
65535	1111 1111 1111 1111	FFFF	Maximum 16-bit count.

The rules for counting in any number system are as follows:

1. Start with all digits set to zero.
2. In the right digit position (LSB), count up from zero to the maximum digit available (for example, 9 in decimal).
3. If a column value is already at its maximum, reset it to zero, and increment (add 1 to) the next column to the left.

In microprocessors, there is a fixed number of digits in the registers that store binary numbers. If the number storage space has a fixed number of digits, leading zeros can be used to fill the empty positions, because each register bit must be either 1 or 0, and leading zeros do not alter the value.

2.1.5 Bits, Bytes and Words

One binary digit represents a 'bit' of information. A group of 8 bits is called a 'byte', and larger binary codes are called 'words'. This last term is used fairly loosely, but it often refers to a 16-bit code, with a 32-bit code called a 'long word'. As we know, in hexadecimal 4 bits are represented by one hex digit, so a byte is two hex digits, and so on.

2.2 Machine Code Programs

Microprocessors store their program code and data in binary form, typically using voltage levels of +5 and 0 volts to represent binary 1 and 0, respectively. The program is stored in non-volatile ROM (read-only memory) or volatile RAM (read-and-write memory). The program can be created in a host computer and 'blown', via a suitable data link, into a ROM chip placed in a programming unit. The program could also be loaded into RAM from the host, or from a disk drive, but it will be lost when the power is switched off.

2.2.1 Data Words

Conventional microprocessors handle the code in 8, 16, 32 or 64-bit binary words. The data word size has increased with the complexity of the integrated circuits available; some examples are given in Table 2.6.

Table 2.6 Comparison of microprocessors and microcontrollers

Microprocessor/ Microcontroller	*Computer/ Application*	*Address bus (bits)*	*External data bus (bits)*	*Internal CPU data (bits)*
Zilog Z80	Spectrum	16	8	8
Rockwell 6502	Commodore/BBC	16	8	8
Motorola 68000	Atari/Amiga/Mac	24	16	32
Intel 8086/8	IBM PC XT	16 + 4	8/16	16
Intel Pentium	WinTel Pentium PC	32	32	64
Intel 8051*	Industrial/Control	(16)	(8)	8
PIC 16F84	Industrial/Control	NA	NA	8

NA, not applicable. *Optional external memory access via multiplexed address and data bus.

Table 2.7 6502 machine code

	Memory address	Hex code	Meaning
First Instruction	0200 0201	A9 55	Load the main data register A with the number 55
Second Instruction	0202 0203 0204	8D 00 03	Store the contents of register A in the memory location whose address is 0300
Next Instruction	0205 0206	** **	next instruction code... next operand...

The first generation of popular home computers, such as the Commodore, Apple, BBC, and Spectrum machines used 8-bit microprocessors; that is, the program and data words were all 8-bit numbers. Second-generation home games machines such as the Atari and Amiga used the 16-bit 68000 chip, which was also the processor used in the Apple Mac, the first mass produced computer to use a WIMP interface.

The original IBM PC was a business oriented personal computer using the Intel 8088, which handled 16 bits inside the CPU, but only 8 bits externally. The Intel processor then went through a progressive development, leading to the 32-bit Pentium processor. At the same time clock speeds increased, and the processor developed, so that the data processing capability of the current Pentium types is several orders of magnitude greater than the original 8-bit CPUs (one order of magnitude = $\times 10$).

The 8051 is a standard microcontroller, which has been available for some years, and is well established in the industrial control market. The PIC is a more recent challenger for the position of leading microcontroller. The 16F84 has many advantages for smaller applications; for example, it has flash ROM program memory, so it can be reprogrammed easily.

2.2.2 Machine Code

Microprocessor machine code is a list of binary codes which can be represented in the equivalent hexadecimal. An example of 6502 code is listed in Table 2.7.

The program code is a list of 8-bit binary numbers, stored in numbered memory locations, here starting at 0200, forming a list of instructions for the microprocessor to execute. The function of this program is to load a number given in the program (55_{16}) into the main data register (called A), and then store it in a memory location (0300_{16}). The program shows two instructions, each of which starts with the instruction code itself ($A9_{16}$, $8D_{16}$), which are followed by data required by the instruction (a number to load, and a memory address to store it in). These are called the operands. Note that in 6502 programs, the instruction may consist of 1, 2 or 3 bytes.

2.2.3 8086 Machine Code

The Intel 8086 was the CPU originally developed for use in the IBM PC. It is useful to know something about 8086 machine code (see Table 2.8) because this is the native language of the PC, and it can be studied without access to any other hardware system. As with other processor families, the same basic instruction set has been expanded for later processors, but the basic syntax is the same, so 8086 code can be run on a Pentium processor.

Table 2.8 PC machine code

Address Segment:Offset	0	1	2	3	4	5	6	7 8	9	A	B	C	D	E	F
1B85:0100	OF	00	B9	8A	FF	F3	AE	47-61	03	1F	8B	C3	48	12	B1
1B85:0110	04	8B	C6	F7	0A	0A	D0	D3-48	DA	2B	D0	34	00	74	1B
1B85:0120	00	DB	D2	D3	E0	03	F0	8E-DA	8B	C7	16	C2	B6	01	16

Backward code compatibility has always been a major feature of the Intel/MSDOS product line.

The code can be observed by entering the command 'debug' at the DOS prompt on the PC, which in turn can be selected using the icon or menu item on Windows machines. A '-' prompt appears, to indicate that debug commands will be accepted. If 'd' (dump) is entered, a block of memory will be dumped to the screen as two hex-digit codes representing 8-bit binary program and data codes.

The addressing system in the Intel processor is a little more complicated than other processors, with the 20-bit address derived from the combination of a 16-bit 'segment' address and an 'offset'. This was originally done when the 16-bit 8086 was introduced, to maintain compatibility with older 8-bit systems. This was achieved by dividing the memory into 10×64k segments (64k is the maximum memory space addressable with 16-bit addresses). This is why the address is shown in the form 'SSSS:OOOO', where SSSS is the four hex-digit segment address and OOOO is the 4-bit offset. For example, if SSSS = 1B85 and OOOO = 0100, then the actual address will be 1B850 + 0100 = 1B950.

If the debug command 'u' is entered, the source assembly language 'mnemonics' (letter codes) are displayed, and it can be seen that each instruction can contain 1, 2, 3 or 4 bytes. Assembly language is the programming method for writing machine code programs. It will be explained in more detail later, using the PIC instruction set as an example.

2.2.4 PIC Machine Code

The PIC machine code program is easier to interpret than the 8086 code, because it has instructions which are of fixed length, i.e., 14 bits. In hexadecimal, these must be represented with four digits, with the most significant two bits unused. The default program start address (which is used if the programmer does not specify another), is 0000 (zero). A simple PIC machine code program is shown in Table 2.9.

Table 2.9 Simple PIC machine code program BIN1

Program memory address	*Program memory code*	*Meaning of machine code*
0000	**3000**	Move the number 00 into the working register.
0001	**0066**	Copy this code into port B data direction register.
0002	**0186**	Clear port B data register to zero.
0003	**0A86**	Increase the value in port B data register by one.
0004	**2803**	Jump back to address 0003.

Table 2.10 PIC machine code in binary form

Memory address	Binary machine code	Meaning
000	11 0000 0000 0000	Load W with 00000000 (0)
001	00 0000 0101 0110	Copy W to Direction Register of Port 110 (6)
002	00 0001 1000 0110	Clear Data Register of Port 110 (6)
003	00 1100 1000 0110	Increment Register of Port 110 (6)
004	10 1000 0000 0011	Jump to back to Address 0000000011 (3)

The machine code for the first PIC program, BIN1, that we will be studying is listed. It consists of five instructions, stored at addresses 0000–0004 in the program memory. The meaning of each is given, and a fuller explanation will come later. Each instruction is 14 bits long, but the actual operation code and operand length varies within the fixed total, as shown in Table 2.10.

The operation code part of the 14-bit instruction is shown in bold, while the operand is shown in italics. The operands refer to numbered registers and addresses within the PIC. For example, the last instruction operand contains the address of the third instruction, because the program jumps back and repeats from this point.

The PIC machine code can be seen in the programmer software (MPSTART) window prior to downloading, or printed in the source program list file. When the PIC chip is the programmer unit, the binary codes for the program are sent to its program memory, in serial form, one bit at a time. Each 14-bit code is stored at the address location (0000–0004) specified. When the program is later executed, the codes are interpreted by the processor and the action carried out. The meaning and use of the registers will be explained later, when this program will be analysed in more detail.

2.3 ASCII Code

ASCII (American Standard Code for Information Interchange) is a type of binary code for representing alphanumeric characters, as found on your computer keyboard. The basic code consists of 7 bits. For example, capital (or 'upper case') 'A' is represented by code $100\,0001_2$ (65_{10}), 'B' by 66, and so on to 'Z' $= 65 + 25 = 90_{10} = 101\,1010_2$. Lower-case letters and other common keyboard characters such as punctuation, brackets and arithmetic signs, plus some special control characters, also have a code in the range $0–127_{10}$. The numerical characters also have a code, for example '9' = 011 1001, so you sometimes need to make it clear if the code is the binary equivalent (1001) or the ASCII code (011 1001).

We will not be using ASCII codes any further in this book, but we need to know of them, as they are the standard coding method for text files. When a program is typed into the computer to create a 'source code file', this is how the text is stored. Later, the ASCII codes must be converted into corresponding binary machine code instructions. If this is confusing, come back to this point when we have looked at programming in more detail!

Summary

- Programs and data in a microprocessor system are stored in binary form; typically '0' = 0 V and '1' = 5 V.

- The binary codes can be displayed and printed in hexadecimal form, where one hex digit = four binary bits.
- A microprocessor program consists of a sequence of binary codes representing instructions and data which is decoded and executed by the CPU.
- The microprocessor memory contains a set of locations, numbered from zero, where the program is stored.
- Each program instruction consists of an operation code and (often) an operand.
- Each complete instruction may occupy a fixed number of bits or a variable number of bytes.

Questions

A calculator which converts between number systems is required. Do the following calculations manually, and then check the answer on a calculator.

1. Refer to Table 2.5.
   (a) Predict the binary equivalent of 35_{10}, 61_{10}, and 1025_{10}.
   (b) Convert the numbers in (a) from binary to hexadecimal.
   (c) Work out the 8-bit binary code for the 6502 program code in Table 2.7.
   (d) Write down the 16-bit binary code for the hex address 0203_{16}.
2. Write down the hex code, and work out the decimal equivalent number for the binary numbers:
   (a) 101_2
   (b) 1100_2
   (c) $1001\,1110_2$
   (d) $0011\,1010\,1111\,0000_2$
3. Light emitting diodes are often used to display output codes in simple test systems, where a binary '1' lights the LED. For an 8-bit output, work out the binary and hex code required:
   (a) to light all the LEDs;
   (b) to switch them all off;
   (c) to light alternate LEDs, with the LSB = 1.
   Now work out the hex data sequence of 8 two hex-digit numbers which will produce:
   (d) a bar graph effect on a set of eight LEDs (all off, then LSB on, two on, three on and so on until all eight are on);
   (c) a scanning effect (switch on one LED at a time, in order, LSB first).

Answers

1. (a) $35_{10} = 100011_2$, $61_{10} = 111101_2$, $1025_{10} = 10000000001_2$.
   (b) $35_{10} = 23_{16}$, $61_{10} = 3D_{16}$, $1025_{10} = 401_{16}$.
   (c) $A9_{16} = 1010\,1001_2$, $55_{16} = 0101\,0101_2$, $8D_{16} = 1000\,1101_2$, $00_{16} = 0000\,0000_2$, $03_{16} = 0000\,0011_2$.
   (d) $0203_{16} = 0000\,0010\,0000\,0011_2$.

2. (a) $101_2 = 5_{16} = 5_{10}$.
   (b) $1100_2 = C_{16} = 12_{10}$.
   (c) $1001\,1110_2 = 9E_{16} = 158_{10}$.
   (d) $0011\,1010\,1111\,0000_2 = 3AF0_{16} = 15088_{10}$.

3. (a) $1111\,1111_2 = FF_{16}$.
   (b) $0000\,0000_2 = 00_{16}$.
   (c) $0101\,0101_2 = 55_{16}$.
   (d) 00, 01, 03, 07, 0F, 1F, 3F, 7F, FF.
   (e) 00, 01, 02, 04, 08, 10, 20, 40, 80.

Activities

1. The seven-segment display is a device which we will use later as an output device for the PIC chip. Digits are displayed by illuminating selected segments. A diagram showing the connections to the LED segments is given in Fig. 2.1(a). The segments are identified by letter; 'a' for the top segment, 'b' is the next clockwise round the outside, and so on up to 'f' for the top left segment, with the middle segment called 'g'.

   These are connected as shown to a port data register bits 1–7, with the LSB not connected. Work out the binary and hex codes required to obtain the displayed characters 0 to F shown in Fig. 2.1(b), if the display operates 'active high', that is, a '1' in the register switches the corresponding segment on. Assume that bit 0 = '0'.

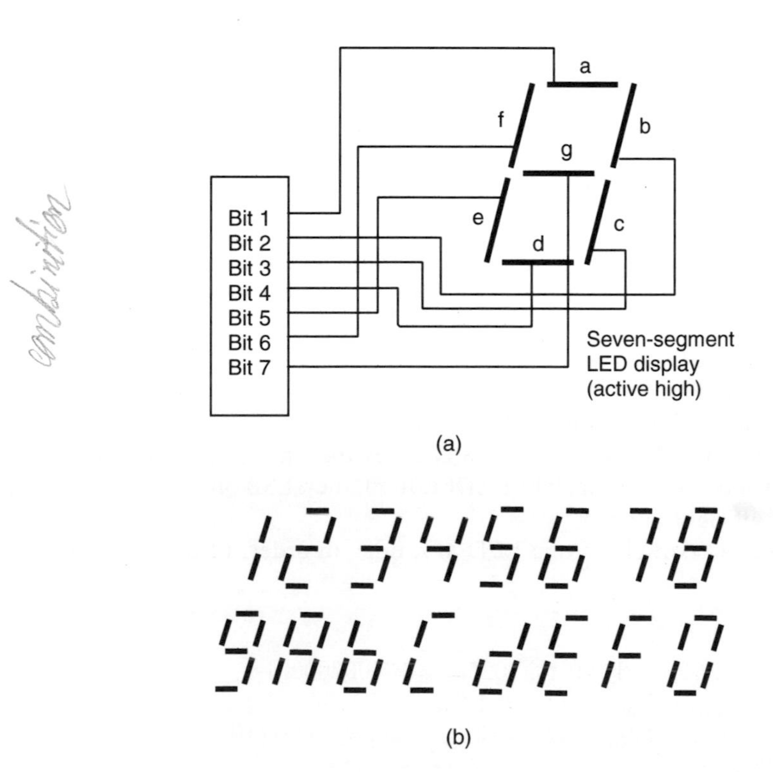

Figure 2.1 Seven-segment display of hex digits.

2. Debug is a DOS utility which allows you to operate at machine code level in the PC system. At the DOS prompt on a PC, enter the command 'debug'; a prompt '-' is obtained.
   (a) Enter '?' and the debug commands are displayed.
   (b) Enter 'd' (dump) and the contents of the current memory range are displayed in hex bytes (two digits), with the ASCII character equivalent at the right. The four-digit codes on the left are the segment address and offset, separated by a colon. The addresses are displayed at intervals of 16 (=10H) locations, since each row shows 16 bytes.
   (c) Enter 'u' (unassemble) and the assembly code is displayed, one instruction per line. Note the presence of instructions such as MOV (move data), ADD (add data), INC (increase value by 1) and so on. Note also the variable instruction length.
   (d) Enter 'r' and the processor registers are displayed. Note that at least one of the segment registers (CS, DS, SS, ES) contains the segment address, and the instruction pointer (IP) contains the offset.
   (e) Enter 'q' to quit.

Chapter 3
Microelectronic Devices

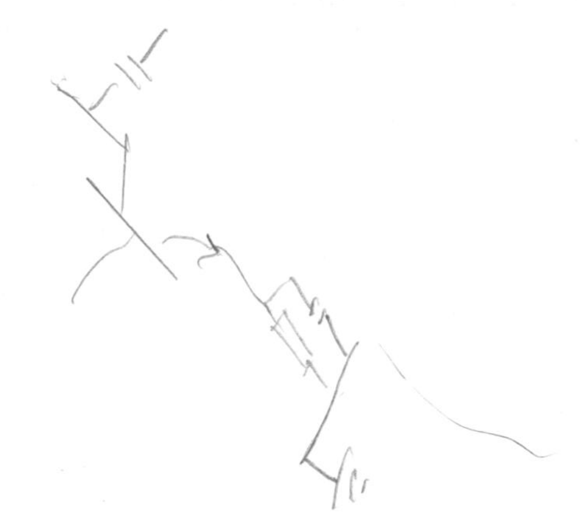

We have seen in Chapter 2 that a microcontroller program consists of a list of binary codes, which itself is a mixture of instructions and data stored in non-volatile memory. The instructions are executed in sequence, and usually process data obtained from the chip registers or inputs. The results are stored back in the registers or sent to an output device. We will now look briefly at the basic circuit elements needed to provide these functions. The intention is to explain logic system components in enough detail to allow the reader to understand the PIC data sheet.

3.1 Digital Devices

The binary codes which make up the program and data in the microcontroller are stored and processed as electronic signals. The binary numbers are typically represented as follows: binary 0 = 0 volts and binary 1 = +5 volts. A +5 V supply, which is usually derived from the mains, is therefore required to power the circuits. It must be able to provide sufficient current for the processor circuits at a voltage which is specified to be between 4.75 V and 5.25 V for standard TTL (Transistor–Transistor Logic). However, microprocessor circuits are now being designed to operate at a lower voltage. For example, a supply of 3.3 V is used to reduce power dissipation in large chips such as the Pentium processor (power dissipation is proportional to the supply voltage squared, so this will produce a heating reduction of greater than 50%). On the other hand, for flexibility of supply, the PIC chip can work within a supply range of 2–6 V. It can thus be battery powered with, for example, 2 × 1.5 V dry cells, and it will tolerate a variation in supply voltage as the batteries discharge.

Digital signals are processed using logic gates, so we now need to look briefly at how these work. The PIC chip itself is a CMOS (complementary metal oxide semiconductor)

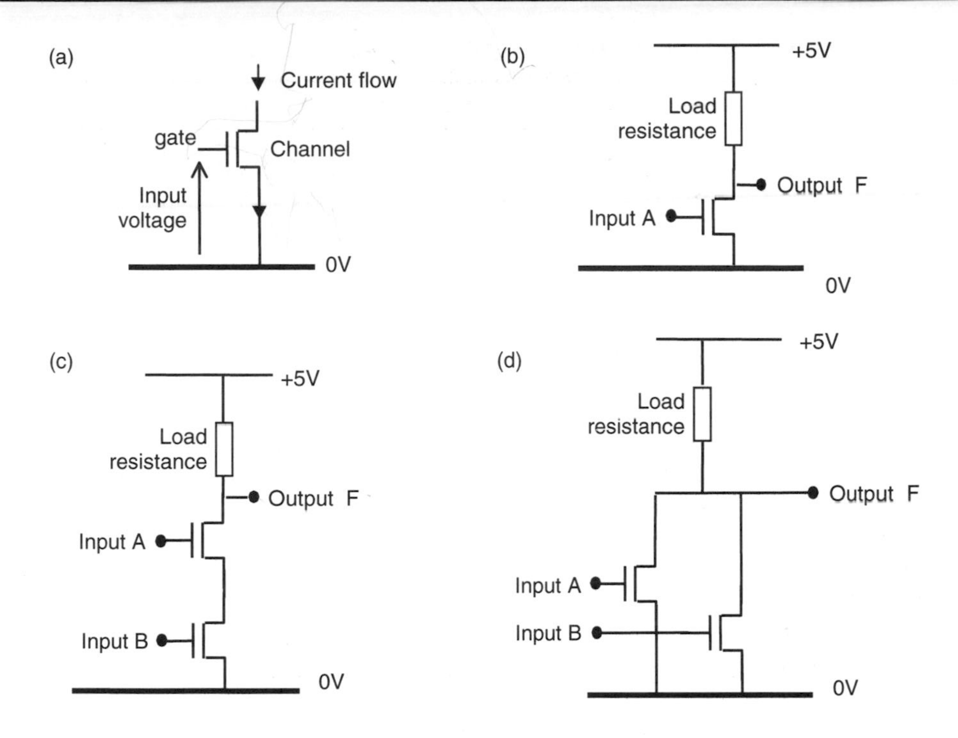

Figure 3.1 Field effect transistor logic gates. (a) Field effect transistor; (b) FET logic inverter; (c) simplified NAND gate; (d) simplified NOR gate.

device, which means that the logic circuits are made from field effect transistors (FETs). These circuits use FETs as digital switches, which, when combined together in various ways, create logic gates that can process binary data. For example, we will see how logic gates can be combined to create a binary adder, which is an essential component of any microprocessor.

3.1.1 FET Gates

The FET is the basic switching device which appears in the PIC data sheet in the equivalent circuits for various functional blocks. It is a transistor which works as a current switch; current flow through a semiconductor 'channel' is controlled by the voltage at the input 'gate'.

A single FET is shown in Fig. 3.1(a), indicating the current flow through the channel when it is switched on by applying a positive voltage between 0 V and the gate. When the input voltage is zero, the channel has a high resistance to current flow, and the device is off. Some FETs operate with a negative voltage at the input to control the current flow.

A logical 'invert' operation is implemented by the FET circuit in Fig. 3.1(b). Assume that the FET is switched on with +5 V at input A. The channel will be conducting current, and this current also flows in the load resistance, R, causing a voltage drop across it. This means that the voltage at F must fall, and for correct operation, F must be near 0 V when the FET is on. Thus the output is near 0 V (logic 0) when the input is +5 V (logic 1). Conversely, the output is 'pulled up' to +5 V (logic 1) by R when the input is low (logic 0), because there is no current flow in the FET channel, and therefore approximately a 0 V drop across the resistor.

Table 3.1 Logic table for one and two input gates

Inputs	Outputs					
	NOT	AND	OR	NAND	NOR	XOR
0	1	–	–	–	–	–
1	0	–	–	–	–	–
00	–	0	0	1	1	0
01	–	0	1	1	0	1
10	–	0	1	1	0	1
11	–	1	1	0	0	0

The logic operation 'AND' requires the output of a gate to be high only when all inputs are high (see Table 3.1). 'NAND', the inverse output, requires that the output is low only when all inputs are high. This operation can be implemented as shown in Fig. 3.1(c). The output F is only low when both transistors are on. The AND function can be obtained by inverting the NAND output; this can be achieved by connecting the inverter circuit to the NAND output.

The logic operation 'OR' requires the output of a gate to be high when either input is high (see Table 3.1). 'NOR', the inverse output, requires that the output is low when either input is high. This operation can be implemented as shown in Fig. 3.1(d). The output F is low when either transistor is on. The OR function can then be obtained by inverting the NOR output, by connecting the inverter circuit.

In real logic gates, the circuits are a little more complex. There are no actual resistors used because they waste too much power; instead, other FETs are used as 'active loads', which reduces the power that would be dissipated as heat in the resistors.

3.1.2 Logic circuits

Digital circuits are based on various combinations of these logic gates, fabricated on a silicon wafer. They can be supplied as discrete gates on small scale ICs, or as complete logic circuits on large scale ICs. Microprocessors are the most complex of all, containing thousands of gates, and sometimes, millions of transistors.

In the original small scale chips, bipolar transistors were used to form TTL gates. However, these have relatively large power dissipation, and run at a correspondingly high temperature. This limits the number of gates that can be operated on one chip, so VLSI (very large scale integrated) circuits normally use FET based logic gates, because of their lower power consumption. CMOS chips, like the PIC, can run from a wider range of supply voltages, so are more suitable for battery powered applications, such as laptop computers. There is continuing development of logic technologies to obtain higher speed, lower cost and lower power dissipation in increasingly complex chips.

3.1.3 Logic Gates

Whichever technology is used to fabricate the gates, the logical operation is the same. The symbols for logic gates used in most data sheets, including the PIC, conform to US standards, because that is where the chips are usually designed. The basic set of logic devices are the AND gate, OR gate and NOT gate (or logic inverter), shown in Fig. 3.2.

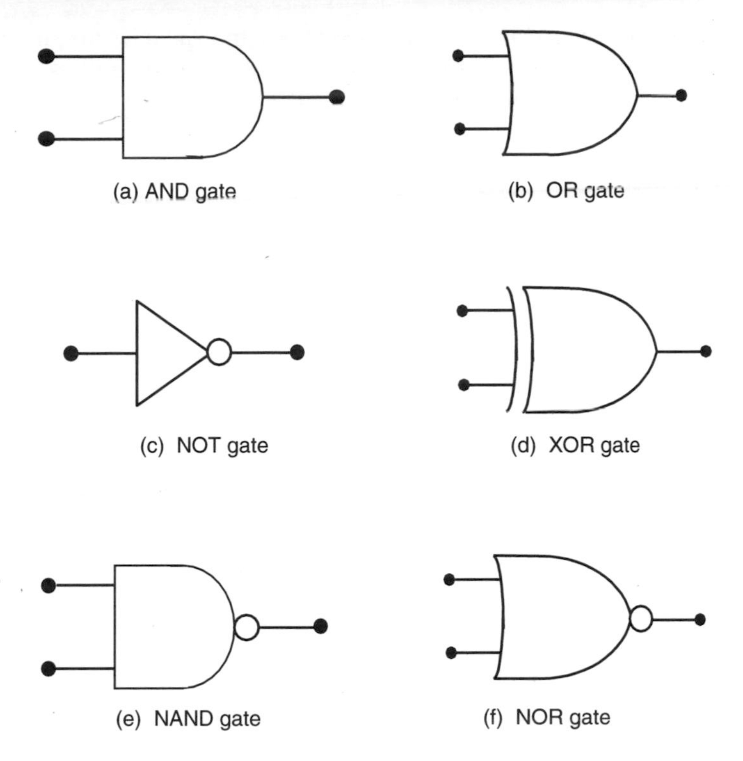

Figure 3.2 Logic gate symbols (US standard).

There are three additional gates which can be made up from the basic set, the NAND gate, NOR gate and XOR (exclusive OR) gate. The NAND is just an AND gate followed by a NOT gate, and a NOR gate is an OR gate followed by a NOT gate. An XOR gate is similar to an OR gate (see Table 3.1). The inputs on the left accept logic (binary) inputs, producing a different function at the output on the right. These logic values are typically represented by +5 V and 0 V in the standard circuit, as we have seen.

These gates, in various combinations, are used to make the control and data processing circuits in a microprocessor, microcontroller and supporting chips. Table 3.1 shows all the possible input combinations for one and two inputs. Obviously, the only possible inputs for the inverter are 1 and 0. The number of different inputs for the two input gates is four, that is, the total number of unique 2-bit codes. When specifying logic gate or circuit operation, all possible input combinations can be generated by counting up from zero in binary to the maximum allowed by the number of inputs.

The resulting output that is obtained from each gate is listed. The operation of logic circuits is shown in this way in IC data sheets, often with 0 represented by L (low) and 1 by H (high). Note that only two inputs to each gate are shown here, but there can be more than two. The logical operation will be similar; for instance, a three-input AND gate requires all inputs to be high to give a high output.

Variations may appear in data sheets. For instance, the circle representing logic inversion may be used at the input to a gate, as well as the output. However, it should always be possible to work out the logical operation from the basic logic symbol set. The more detailed analysis and design of discrete logic circuits is described in standard textbooks, and does not need to be covered here. Such discrete design principles are, in any case, less important than previously, partly due to the availability of microcontrollers such as the PIC which provide an alternative to pure hardware logic.

3.2 Combinational Logic

Logic circuits can be divided into two categories, combinational and sequential. Combinational logic describes circuits in which the output is determined only by the current inputs, and not by the inputs at some previous point in time. Circuits for binary addition will be outlined as examples of simple combinational logic. Binary addition is a basic function of the arithmetic and logic unit (ALU) in any microprocessor. A 4-bit binary addition is shown in Fig. 3.3, to illustrate the process required.

The process of binary addition is carried out in a similar way to decimal addition. The digits in the least significant column are added first, and the result 1 or 0 inserted in the 'Sum' row. If the sum is two (10_2), the result is zero with a carry into the next column. The carry is then added to the sum of the next column, and so on, until the last carry out is written down as the most significant bit of the result. The result can therefore have an extra digit, as in our example.

Having specified the process required, we can now design a logic circuit to implement this process. We will use a binary adder circuit for each column, feeding the carry bits forward as required.

```
             1 1 1 1    (A)
      +      0 1 1 0    (B)
      =    1 0 1 0 1    (Sum)
 Carry:    1 1 1
```

Figure 3.3 Binary addition.

3.2.1 Simple Binary Adder

The basic operation can be implemented using logic gates as shown in Fig. 3.4.

The two binary bits are applied at A and B, giving the result at F. Obviously, some additional mechanism is needed to store and present this data to the inputs, but this will

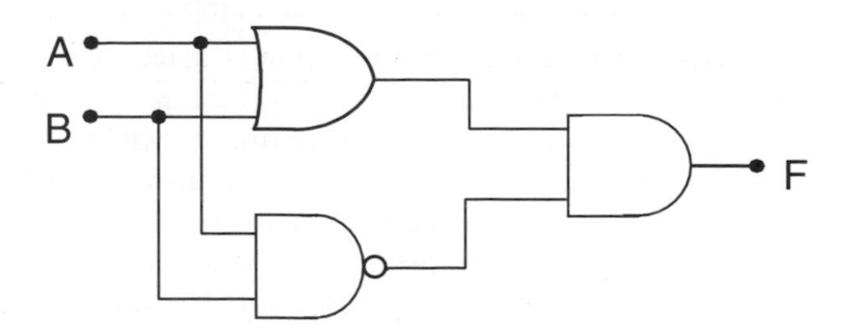

Figure 3.4 Binary adding circuit.

be explained later. This circuit is equivalent to a single XOR gate, which can thus be used as our basic binary adder.

3.2.2 *Full Adder*

To add complete binary numbers, a carry bit must be generated from each bit adder, and added to the next significant bit in the result. This can be done by elaborating the basic adder circuit as shown in Fig. 3.5(a).

The required function of the circuit can be specified with a logic table, as shown in Fig. 3.5(b). To implement this logic function, the carry out (Co) from each stage must be connected to the carry in (Ci) of the next, so that we end up with four full adders cascaded together. The overall carry in must be applied to the Ci of stage 1 and the carry out will then be obtained from Co of stage 4.

3.2.3 *4-Bit Adder*

A set of four full adders can be used to produce a 4-bit adder, or any other number of bits, by cascading one adder into the next. The PIC 16F84 ALU, for example, processes 8-bit data. As we are not particularly concerned with exactly how the logic is designed, we can

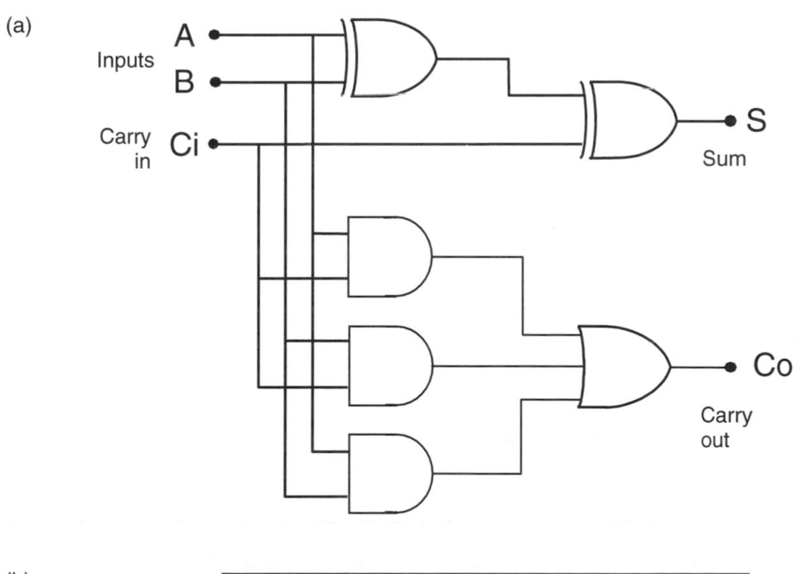

(b)

Input A	Input B	Carry In Ci	Carry Out Co	Sum S
0	0	0	0	0
0	0	1	0	1
0	1	0	0	1
0	1	1	1	0
1	0	0	0	1
1	0	1	1	0
1	1	0	1	0
1	1	1	1	1

Figure 3.5 (a) Full adder logic circuit; (b) full adder logic table.

(a)

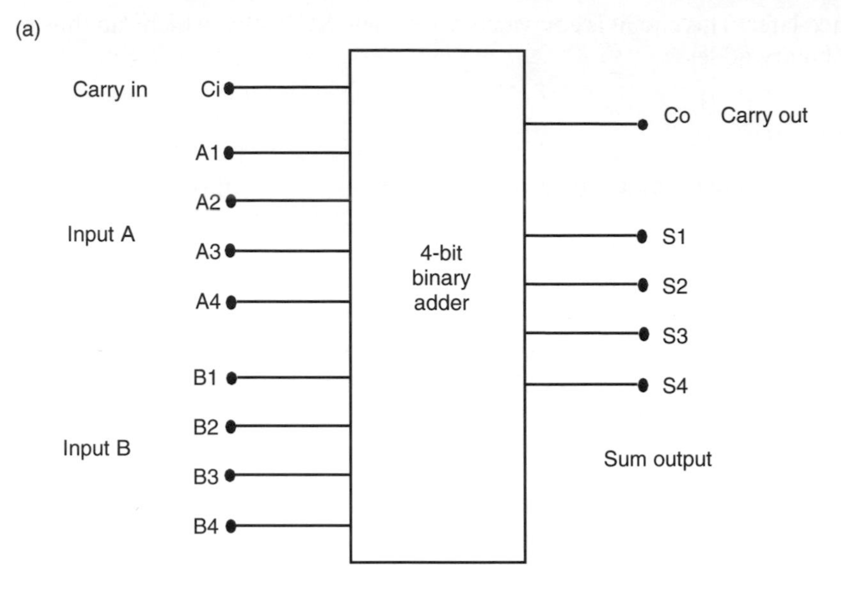

(b)

Row	Input									Output					
	Input A				Input B						Output sum				
	A4	A3	A2	A1	B4	B3	B2	B1	Ci	Co	S4	S3	S2	S1	Dec
0	0	0	0	0	0	0	0	0	0	0	0	0	0	0	0
1	0	0	0	0	0	0	0	0	1	0	0	0	0	1	1
2	0	0	0	0	0	0	0	1	0	0	0	0	0	1	1
3	0	0	0	0	0	0	0	1	1	0	0	0	1	0	2
4	0	0	0	0	0	0	1	0	0	0	0	0	1	0	2
5	0	0	0	0	0	0	1	0	1	0	0	0	1	1	3
6	0	0	0	0	0	0	1	1	0	0	0	0	1	1	3
.	.	.	.	.	.	.	.	.	.	.	.	.	.	.	
etc															etc
.	.	.	.	.	.	.	.	.	.	.	.	.	.	.	
509	1	1	1	1	1	1	1	0	1	1	1	1	0	1	30
510	1	1	1	1	1	1	1	1	0	1	1	1	1	0	30
511	1	1	1	1	1	1	1	1	1	1	1	1	1	1	31

Figure 3.6 (a) 4-bit adder block; (b) logic table for 4-bit adder.

hide it inside a block, and then define the required logical inputs and the resulting outputs (see Fig. 3.6).

All possible input combinations are required, and these can be generated by using a binary count in the input columns. The state of the output for each possible input code is then defined. With 2×4-bit inputs, plus the carry in, there are 512 possible input combinations in all, so the logic table only shows the first few and last rows.

In the past, logic circuits had to be designed using Boolean mathematics and built from discrete chips. Now, programmable logic devices (PLDs) make the job easier, as the required operation can be defined with a logic table or function on a PC host and programmed directly into a PLD chip, in much the same way that the PIC itself can be programmed.

3.3 Sequential Logic

Sequential logic refers to digital circuits whose outputs are determined by the current inputs *and* the inputs which were present at an earlier point in time. That is, the *sequence* of inputs determines the output. Such circuits are used to make data storage cells in registers and memory, and counters and control logic in the processor.

3.3.1 Basic Latch

Sequential circuits are made from the same basic set of logic gates shown in Fig. 3.2. They are all based on a simple latching circuit made with two gates, where the output of one gate is connected to an input of the other, as shown in Fig. 3.7.

This circuit uses NAND gates, but NOR gates will work in a similar way. When both inputs, A and B, are low, both outputs must be high. This state is not useful here, so is called 'invalid'. When one input is taken high, the output of that gate is forced low, and the other output high. Thc latch is now set, or reset, depending on which output is being

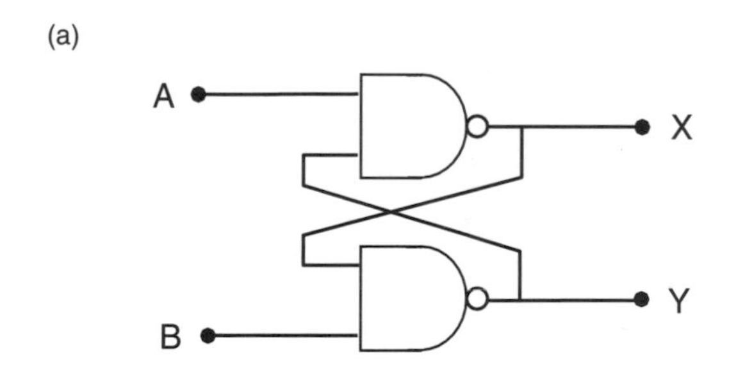

(b)

Time	Inputs A	B	Outputs X	Y	Comment
1	0	0	1	1	Invalid
2	0	1	1	0	X = 1
3	1	1	1	0	Hold X = 1
4	1	0	0	1	Reset X = 0
5	1	1	0	1	Hold X = 0
6	0	1	1	0	Set X = 1
7	1	1	1	0	Hold X = 1

Figure 3.7 (a) Basic latch circuit; (b) sequential logic table for basic latch.

used. In Fig. 3.7, X is taken as the output, and is set high. This state is 'held' when the other input is taken high, and this gives us the data storage operation required.

The output X can now be reset to zero by taking input B low. This state is held when B is returned high. Thus at time slot 3 a data bit '1' is stored at X, while at time slot 7 data bit '0' is stored. Note that in the time slots when both inputs are high, output X can be high or low, depending on the sequence of inputs before that step was reached.

With additional control logic, the basic latch can be elaborated to give two main types of circuit: the D-type ('data') bistable or latch which acts as a 1-bit data store, and T-type ('toggle') bistable which is used in counters. Bistable devices are frequently called as 'flip-flops', and different types of sequential circuit including counters and registers can be constructed from a general purpose device called a 'J–K flip-flop', if necessary.

3.3.2 Data Latch

A basic sequential circuit block is a data latch, which is shown as a block in Fig. 3.8(a). The inputs and output sequence can be represented on a logic table, as shown in Fig. 3.8(b). When the enable (EN) input is high, the output (Q) follows the state of the input (D). When the enable is taken low, the output state is held. The output does not change until the enable is taken high again. This is called a transparent latch, because the data goes straight through when the enable is high. There are other types of latches, called edge-triggered, which latch

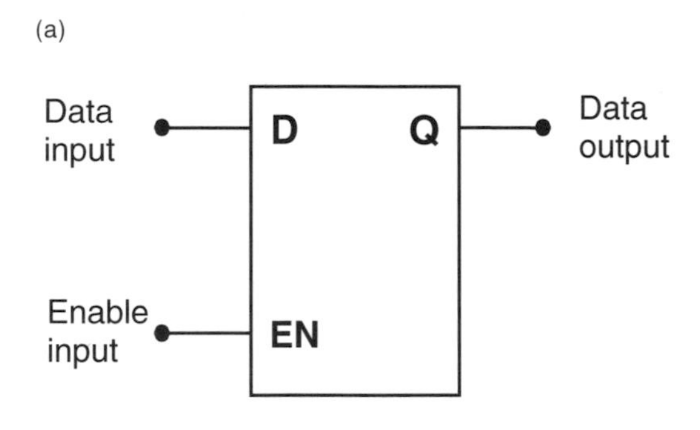

(b)

Time	Inputs D	Inputs EN	Output Q	Comments
1	0	0	x	Output unknown
2	0	1	0	Output = Input 0
3	0	0	0	Data 0 Latched
4	1	0	0	Data 0 Held
5	1	1	1	Output = Input 1
6	1	0	1	Data 1 Latched
7	0	0	1	Data 1 Held
8	0	1	0	Output = Input 0

Figure 3.8 (a) Transparent data latch; (b) Sequential logic table for data latch.

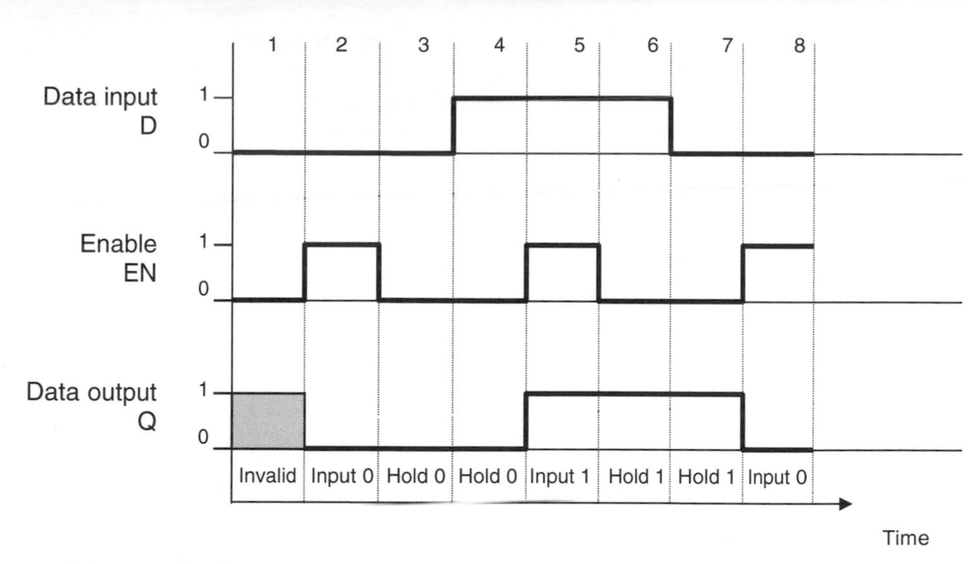

Figure 3.9 Data latch timing diagram.

the data input at a specific point in time, when the enable (or 'clock') signal changes. This type of circuit block is used in registers and static RAM to store groups of, typically, 8 bits.

A timing diagram (Fig. 3.9) can provide more complete information about the precise timing of the signals. Any delays between signals caused by the time taken for the gates to switch state can be shown, and the timing diagram may be easier to interpret. The same signals can be displayed in the actual circuit using an oscilloscope or logic analyser, which operates like a multichannel digital oscilloscope.

3.4 Data Devices

All data processing or digital control systems have circuits to carry out the following operations: data input, data storage, data processing, data output, control and timing.

Data processing devices must be controlled in sequence to carry out useful work. In a microprocessor system, most of this control logic is built into the CPU and its support chips, but additional control circuits usually need to be designed for each specific system. In order to illustrate the principles of operation of microprocessor and microcontroller systems, a set of basic logic devices will be used to make up a simple data processing circuit; they are shown in Fig. 3.10.

3.4.1 Data Input Switch

In Fig. 3.10(a), a switch (S) and resistor (R) are connected across a 5 V supply. If the switch is open, the output is 'pulled up' to +5 V, via the resistor. This will only work if there is negligible current flow in the resistor, which means that a high resistance (that is, a small load) must be connected to the data output. In practice, this limits the number and type of circuits that can be connected to the switch circuit. The same effect limits the number of logic gate inputs that can be connected to a particular gate output.

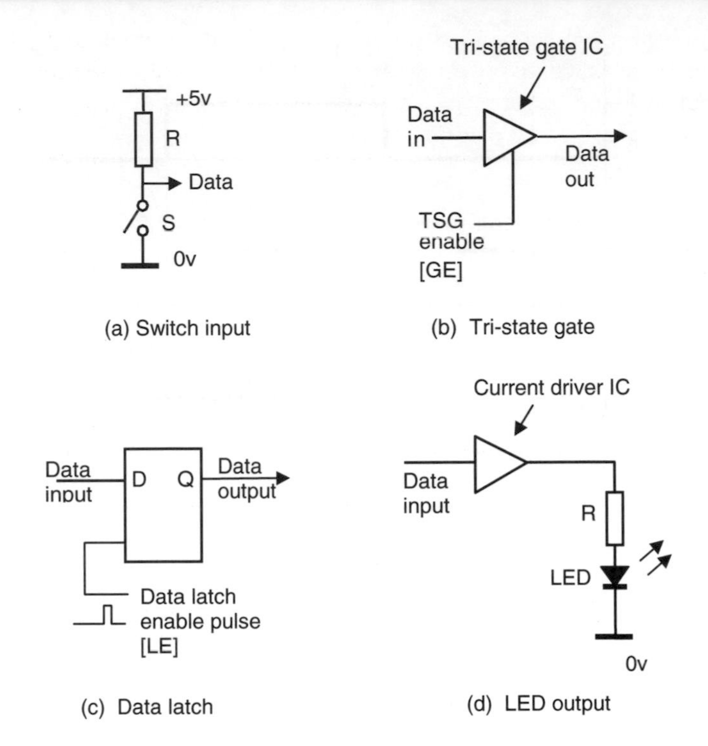

Figure 3.10 Data circuit elements.

If the switch is closed, the logic level must be 0, as it is connected directly to ground. The resistor is required to prevent a short circuit between the +5 V and 0 V supplies. A capacitor may be connected across the switch to 'debounce' it, that is, ensure a smooth change from high to low, and back.

3.4.2 Tri-State Gate

The tri-state gate (TSG) (Fig. 3.10(b)) is a digital device which allows electronic switching and routing of signals through a data processing system. It is controlled by the gate enable input (GE). When GE is active (in this example high), the gate is switched on, and all data is allowed through, 1 or 0. When GE is inactive (low), the data is blocked, and the output goes into a high impedance (HiZ) state, which effectively disconnects it from the input of the following stage. The TSG may have an active low input, in which case the control input has a circular invert symbol.

TSGs can be obtained as individual gates in a small scale integrated (SSI) circuit chip, and are used as basic circuit building blocks within large scale integrated (LSI) circuits, such as the PIC microcontroller.

3.4.3 Data Latch

A data latch (Fig. 3.10(c)) is a circuit block which stores 1 bit of data. A transparent latch has already been described. If a data bit is presented at the input D (1 or 0), and the latch 'clocked' by pulsing the latch enable input (0,1,0), the data appears at the output Q. It remains stored there when the input is removed or changed, until the latch is clocked again. The data can then be retrieved at a later time in the data process.

3.4.4 LED Data Display

A light emitting diode (LED) can provide a simple data display device; a set of eight can display a binary byte if the data does not change too quickly to be viewed. In Fig. 3.10(d) the logic level to be displayed (1 or 0) is fed to the current driver, which operates as a current amplifier and provides enough current to make the LED light up (typically about 10 mA) when the data is '1'. The resistor value controls the size of the current. Seven-segment, and other, matrix displays, use LEDs to display decimal or hexadecimal digits by lighting up suitably arranged LED segments or dots.

3.5 Simple Data System

The way that data is transferred through a digital system using the devices described above is illustrated in Fig. 3.11. The circuit allows one data bit to be input at the switch (0 or 1), stored at the output of the latch, and displayed on the LED. The operational steps are as follows.

1. The data at D1 is generated manually at the switch ('0' = 0 V and '1' = +5 V).

2. When the tri-state gate is enabled, the data is available at D2. When the gate is disabled, the line D2 is floating, or indeterminate (HiZ).

3. When the data latch is pulsed, level D2 is stored at its output, D3. D3 remains stored until new data is latched, or the system powered down.

4. While latched, the data at D3 is displayed by the LED (ON = '1'), via the current driver stage.

Note that all active devices (gate, latch and driver) must be connected to the +5 V power supply, but these supply connections do not have to be shown in a block diagram or

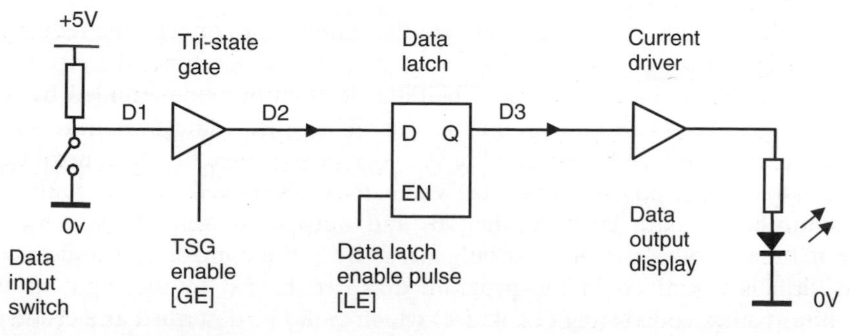

Figure 3.11 1-bit data system.

Table 3.2 1-Bit system operating sequence

Operation	*Switch*	*D1*	*GE*	*D2*	*LE*	*D3*
Data input 1	Open	1	0	x	0	x
Input enable	Open	1	1	1	0	x
Latch data	Open	1	1	1	0-1-0	1
Input disable	Open	1	0	x	0	1
Data input 0	Closed	0	0	x	0	1
Input enable	Closed	0	1	0	0	1
Latch data	Closed	0	1	0	0-1-0	0
Input disable	Closed	0	0	x	0	0

logic circuit. Table 3.2 details the control sequence, with the data states which exist after each operation. Note that 'x' represents 'don't know' or 'don't care' (it could be 1, 0 or floating).

3.6 4-Bit Data System

Data is often moved and processed in parallel form in a microprocessor system. The circuit shown in Fig. 3.12 illustrates this in a simplified way. The function of the 4-bit system is to add two numbers that are input at the switches. The two numbers A and B will be stored, processed and output on a seven-segment display which shows the output value in the range 0–F. The display has a built-in decoder which converts the 4-bit binary input into the corresponding digit pattern on the segments. To obtain a correct answer, the two numbers that are input must add up to 15 (decimal) or less. The system could be modified to carry out any arithmetic or logical operation, as specified by a control program, if an ALU were substituted for the adder, and some additional control circuitry was added.

The common data bus is used to minimize the number of connections required, but it means that only one set of data can be on the bus at any one time, therefore, only one set of gates must be enabled at a time. The data destination is determined by which set of latches is operated when data is on the bus. The gates (data switches) and latches (data stores) must therefore be operated in the correct sequence by the control unit.

Table 3.3 shows the control sequence which must be issued by the control unit to read in a number to data latch A, then B, and output the sum. If we now take the sequence of binary codes which must be generated by the control unit and assume that the input data is contained in the program and can be fed to the input at the right time, we obtain a hex code listing (Table 3.4) which could be described as a crude machine code program.

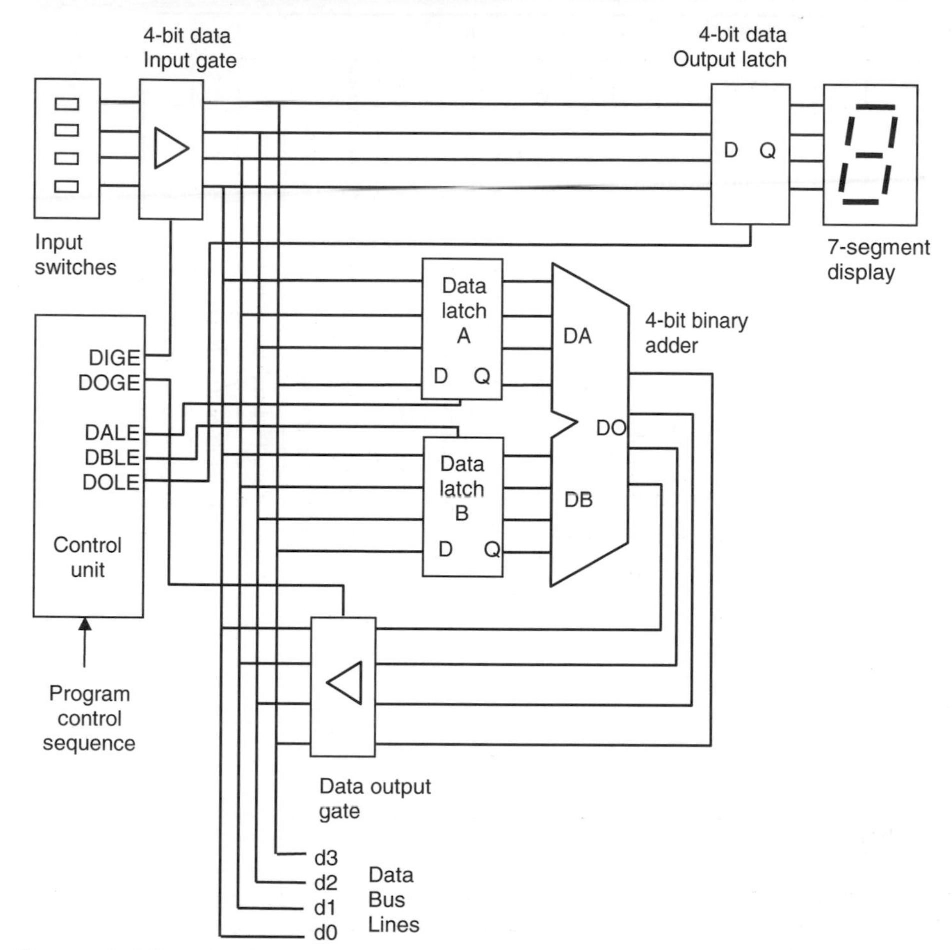

Figure 3.12 4-bit data system.

The 'program' for the 4-bit adder works as follows. The 'instruction' code is the binary number required from the control unit to set up the gates and latches for the operation. The first instruction (14) enables the data input gate and operates the data A latch. The data ('operand') must in some way be fed to the four input lines, bypassing the manual switch inputs, at the appropriate time. We must also assume that this happens before the latch is operated, in order to store valid data. The second instruction then does the same thing with data B. The third instruction needs no operand data, and allows the result from the binary adder to be stored and displayed at the output. We simply place zeros in the unused digits. Again, the data output gate enable must operate before the latch.

Thus a sequence can be pre-programmed, if the control circuit can be designed to provide the control signals in the right order, and additional hardware is provided to store the program and feed the operands to the inputs at the right time. A ROM device can be used to store such a control sequence; the operation of memory devices will be covered in the next chapter.

The program described above has three instructions, of 9 bits in length – and we now have a simple microprocessor system with a machine code program!

Table 3.3 4-bit system operating sequence

	Input switches (4-bit binary)	DIGE *(Data input gate enable)*	*DOGE (Data output gate enable)*	*DALE (Data A latch enable)*	*DBLE (Data B latch enable)*	*DOLE (Data output latch enable)*	*Hex display 0–F*	*Data bus (4-bit binary)*	*Operation*
0	xxxx	0	0	0	0	1	X	xxxx	Ready for input
1	0110	0	0	0	0	1	X	xxxx	Set data input number A on switches
2	0110	1	0	0	0	1	6	0110	Enable data A onto bus by switching on input gates
3	0110	1	0	0–1–0	0	1	6	0110	Store data A in latch A by clocking it with a pulse
4	0110	0	0	0	0	1	X	xxxx	Disable input gates – no valid data on bus
5	0101	0	0	0	0	1	X	xxxx	Set data input number B on switches
6	0101	1	0	0	0	1	5	0101	Enable data B onto bus by switching on input gates
7	0101	1	0	0	0–1–0	1	5	0101	Store data B in latch B by clocking it with a pulse
8	0101	0	0	0	0	1	X	xxxx	Disable input gates – no valid data on bus
9	xxxx	0	1	0	0	1	B	1011	Enable result from ALU onto bus
10	xxxx	0	1	0	0	0	B	1011	Store result in output latch
11	xxxx	0	0	0	0	0	B	xxxx	Result displayed – ready for next input

Table 3.4 4-bit system 'machine code program'

'Instruction' code	Operand	Hex 'program'	Operation
1 0101	0110	15 6	Input and latch data A
1 0011	0101	13 5	Input and latch data B
0 1001	0000	09 0	Latch and display result

Summary

- MOS digital circuits are based on the field effect transistor acting as a current switch. These are combined to form logic gates on integrated circuits.
- The basic set of logic gates is AND, OR and NOT, from which all logic functions can be implemented. NAND, NOR and XOR form a useful additional set.
- Combinational logic gives outputs which depend only on the current input combination. Sequential logic outputs additionally depend on the prior sequence of inputs.
- The data latch is used for storage and the tri-state gate for routing data. Input and output devices are also needed. Combinational logic provides processing and control.

Questions

1. Why is it necessary for battery powered digital circuits to operate at a wide range of voltages?
2. Draw a simple logic inverter using a FET and a resistor, and then elaborate the circuit to provide the AND and OR logic functions.
3. Describe in one sentence the operation of (a) an OR gate, and (b) an AND gate.
4. Representing a 1-bit full adder as a single block, with inputs A,B and Ci, and outputs S and Co, draw a 4-bit adder consisting of four of these blocks, with the inputs and outputs shown in Fig. 3.6(a).
5. Construct a timing diagram for the sequential logic table shown in Fig. 3.7(b), to show how a basic latch works.
6. Draw a circuit with two input logic switches whose data can be stored in one of three different D-type latches. Describe how the control logic must work to allow either input to be stored in any of the latches.
7. Modify the 4-bit system operating sequence so that only the final result is displayed.

Activities

1. Construct the full adder circuit using the necessary logic chips and check that it works as described.
2. Investigate the operation of a 'programmable logic device', and work out how it would be programmed to create a 4-bit adder.
3. Investigate how the basic latch circuit can be used to 'debounce' our data input switch.
4. Using a suitable circuit simulator, such as Electronic Workbench™, test the 4-bit system operation.

Chapter 4
Digital Systems

The basic digital devices described in Chapter 3 are enough to build working data circuits, but they can be combined into useful building blocks for designing more complex digital systems. This chapter outlines some of the more common sub-circuits, which may still be available as small scale ICs, or are built into larger scale devices such as the PIC microcontroller.

4.1 Encoder and Decoder

A digital encoder is a device which has a number of separate inputs and a binary output. An output binary number is generated corresponding to the numbered input which is 'active'. A decoder is a device which, unsurprisingly, carries out the inverse logical operation: a binary input code activates the corresponding output. Thus, if the binary code for 5 (101) is input, output 5 goes active (usually low). An example of encoder and decoder operation is described below, where they are used to operate a keypad (Fig. 4.1).

A set of switches can be combined in a two-dimensional array to form a simple keyboard. These may have 12 keys (decimal) or 16 keys (hexadecimal). The decimal pad has digits 0–9, hash (#) and star (*), while the hex keypad has digits 0–F.

To read the hex keypad, one of 16 push buttons must be detected when pressed. There are four rows of select lines output from the row decoder which are normally high (pull-up resistors can be attached to each line if necessary). When a binary input code is applied, the corresponding row select line goes low. A binary counter can be used to drive the row decoder

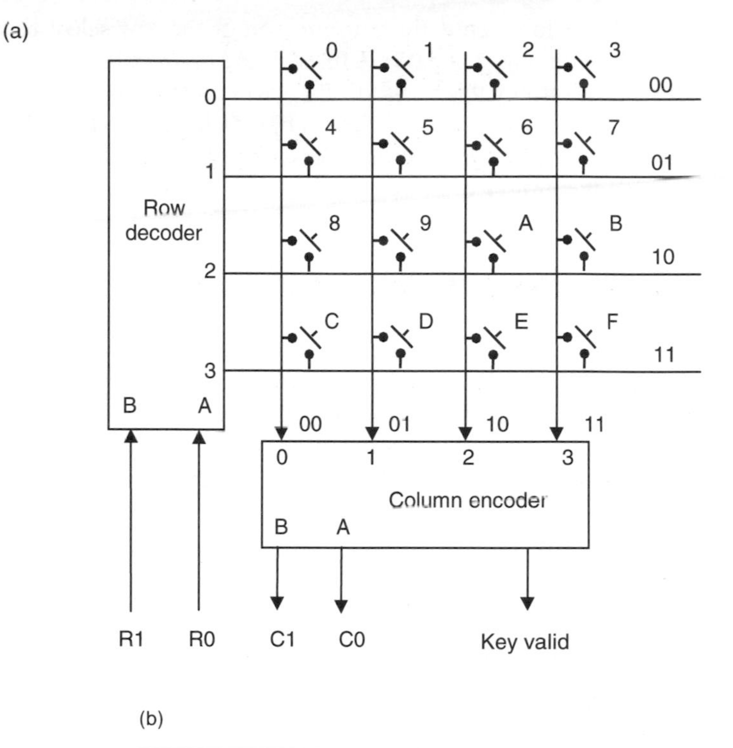

(b)

Inputs		Outputs			
B	A	0	1	2	3
0	0	0	1	1	1
0	1	1	0	1	1
1	0	1	1	0	1
1	1	1	1	1	0

(c)

Inputs				Outputs	
0	1	2	3	B	A
0	1	1	1	0	0
1	0	1	1	0	1
1	1	0	1	1	0
1	1	1	0	1	1

Figure 4.1 (a) Keypad scanning using an encoder and decoder; (b) Row decoder logic table; (c) Column encoder logic table.

(see below for counters), which will generate each row code in turn, continuously. If a switch on the active row is pressed, this low bit can be detected on the column line.

The column lines, which are also normally high, are connected to a column encoder. This generates a binary code that corresponds to the input that has been taken low by connection to a row that is low. Thus the combination of the row select binary code (R1, R0) and the column detect binary code (C1, C0), will give the number of the key which has been pressed. For instance, if key 9 is pressed, row 2 will go low when the input code is 10. This will take column 1 low, which will give the column code 01 out. The whole code is then 1001, which is 9 in binary. Encoders and decoders are combinational logic circuits which can be designed with any number of code bits, n, giving 2^n select lines.

4.2 Multiplexer, Demultiplexer and Buffer

These devices can be constructed from the same set of gates: two tri-state gates and a logic inverter, Fig. 4.2. A multiplexer is basically an electronic changeover switch, which can select data in from alternative sources within the data system. A typical application is to allow two different signal sources to use a common signal path at different times. This is achieved for one bit of the data path as shown in Fig. 4.2(a). Input 1 or 2 is selected by

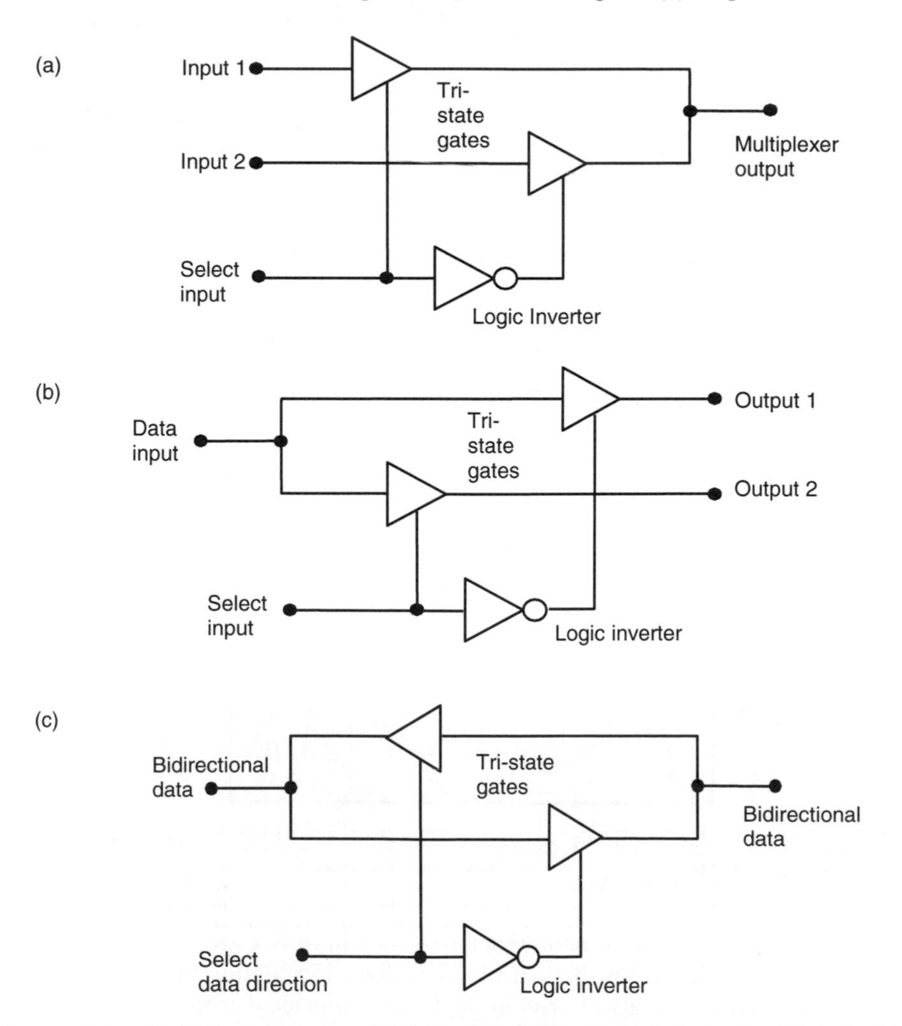

Figure 4.2 (a) 1-bit multiplexer; (b) 1-bit demultiplexer; (c) bidirectional data buffer.

the logic state of the select input. The logic inverter ensures that only one of the TSGs is enabled at a time. Conversely, a demultiplexer (Fig. 4.2(b)) splits the signal using the same basic devices. That is, it can pass data to alternative destinations.

The bidirectional buffer (Fig. 4.2(c)) is used to allow data to pass in one direction at a time along a data path, for example, on a bidirectional data bus. To achieve this, the TSGs are connected nose to tail, and operate alternately as in the multiplexer. When the control input is low, the data is enabled through from left the right, and when high, from right to left.

4.3 Registers and Memory

We have seen previously how a 1-bit data latch works. If some bidirectional control logic is added, data can be read from a data line into the latch, or written to the data line from it. We then have a register bit store. In Fig. 4.3(a), the data in/out line can be connected to the D input or Q output, depending on the state of the data direction select. If data is to be stored by the latch from the data line, latch enable is activated at the appropriate time.

If a set of these register elements are used together, a data word can be stored. A common data word size is 8 bits (1 byte), and most systems handle data in multiples of 8 bits. An 8-bit register, consisting of eight data latches, can be represented as shown in Fig. 4.3(b). The

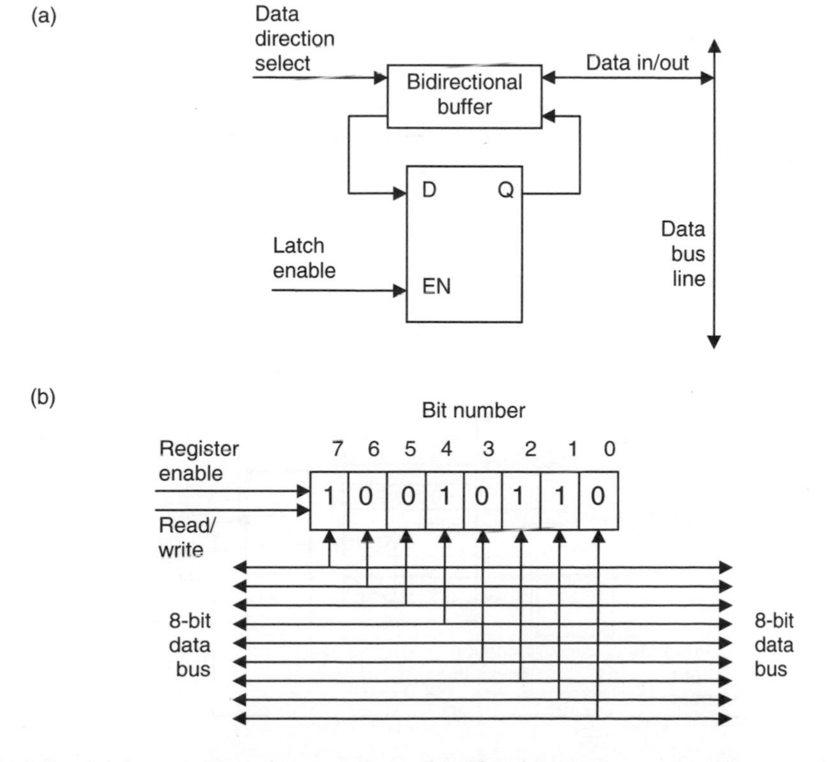

Figure 4.3 (a) Data register bit operation; (b) 8-bit data register operation.

register enable and read/write (data direction select) lines are connected to all the register bits, which operate simultaneously, to read and write data to and from the 8-bit data bus.

4.4 Memory Address Decoding

A static RAM location operates in a similar way to a register. The memory device typically stores a block of 8-bit data bytes which are accessed as numbered locations (Fig. 4.4). Each location consists of eight data latches that are loaded and read together. A read operation is illustrated; the data is being output from the selected location. A 3-bit code is needed to select one of eight locations in the memory block, using an internal address decoder to generate the location select signal. The selected data byte is enabled out via an output buffer, which allows the memory device to be electrically disconnected from the external data bus when the memory chip is not required.

The number of locations in a memory device can be calculated from the number of address pins on the chip. In Figure 4.4, a 3-bit address provides eight unique location addresses (000_2 to 111_2). This number of locations can be calculated directly as $2^3 = (2 \times 2 \times 2) = 8$. Thus, the number of locations is calculated as 2 raised to the power of the number of address lines. Some useful values are listed below:

$$2^8 = 256 \text{ bytes}$$

$$2^{10} = 1024 = 1 \text{ kb (kilobyte)}$$

$$2^{16} = 65536 = 64 \text{ kb}$$

$$2^{20} = 1048576 = 1 \text{ Mb (megabyte)}$$

In a microprocessor system, program and data words (binary numbers) are stored in memory, ready for processing in the CPU. The data transfer is usually implemented

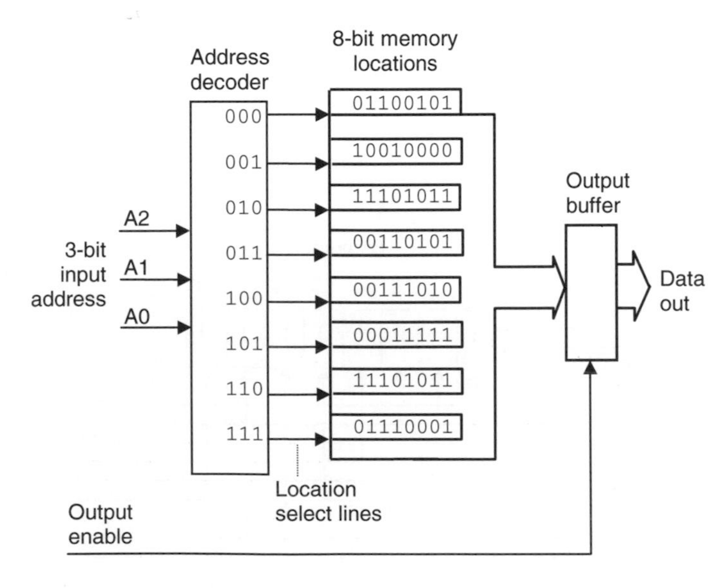

Figure 4.4 Memory device operation.

using a bus, which is a common set of lines which pass data between memory and registers, with logic gates directing the data to the correct destination.

4.5 System Address Decoding

There are usually several memory and input/output devices connected to a common data bus in the typical microprocessor system. Only one can use the data bus at any one time, so a system of chip selection is needed, so that the processor can 'talk to' the required chip. A system address decoder is used as illustrated in Fig. 4.5.

Although we are mainly concerned with microcontroller architecture, it is worth looking briefly at memory and I/O operation in a conventional system: first, because it explains the process which occurs within the microcontroller chip; second, because it is important for an overview of microprocessor systems; and third, because it is a logical extension of address decoding within each memory chip.

Figure 4.5 shows the basic connections in a microprocessor system that allow the CPU to read and write data to and from the memory and I/O devices. A binary address is output on the address bus by the CPU from its program counter that specifies a memory location. The system address decoder takes, in this case, the two most significant address lines and sets one of the four chip select lines active accordingly, which selects the chip to be accessed. The low order address lines are used, as described above, to select the required location within the chip. Thus, the location select is a two-stage process, with external (system) and internal (chip) decoding of the address.

When the location has been selected, the data stored in it can be read or written via the data bus according to the setting of the read/write line from the CPU. Note that ROM cannot be written and therefore does not need the read/write line connected. The I/O port only has a few addressable locations, its registers, so only a few of the address lines are needed for this device.

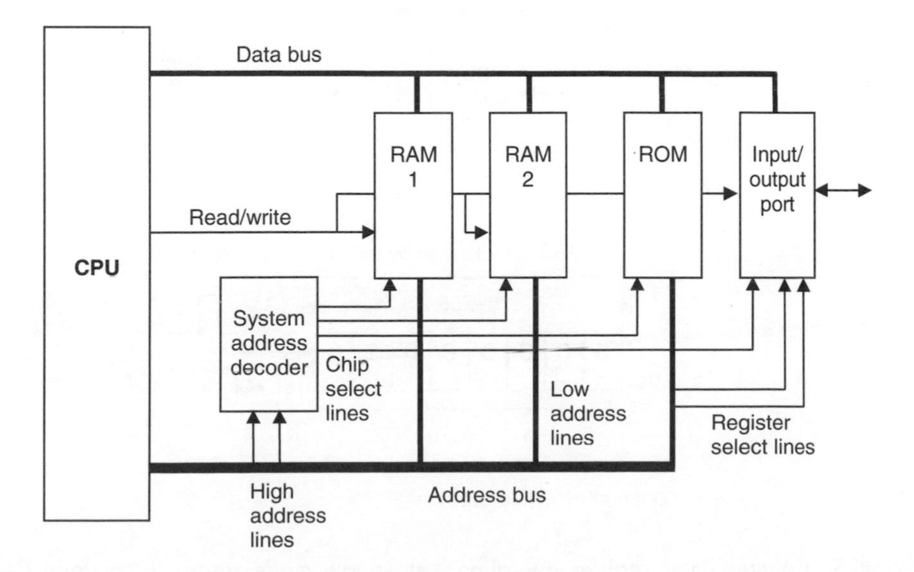

Figure 4.5 Microprocessor system addressing.

4.6 Counters and Timers

A counter/timer register can count the number of digital pulses applied to its input. If a clock signal of known frequency is used, it becomes a timer, because the duration of a count is equal to the count value multiplied by the clock period. Like the data register, the counter/timer register is made from bistable units, but connected in toggle mode, so that each stage drives the next. Each stage outputs one pulse for every two pulses which are input, so the output pulse frequency is half the input frequency. The counter/timer register can therefore be viewed as a binary counter or frequency divider, depending on the application.

Figure 4.6 shows an 8-bit counter/timer, with the input to the LSB at the right. The binary count stored is incremented each time the LSB is pulsed. Two pulses have been applied, so the counter shows binary 2. After 255 pulses have been applied, the counter will 'roll over' from 11111111 to 00000000 on the next pulse. A signal may be output to indicate this, which can be used as a 'carry out' in counting operations or 'time-out' in timing operations. In a microprocessor system, the 'time out' signal typically sets a bit in a 'status' register to record this event, and an 'interrupt' signal may be generated each time. There will be more on interrupts later.

If the pulse frequency is 1 MHz, the period will be 1 μs, and the counter will generate a time-out signal every 256 μs. If the counter can be pre-loaded, we can make it time out after some other number of input pulses. For example, if pre-loaded with a count of 56, it will time out after 200 μs. In this way, known time intervals can be generated in microprocessor systems. In conventional microprocessor systems, the I/O ports often contain timers that the processor uses for timing operations. The PIC 16F84 has one 8-bit counter/timer, with a 'prescaler' that divides the input frequency by a factor of between 2 and 256 in order to extend its range. The program counter is another example of a binary counter, which is incremented to keep track of the current address as a program is executed.

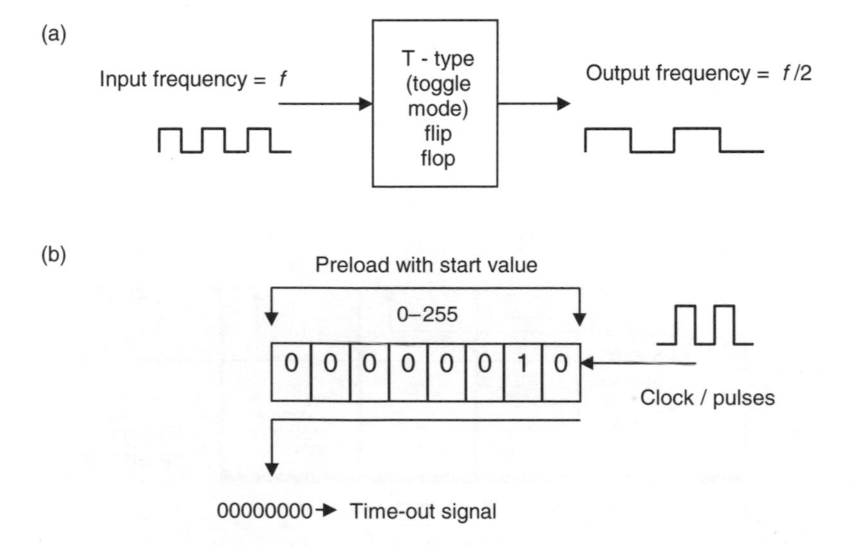

Figure 4.6 Counter/Timer register operation. (a) Toggle mode stage; (b) 8-stage Counter register.

4.7 Serial and Shift Registers

The general purpose data register, as described above, is loaded and read in parallel. A shift register consists of a set of data latches that are connected so that a data bit fed into one end can be moved from one stage to the next, under the control of a clock signal. An 8-bit shift register can therefore store a data byte which is read in 1 bit at a time from a single data line. The data can then be shifted out again, 1 bit at a time, or read in parallel. Alternatively, the register could be loaded in parallel and the data shifted out onto a serial output line.

In Fig. 4.7(a), the 8-bit shift register is fed data from the right. The shift clock has to operate at the same rate as the data, so that the register samples the data at the serial data input at the right time. As each bit is read in, the preceding bits are shifted left to allow the next bit into the LSB. The timing diagram shows the data being sampled and shifted on to the falling clock edge; note that only the state of the input at the sampling instant is registered, so the short pulse is ignored.

This type of register is used in a microprocessor serial port, where data is sent or received in serial form. In the PC, this could be the modem or network port, the keyboard input or VDU output.

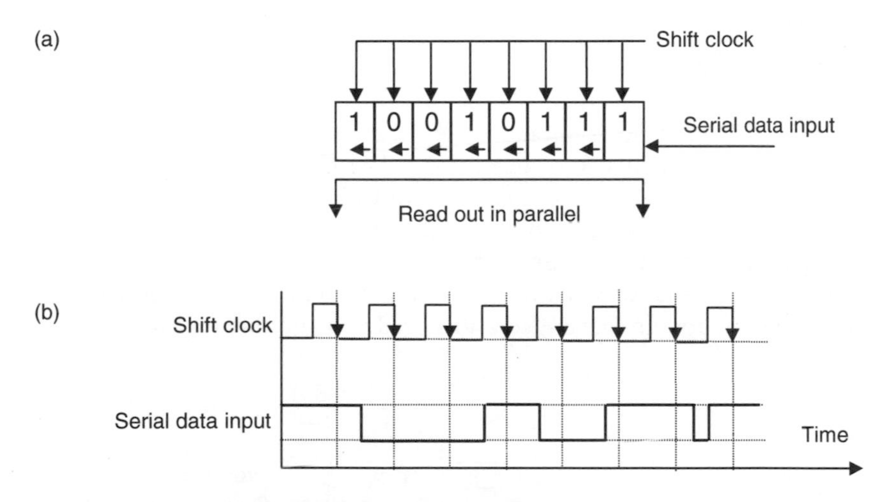

Figure 4.7 (a) Shift register operation; (b) Timing diagram.

4.8 Arithmetic and Logic Unit (ALU)

The main function of any processor system is to process data, for example, to add two numbers together. The arithmetic and logic unit (ALU) (Fig. 4.8) is therefore an essential feature of any microprocessor or microcontroller.

A binary adder block has already been described, and this would be just one of the functions of an ALU. The ALU takes two data words as inputs and combines them together by adding, subtracting, comparing or carrying out logical operations such as AND, OR, NOT, XOR on the corresponding pairs of bits. The operation to be carried out is determined by function select inputs. These in turn are derived from the instruction code in the program being executed. The broad arrows used in the diagram indicate the

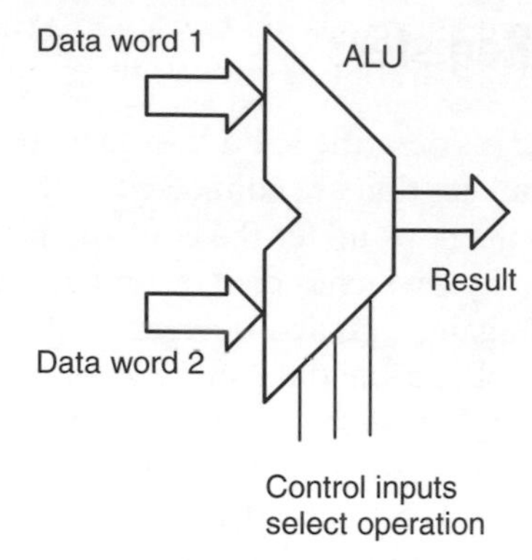

Figure 4.8 Arithmetic and logic unit.

parallel data paths, which carry the operands to the ALU, and the result away. A set of data registers that store the operands is usually associated with the ALU, as seen in the 4-bit data system.

4.9 Processor Control

The instruction decoder is a logic circuit in the CPU that uses the instruction codes from the program to determine the sequence of operations. The decoder output lines, which are connected to the registers, ALU, gates and other control logic, are set up for a particular instruction to be carried out (e.g. add two data bytes). The processor control block (Fig. 4.9) also includes timing control and other logic to manage the processor operations. The clock signal drives the sequence of events, so that after a certain number of clock cycles, the results of the instruction are generated and stored in a suitable register or back in the memory.

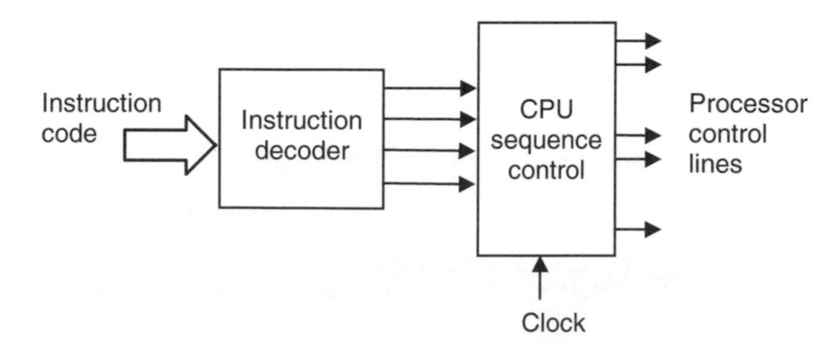

Figure 4.9 CPU control logic.

Summary

- An encoder generates a binary code corresponding to the active numbered input; the decoder carries out the inverse operation, generating a selected output from a binary input.

- The multiplexer allows a selected input to be connected to a single output line; the demultiplexer carries out the inverse operation, connecting a single input line to a selected output line.
- A register or memory cell stores one bit of data, using a data latch and bidirectional buffer.
- Numbered memory locations are accessed by decoding the address to generate a system chip select and chip location select.
- Counters and timers use a counting register to count digital pulses, or measure time intervals using a clock input.
- Shift registers convert parallel to serial data, and back.
- The ALU provides the data processing operations.
- The processor control signals are generated by the instruction decoder and timing circuits.
- The clock signal provides the timing reference signal for all processor operations.

Questions

1. Describe the process whereby an encoder and decoder could be used to scan a 4 × 32 key computer keyboard.

2. Two 8-bit registers, A and B, are connected to an 8-bit data bus via bidirectional buffers, so as to allow data to be stored and retrieved. Draw a block diagram, and explain the sequence of signals required from the controller circuits to transfer the contents from A to B, using data direction select and data latching signals.

3. A minimal microprocessor system, configured as shown in Fig. 4.5, has a 16-bit address bus. The two most significant lines, A14 and A15 are connected to the 2-bit decoder, which operates as specified in Fig. 4.1(b). The four select outputs are connected to memory and I/O chip select inputs as follows:

Output	Chip
0	RAM1
1	RAM2
2	ROM
3	I/O

   The RAM1 chip is selected in the range of addresses from 0000 to 3FFF (hex). Work out the lowest address where each of the three remaining chips are selected.

4. Calculate the number of locations in a memory chip which has 12 address pins.

5. Calculate the time interval generated by an 8-bit timer pre-loaded with the value 11001110 and clocked at 125 kHz.

Answers

3. 4000, 8000, C000.

4. 4096.

5. 400 μs.

Activities

1. In a suitable TTL logic device data book or suppliers catalogue, look up the chip numbers and internal configuration of the medium scale ICs: 3- to 8-line Decoder, Octal D Latch, Octal Bus Transceiver, 8-bit Shift Register, 4-bit Binary Counter; also identify the largest capacity RAM chip listed in your source.

2. Refer to the PIC 16F8X block diagram, Fig. A.1 in Appendix A. State the function of the following features: ROM program memory, RAM file registers 68 × 8, Program Counter, Instruction register, Instruction Decode and Control, MUX, ALU, W reg, I/O ports, TMR0.

3. Refer to the PIC 16F8X data sheet, Fig. A.10. Identify a: FET, OR gate, tri-state gate, transparent data latch, edge-triggered data latch. Work out how input data would be transferred onto the internal data bus line from the I/O pin.

Chapter 5
Microcontroller Operation

To understand the operation of a microcontroller requires some knowledge of both the internal hardware arrangement and the instruction set that it uses to carry out the program operations. In this chapter we will look at a general hardware configuration and some basic features of machine code programs.

5.1 Microcontroller Architecture

The architecture (internal hardware arrangement) of a complex chip is best represented as a block diagram. This allows the overall operation to be described without having to analyse the circuit, which will be very complex, at component level. The PIC data sheet contains the definitive block diagram of the PIC 16F84 (Fig. A.1), but simplified versions will be used later to help explain particular aspects of PIC 16F84 operation. First, however, we will look at a general block diagram that shows some of the common features of microcontrollers (Fig. 5.1).

The block diagram shows a general microcontroller that can be considered in two parts, the program execution section and register processing section. This division reflects the PIC architecture, where the program and data are accessed separately. This arrangement increases overall program execution speed, and is known as a Harvard architecture.

The program execution section contains the program memory, instruction register and control logic which store, decode and execute the program. The register processing section has special registers used to set up the processor options, data registers to store the current data, port registers for input and output, and the ALU to process the data. The timing and control block co-ordinates the operation of the two parts as determined by the program instructions, and responds to external control inputs, such as the reset.

5.1.1 ROM Program Memory

The control program is normally stored in non-volatile read-only memory. Microcontrollers that are designed for prototyping and short production runs typically have

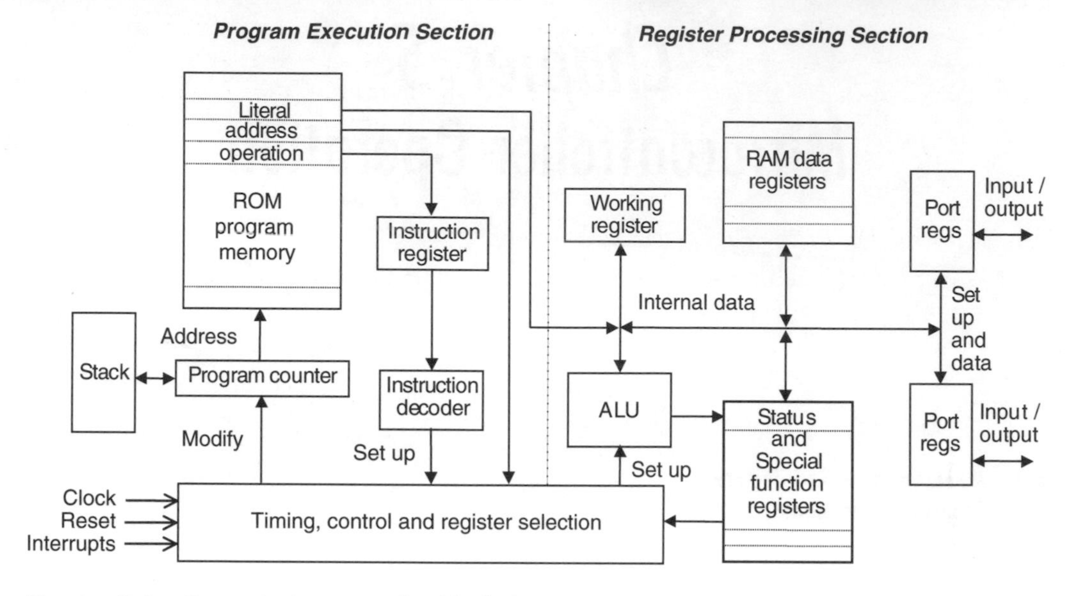

Figure 5.1 General microcontroller block diagram.

erasable programmable ROM (EPROM) into which the program can be 'blown' using a suitable programming unit. EPROM can be erased and reprogrammed, but the chip must be removed and placed under an ultraviolet lamp to clear an existing program.

The PIC 16F84 is convenient for learning programming and prototyping applications because it has flash ROM program memory. This type of memory can be reprogrammed without any special erasing process, and this can even be done while the chip is still in the application circuit. Usually, however, the chip is placed in a programming unit attached to the host computer for program downloading, prior to fitting it in the target board. The program is stored in a set of numbered locations (addresses), starting at zero. The PIC program has one complete instruction in each location.

5.1.2 Program Counter (PC)

The program counter is a register which keeps track of the program sequence, by storing the address of the instruction currently being executed. The default start address of the program is usually zero; this is where the first instruction in the program will be stored unless the programmer specifies otherwise. The program counter is automatically loaded with zero when the chip is powered up, or reset. Sometimes, it is necessary to jump from address zero to the start of the actual program at a higher address, because control words must be stored in specific low addresses. For instance, the PIC 16F84 uses address 004 to store the 'interrupt vector' (we will see how interrupts work later). However, this problem can be ignored for programs which do not use interrupts, and therefore our simple programs can be located at address zero.

As each instruction is executed, the program counter is incremented (increased by one) to point to the next instruction. Program jumps are achieved by changing the program counter to point to an instruction other than the next in sequence. For instance, if a branch back by three instructions is required, 3 is subtracted from the contents of the PC.

Associated with the program counter is the 'stack'. This is a temporary program counter store. When a subroutine is executed (see Section 5.2.3), this set of registers stores the current address, so that it can be recovered at a later point in the program. It is called a stack, because the addresses are restored to the PC in the reverse order to which they were stored, that is, 'last in, first out' (LIFO), like a stack of plates.

5.1.3 *Instruction Register (IR) and Decoder*

To execute an instruction, the processor copies the instruction code from the program memory into the instruction register. It can then be decoded by the instruction decoder, which is a combinational logic block which sets up the processor control lines as required. These control lines are not shown explicitly in the block diagram, as they go to all parts of the chip.

In the PIC, the instruction code includes the operand, which may be a literal value or a register address. For example, if a literal given in the instruction is to be loaded into the Working (W) register, it is placed on an internal data bus and the W register latch enable lines are activated by the timing and control logic. The internal data bus can be seen in the manufacturer's block diagram (Fig. A.1) in the data sheet.

5.1.4 *Timing and Control*

This sequential logic block provides overall control of the chip, and from it control signals go to all parts of the chip to move the data around and carry out logical operations and calculations. A clock signal is needed to drive the program sequence; it is usually derived from a crystal oscillator, which provides an accurate, fixed frequency signal. A maximum frequency of operations is always specified, e.g. 10 MHz. The PIC 16F84 can operate at any frequency below this maximum, down to zero.

The reset input can restart the program at any time by clearing the program counter to zero. If the program runs in a continuous loop, and there is no instruction to exit the loop, the reset may be needed. However, it is not essential to connect an active reset input because the program will start automatically at program ROM address zero, as long as the reset input is connected in its active state, which is usually high (MCLR = 'high' in the PIC). In most of the sample programs in this book, it is assumed that the chip will be switched off to restart the program, and reset switch is not required.

The only other way to stop or redirect a continuous loop is via an 'interrupt'. Interrupts are signals generated externally or internally, which force a change the sequence of operations. If an interrupt source goes active in the PIC 16F84, the program will restart at address 004, where the sequence known as the 'interrupt service routine' is stored. More details are provided in Chapter 9.

5.1.5 *Working (W) Register*

In some microcontrollers and microprocessors this is called the Accumulator (A), but the name Working register used in the PIC system is a better description. It holds the data that the processor is working on at the current time, and most data has to pass through it. In the PIC, for instance, if a data byte is to be transferred from the port register to a RAM data register, it must be moved into W first. The Working register or Accumulator works closely with the ALU in the data processing operations.

5.1.6 Arithmetic and Logic Unit (ALU)

This is a combinational logic block that takes one or two input binary words and combines them to produce an arithmetic or logical result. In the PIC, it can operate directly on the contents of a register, but if a pair of data bytes are being processed (for instance, added together), one must be in W. The ALU is set up according to the requirements of the instruction being executed by the timing and control block. Typical ALU/register operations are described later in this chapter.

5.1.7 Port Registers

Input and output in a microcontroller are achieved by simply reading or writing a port data register. If a binary code is presented to the input pins of the microcontroller by an external device (e.g. a set of switches), the data immediately becomes the content of the register allocated to that port. This input data can then be moved into another register for processing. If a port register is initialized for output, the data moved to that register is immediately available at the pins of the chip, and can be displayed, for example, on a set of LEDs. Each port has a 'data direction' register associated with its data register. This allows each pin to be set individually as an input or output before the data is read from or written to the port data register.

In the PIC 16F84, there are two ports, A and B. Port A has five pins and Port B has eight. A '0' in the data direction register sets the port bit as an output, and a '1' sets it as an input. These port registers are mapped (addressed) as special function registers.

5.1.8 Special Function Registers

These registers provide dedicated program control registers, such as stack pointer, timers, interrupt and index registers. In the PIC, the program counter, port registers and spare registers are mapped as part of this block. The Working register is the only one that is not numbered in the main register block and is accessed by name, 'W'.

The processor will also contain control registers whose bits are used individually to set up the processor operating mode or record significant results obtained from those operations. For example, microprocessors invariably have a Status Register, which contains a Zero (Z) Flag. This bit is automatically set to '1' if the result of any operation on a register is zero. The Carry (C) Flag is another standard status flag; it is set if there is a carry out of the most significant bit of a register as a result of an arithmetic operation. The status register bits are often used to control the program sequence via conditional branching (see Section 5.2, Program Operations). An example of a control bit is the Interrupt Enable flag. In the PIC, Global Interrupt Enable (GIE) must be set as part of the program initialization if interrupts are to be used. Interrupts allow external events to change the program execution sequence, and are explained more fully in Chapter 9, Section 9.3.

5.2 Program Operations

We have seen in Chapter 2 that a machine code program consists of a list of binary codes stored in the microcontroller memory. They are decoded in sequence by the processor, which generates control signals that set up the microcontroller to carry out the instruction. Typical operations are

- load a register with a given number (a 'literal');
- move (i.e. copy) data between registers;
- carry out an arithmetic or logic operation on a data word, or pair of data words;
- test a bit or whole data word in a register;
- jump to another point in the program;
- test and jump conditionally, depending on the result of the test;
- jump to a subroutine, storing the return address;
- carry out a special control operation.

The machine code program must be made up only from those binary codes which the instruction decoder will recognize as meaningful. These codes could be worked out manually from the instruction set given in the data sheet (Table A.4). When computers were first developed, this was indeed how the program was entered, using a set of switches. This is obviously time consuming and inefficient, and it was soon realized that it would be useful to have a software tool that would generate the machine code automatically from a program that was written in a more user-friendly form. Assembly language programming was therefore developed, when hardware had moved on enough to make it practicable.

Assembly language allows the program to be written using mnemonic ('designed to aid the memory') code words. Each processor has its own set of instruction codes and corresponding mnemonics. For example, a commonly used instruction mnemonic in PIC programs is 'MOVWF', which means move (actually copy) the contents of the Working register (W) to a file register which is specified as the 'operand'. The destination register is specified by number (file register address), such as 0C (the first general purpose register in the PIC 16F84). The complete instruction is:

```
MOVWF 0C
```

There are two main types of instruction: data processing operations and program sequence operations. Data processing operations include:

MOVE	copy data between registers
REGISTER	manipulate data in a register
ARITHMETIC	combine register pairs arithmetically
LOGIC	combine register pairs logically.

Program sequence control operations include:

UNCONDITIONAL JUMP	jump to a specified destination
CONDITIONAL JUMP	jump, or not, depending on a test
CALL	jump to a subroutine and return afterwards
CONTROL	miscellaneous operations.

Together, these types of operations allow inputs to be read and processed, and the results stored or output, or used to determine the subsequent program sequence.

Table 5.1 Single register operations

Operation	Before		After	Comment
CLEAR	0101 1101	→	0000 0000	Reset all bits to zero
INCREMENT	0101 1101	→	0101 1110	Increase binary value by one
DECREMENT	0101 1101	→	0101 1100	Decrease binary value by one
COMPLEMENT	0101 1101	→	1010 0010	Invert all bits
ROTATE LEFT	0101 1101	→	1011 1010	Shift all bits left by one place, replace MSB in LSB
SHIFT RIGHT	0101 1101	→	0010 1110	Shift all bits right by one, losing the LSB
CLEAR BIT	0101 1101	→	0101 0101	Reset bit (3) to 0
SET BIT	0101 1101	→	1101 1101	Set bit (7) to 1

5.2.1 Single Register Operations

The processor operates on data stored in registers, which typically contain 8 bits. The data can originate in three ways:

1. a literal (numerical value) provided in the program;
2. an input via a port data register; or
3. the result of a previous operation.

This data can be processed using the set of instructions defined for that processor. Table 5.1 shows a basic set of operations which can be applied to a single register. The same binary number is shown before processing, and then after the operation has been applied to the register.

As an example of how these operations are specified in mnemonic form in the program, the assembler code to increment a PIC register is:

```
INCF 06.
```

Register number 06 happens to be port B data register, so the effect of this instruction can be seen immediately at the I/O pins of the chip. The corresponding machine code instruction is '0A86' in hexadecimal, or '00 1010 1000 0110' in binary (14 bits). As you can see, it is easier to recognize the mnemonic form!

5.2.2 Register Pair Operations

Table 5.2 shows the basic operations that can be applied to pairs of registers. Normally, the result is retained in one of the registers, which is referred to as the destination register. A binary code to be combined with the contents of the destination register is obtained from the source register. The source register contents remain unchanged after the operation. The meaning of each type instruction is explained below, with an example from the PIC instruction set.

Table 5.2 Operations on register pairs

Operation		Register before	Register after	Comment
MOVE				Copy operation.
	Source	0101 1100	101 1100	Overwrite destination with source,
	Destination	xxxx xxxx	101 1100	leaving source unchanged.
ADD				Arithmetic operation.
	Destination	001 0010	110 1110	leaving source unchanged,
	Source	101 1100	101 1100	Add source to destination.
SUB				Arithmetic operation.
	Destination	101 1100	100 1010	Subtract source from destination,
	Source	001 0010	001 0010	leaving source unchanged.
AND				Logical operation.
	Destination	101 1100	001 0000	AND source and destination bits
	Source	001 0010	001 0010	leaving source unchanged.
OR				Logical operation.
	Destination	101 1100	101 1110	OR source and destination bits
	Source	001 0010	001 0010	leaving source unchanged.
XOR				Logical operation.
	Destination	101 1100	100 1110	Exclusive OR source and destination bits
	Source	001 0010	001 0010	leaving source unchanged.

Move Operation

The most commonly used instruction in any program simply moves data from one register to another. It is actually a copy operation, as the data in the source register remains unchanged until overwritten or the processor is reset.

```
MOVF 0C,W
```

This moves the contents of register 0C (12) into the Working register.

Arithmetic Operation

Add and subtract are the basic arithmetic operations, carried out on binary numbers. Some processors also provide multiply and divide in their instruction set, but these can be created if necessary by using shift, add and subtract operations in a suitable sequence.

```
ADDWF 0C
```

This instruction adds the contents of W to register 0C.

Logical Operations

Logical operations act on the corresponding pairs of bits in the destination and source registers, with the result normally being retained in the destination, leaving the source unchanged. The result in each bit position is obtained as if the bits had been fed through the equivalent logical gate (see Chapter 3).

```
ANDLW 001
```

This carries out an AND operation on the corresponding pairs of bits in the binary number in W and the binary number 00000001, leaving the result in W. The result is zero if the LSB in W is zero. This type of operation can be used for bit testing if the processor does not have a bit test instruction.

5.2.3 Program Control

As we have already seen, the microcontroller program is a list of binary codes in the program memory that are executed in sequence. The sequence is controlled by the program counter. Most of the time, the PC is simply incremented by one to execute the next instruction. However, if a program jump (branch) is needed, the PC must be modified, that is, the address of the destination instruction must be loaded into the PC, replacing the existing value. This new address can be given as a jump instruction operand, or calculated for the current address, for example, by adding a number to the current PC value.

The program counter is cleared to zero when the chip is reset or powered up for the first time, so program execution starts at address 0000. If the clock is running, the first instruction will be executed. During the execution cycle, the PC is incremented to 0001, so that the processor is ready to execute the next instruction. This process is repeated unless there is a jump instruction.

The jump instructions must have a destination address or offset as the operand. This can be given a numerical address, but this would mean that the instructions would have to be counted up by the programmer to work out this address. So, as we will see later in the program examples, a destination address is usually specified in the program source code by using a recognizable label, such as 'again', 'start' or 'wait', in the same way that mnemonics are used to represent the binary machine code instructions. The assembler program then replaces the label with the actual address when the assembler code is converted to machine code.

Program sequence control operations are illustrated in Figs 5.2, 5.3 and 5.4. The diagrams show the program memory from address zero, with jump instructions at address 0002.

Unconditional Jump

The unconditional jump (Fig. 5.2) forces a jump to another point in the program every time it is executed. This is carried out by replacing the contents of the program counter with the address of the destination instruction, 0005; execution continues from the new address.

The unconditional jump is often used at the very end of a program to go back to the beginning of the sequence, and keep repeating it. It may also be used to jump over subroutines that are placed at the beginning of a program. In the PIC, we will see that the unconditional jump instruction 'Goto address/label' is used with 'Bit Test and Skip' instructions to create conditional jumps.

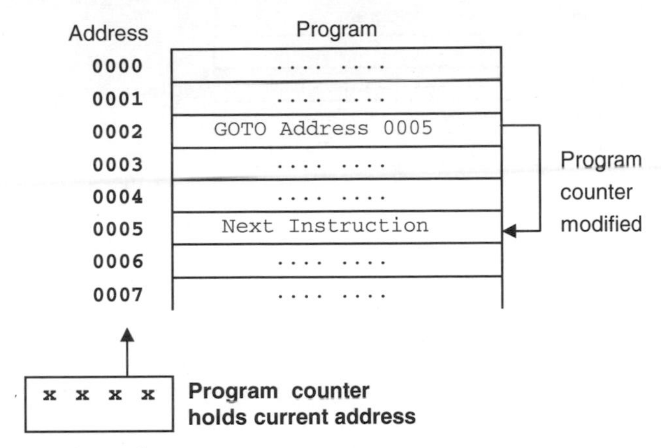

Figure 5.2 Unconditional jump.

On the question of terminology, 'jump' and 'branch' are two terms for describing sequence control operations, but the ways in which they work are slightly different. A 'branch' is made relative to the current address, by adding to the current value in the PC. A 'jump' uses absolute addressing, that is, the contents of the PC are replaced with the destination address. The PIC is a RISC (reduced instruction set) processor and does not provide branch instructions in its basic instruction set, but such operations can be created from the available instructions, if required. In the PIC, for example, the program counter can be modified directly using a register processing operation to create a relative jump, or branch.

The unconditional jump may be illustrated by a program sequence description:

```
start First program instruction
      Next program instruction
      etc....
      ....
      ....
      Last program instruction
      GOTO start
```

Note the use of the label '`start`' in the first column to identify the jump destination. This is the technique normally used in assembler programming. 'GOTO' is the actual mnemonic used in PIC assembler for the unconditional jump. The program will execute continuously until the processor is reset or switched off. The program description used above is a form of 'pseudocode', which shows the program operations and sequence without using any specific programming language.

Conditional Jump

The conditional jump instruction (Fig. 5.3) is required for making decisions in the program. Instructions to change the program sequence depending on, for instance, the result of a calculation or a test on an input, are an essential feature of any microprocessor.

In complex instruction-set processors, conditional jump instructions that test specific bits in the CPU status register are normally available. When the CPU operates on some data in a register, the status register bits record certain results, such as whether the result was zero

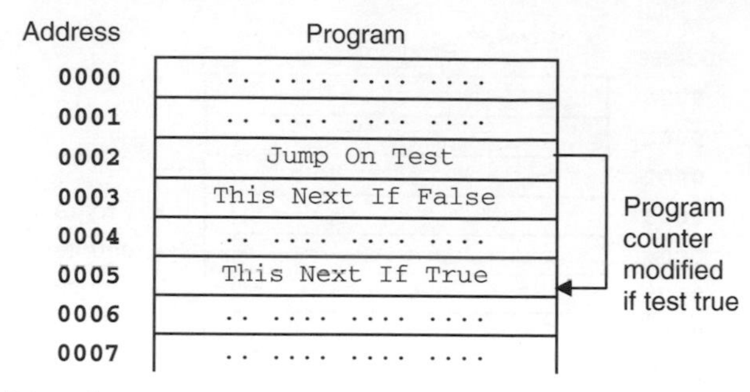

Figure 5.3 Conditional jump.

or not. In this case, if the result of a previous instruction was zero, a 'zero flag' is set to 1 in the status register.

Since the PIC is a RISC processor, designed with a minimal number of instructions, the conditional branch has to be made up from two simpler instructions. The first instruction tests a bit in a register and then skips (misses out) the next instruction, or not, depending on the result. This next instruction is a jump instruction (GOTO or CALL). Thus, program execution continues either at the instruction following the jump, if the jump is skipped, or at the jump destination. Pseudocode for a typical use of the conditional jump, a delay routine, would look like this:

```
        Allocate 'Count' Register
        ....
        ....
        Load 'Count' register with literal XX
again   Decrement 'Count' register
        Test 'Count' register for zero
        If not zero, jump to label 'again'
        Next Instruction
        ....
```

This software timing loop simply causes a time delay in the program, which is useful, for example, for outputting signals at specific intervals. A register is used as a down counter, by loading it with a number, XX, and decremented repeatedly until it is zero. A test instruction then detects that the zero flag has gone active, and the loop is terminated. Each instruction takes a known time to execute, therefore the delay can be calculated.

Subroutine CALL

Subroutines are used to carry out discrete program functions. They allow programs to be written in manageable, self-contained blocks, which can then be executed as required. The instruction CALL is used to jump to a subroutine, which must be terminated with the instruction RETURN.

CALL has the address of the first instruction in the subroutine as its operand. When the CALL instruction is decoded, the destination address is copied to the PC, as for the GOTO instruction. In addition, the address of the next instruction in the main program is saved in the stack. In the PIC, this is a special block of RAM used only for this purpose, but in conventional processors, a part of the main RAM may be set aside for this purpose. The return

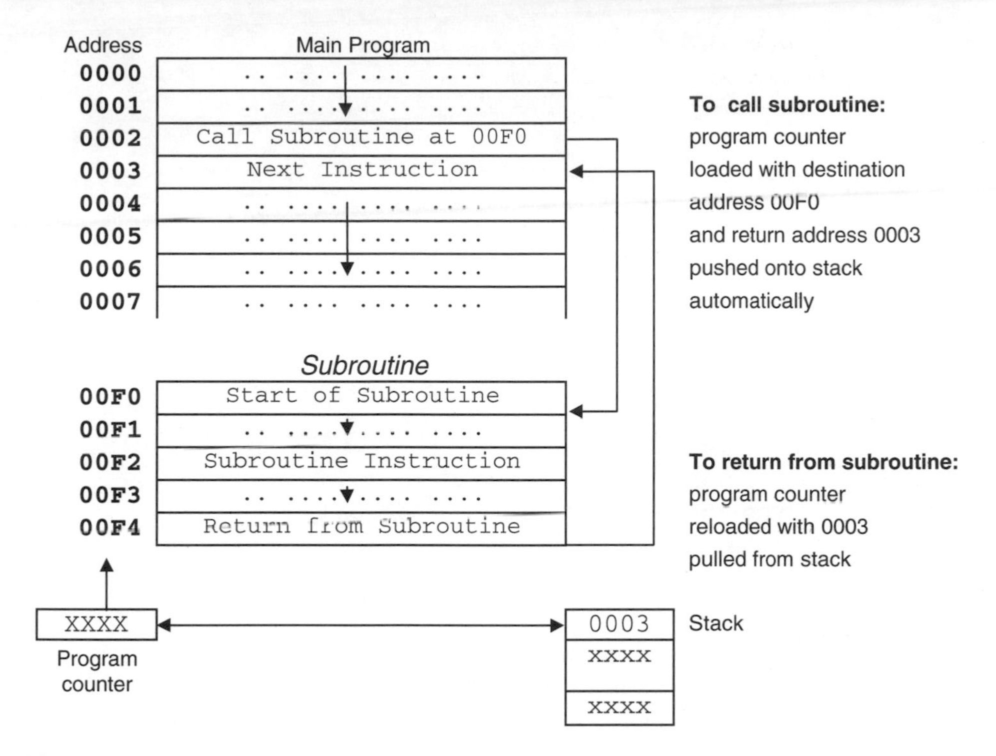

Figure 5.4 Subroutine call.

address is 'pushed' onto the stack when the subroutine is called, and 'popped' back into the program counter at the end of the routine, when the RETURN instruction is executed.

In Fig. 5.4, the subroutine is a block of code which whose start address has been defined by label 00F0. The CALL instruction at address 0002 contains the destination address as its operand. When this instruction is encountered, the processor carries out the jump by copying the destination address (00F0) into the program counter. Prior to this, the address of the next instruction in the main program (0003) is pushed onto the stack, so that the program can come back to the original point after the subroutine has been executed.

One advantage of using subroutines is that the block of code can be used more than once in the program, but only needs to be typed in once. In the PIC program SCALE (see Chapter 12), a delay loop is written as a subroutine. It is a counting loop which uses up time to give a delay between output changes, which is 'called' twice within a loop which sets an output high, delays, sets the output low, and delays again before repeating the whole process. The same program also contains an example of direct modification of the program counter (labelled PCL) to create a data table.

Pseudocode for a delay loop written as a subroutine would be as follows:

```
; Program DELTWICE ********************************

        Allocate 'Count' Register
        ....
        ....
        Load 'Count' register with value XX
```

```
            CALL 'delay'
            Next Instruction
            ....
            ....
            Load 'Count' register with value YY
            CALL 'delay'
            Next Instruction
            ....
            ....
            END of Program

; Subroutine DELAY *******************************

      delay Decrement 'Count' register
            Test 'Count' register for zero
            If not zero, jump to label 'delay'
            RETURN from subroutine

; End of code ************************************
```

Note that the 'delay' routine is called twice, but using a different delay value in the 'Count' register. Thus, the same code can be used to give different delay times. Notice also that we have started using comments in our pseudocode to identify the functional blocks as the pseudocode gets more complex.

Summary

- The typical microcontroller contains a program execution section, and a register processing section.
- The program counter steps through the program ROM addresses, and the instructions are decoded and executed.
- Data is transferred via port registers, stored in RAM/registers and processed in the ALU.
- Special function registers hold control, set-up and status information.
- Instructions move or process data, or control the execution sequence.
- The content of the data registers is manipulated as single data words, or using register pairs.
- Program jumps can be unconditional or conditional, using bit testing or status bits to determine the sequence.
- Subroutines are distinct program blocks which operate using call, execute and return.

Questions

1. Outline the sequence of program execution in a microcontroller, describing the role of the program ROM, program counter, instruction register, instruction decoder, and timing and control block.

2. A register is loaded with the binary code 01101010. State the contents of the register after the following operations on this data: (a) clear, (b) increment, (c) decrement, (d) complement, (e) rotate right, (f) shift left, (g) clear bit 5, (h) set bit 0.

3. A source register is loaded with the binary code 01001011, and a destination register loaded with the 01100010. State the contents of the destination register after the following operations: (a) Move, (b) Add, (c) AND, (d) OR, (e) XOR.

4. In a microcontroller program, a subroutine starts at address 016F and ends with a 'return' instruction at address 0172. A 'call subroutine' instruction is located at address 02F3. Assuming that the microcontroller has one complete instruction in each address, list the changes in the contents of the program counter and stack between the time of execution of the instruction before the call and the instruction following the call.

5. Write a pseudocode program for the process by which two numbers, say 4 and 3, could be multiplied by successive addition. Use the register instructions Clear, Move, Add, Decrement, Test for Zero and Jump if Zero to Label.

Answers

2. (a) 00000000, (b) 01101011, (c) 01101001, (d) 10010101, (e) 00110101, (f) 11010100, (g) 01001010, (h) 01101011

3. (a) 01001011, (b) 10101101, (c) 01000010, (d) 01101011, (e) 00101001

4.

PC	Stack	
....	....	
02F2	XXXX	Instructions
02F3	XXXX	before Call
016F	02F4	Subroutine Start
0170	02F4	Subroutine
0171	02F4	instructions
0172	02F4	Return
02F4	XXXX	Instructions
02F5	XXXX	after Call
....	....	

5.

```
Allocate registers A,B,C
Clear register A
Move 4 into register B
Move 3 into register C
Loop1      Add B to A
           Decrement C
           Test C for zero
           Jump back to 'Loop1' if C not zero
Finished with product in A
```

Activities

1. Study the PIC 16F8X block diagram (Appendix A, Fig. A.1), and identify the features described in Section 5.1.

2. Study the PIC instruction set (Appendix A, Table A.4) and try to allocate the instructions to the following categories: Move, Arithmetic, Logic, Jump and other Control.

Part B
PIC Microcontroller

Chapter 6
A Simple PIC Application

A very simple machine code program for the PIC will now be developed to show how the chip is used, avoiding complicating factors as far as possible. The program, called BIN1, will output a binary count at Port B of the PIC 16F84. A simplified internal architecture will be used to explain the execution of the program, and the program will then be developed further as BIN2, BIN3 and BIN4, with new programming techniques being added at each step.

6.1 BIN Hardware

The first step in developing our first application is to note the pin-out of the microcontroller, so that the connection of the external components can be worked out. We are going to connect LEDs to Port B, which must therefore operate as outputs, and switches to Port A to provide manual inputs. The only other components needed are a capacitor and resistor for the clock circuit.

6.1.1 PIC 16F84 Pin-out

The PIC 16F84 microcontroller is supplied in an 18-pin DIL (dual in line) chip. Simplified pin labelling, taken from the PIC data sheet, is shown in Fig. 6.1 (some of the pins have dual functions which will be discussed later).

The chip has two ports, A and B. The port pins allow data to be input and output as digital signals, at the same voltage levels as the supply that is connected to Vdd and Vss. CLKIN and CLKOUT are used to connect clock circuit components, and the chip then generates a fixed frequency clock signal that drives all its operations. !MCLR ('NOT Master CLeaR') is a reset input, which can (optionally) be used to restart the program. Note that the active low operation of this input is indicated by a bar over the pin label. An exclamation mark at the beginning of the pin label means the same thing. In many applications, this input does not need to be used, but it *must* be connected to the positive supply rail to allow the chip to run. A summary of the pin functions is provided in Table 6.1.

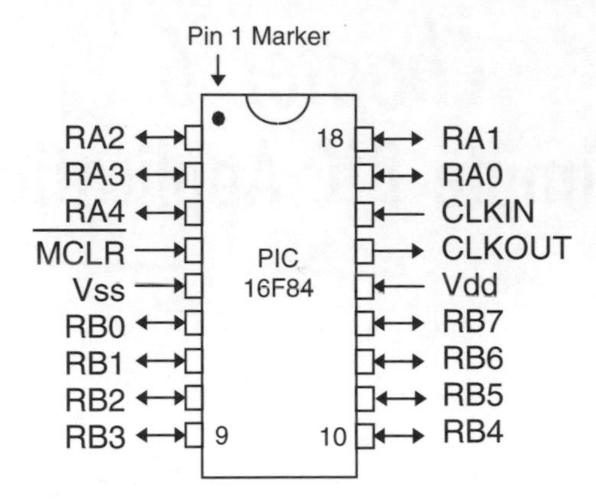

Figure 6.1 Pin-out of PIC 16F84.

6.1.2 BIN Hardware Block Diagram

The simple demonstration application will be developed in stages to run on a minimal hardware system, consisting of the PIC 16F84 chip and a few external components. The hardware arrangement required for any given application can be represented in a simplified form as a block diagram, in this case, Fig. 6.2. The main parts of the hardware and relevant inputs and outputs should be identified in the block diagram, together with the direction of signal flow. The nature of the signals may be described with labels or illustrated with simple

Table 6.1 PIC 16F84 pins arranged by function

Pin	*Label*	*Function*	*Comment*
14	Vdd	Positive Supply	+5 V nominal, 3 V–6 V allowed
5	Vss	Ground Supply	0 volts
4	!MCLR	Master Clear	Active Low Reset Input
16	CLKIN	Clock Input	Connect RC Clock Components to 16
15	CLKOUT	Clock Output	Connect Crystal Oscillator to 15 and 16
17	RA0	Port A, Bit 0	Bidirectional Input/Output
18	RA1	Port A, Bit 1	Bidirectional Input/Output
1	RA2	Port A, Bit 2	Bidirectional Input/Output
2	RA3	Port A, Bit 3	Bidirectional Input/Output
3	RA4	Port A, Bit 4	Bidirectional Input/Output + TMR0 Input
6	RB0	Port B, Bit 0	Bidirectional Input/Output + Interrupt Input
7	RB1	Port B, Bit 1	Bidirectional Input/Output
8	RB2	Port B, Bit 2	Bidirectional Input/Output
9	RB3	Port B, Bit 3	Bidirectional Input/Output
10	RB4	Port B, Bit 4	Bidirectional Input/Output + Interrupt Input
11	RB5	Port B, Bit 5	Bidirectional Input/Output + Interrupt Input
12	RB6	Port B, Bit 6	Bidirectional Input/Output + Interrupt Input
13	RB7	Port B, Bit 7	Bidirectional Input/Output + Interrupt Input

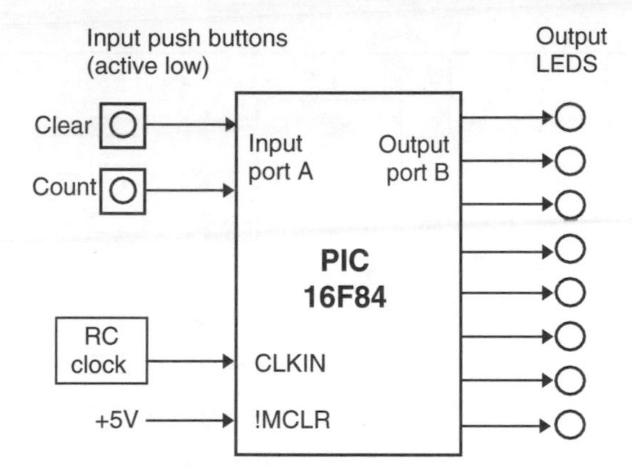

Figure 6.2 Block diagram of BIN hardware.

diagrams. The power supplies need not be shown. The idea is to outline the basic hardware arrangement without having to design the circuit in detail at this stage.

Port A (5 bits) and Port B (8 bits) give access to the data registers of the ports, the pins being labelled RA0–RA4, and RB0–RB7, respectively. A pair of push button switches will be connected to RA0 and RA1, and a set of LEDs connected to RB0–RB7. The switches will be used later in BIN3 to control the output sequence. RA1 will be programmed to act as a 'run' input, enabling the binary count, while RA0 will provide a 'reset' input to restart the output sequence. However, these inputs will not be used in the first program, BIN1. The connections required are shown in Table 6.2.

Table 6.2 PIC 16F84 pin allocation for BIN application

Pin	*Connection*
Vss	0V
Vdd	+5V
!MCLR	+5V
CLKIN	CR circuit
CLKOUT	not connected (n/c)
RA0	Reset Switch
RA1	Count Switch
RA2	n/c
RA3	n/c
RA4	n/c
RB0	LED Bit 0
RB1	LED Bit 1
RB2	LED Bit 2
RB3	LED Bit 3
RB4	LED Bit 4
RB5	LED Bit 5
RB6	LED Bit 6
RB7	LED Bit 7

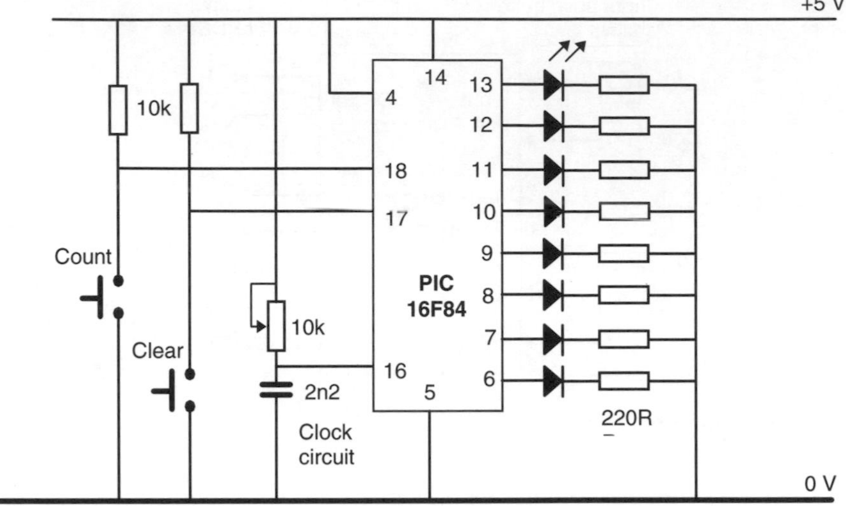

Figure 6.3 Circuit diagram of BIN hardware.

The block diagram can now be converted into a circuit diagram. The input and output circuits have already been introduced in Section 3.4. A simple RC clock circuit will be used, as the frequency of operation is not critical. The circuit diagram is shown in Fig. 6.3.

6.1.3 BIN Circuit Operation

Active low switch circuits, consisting of normally open push buttons and pull-up resistors, are connected to the control inputs. The resistors ensure that the inputs are high when the buttons are not pressed. The outputs are connected to LEDs in series with current limiting resistors. The PIC outputs are capable of supplying enough current to drive LEDs directly, unlike standard microprocessor port chips, making the circuit relatively simple. The external clock circuit consists of a capacitor (C) and resistor (R) in series. The value of C and R multiplied together will determine the chip clock rate. The resistance in this circuit has been made variable, and the values shown should allow the clock frequency to be adjusted to around 100 kHz. The reset input (!MCLR) must be connected to the positive supply (+5 V) to allow the chip to run. Other unused pins can be left open circuit.

6.2 Program Execution

A block diagram showing a simplified program execution model for the PIC 16F84 is shown in Fig. 6.4. The hexadecimal program is stored in ROM. The instructions are decoded one at a time by the instruction decoder, and the required operations set up in the registers by the control logic. The file registers are numbered from 00–4F, with the first 12 registers (00–0B) being reserved for specific purposes. These are called the special function registers (SFRs). The rest may be used for temporary data storage, and are called the general purpose registers (GPRs). Only GPR1 is shown in Fig. 6.4.

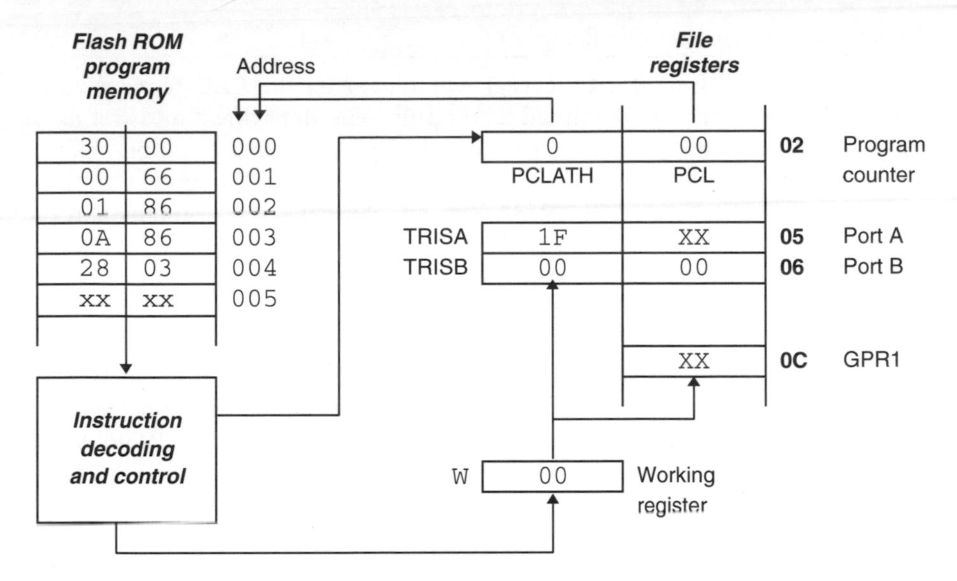

Figure 6.4 PIC 16F84 simple program execution model.

6.2.1 Program Memory

This is a block of flash ROM, which means it is non-volatile, but can be easily re-programmed. The program is created in a host computer and can be downloaded via selected port register pins when the chip is placed in its programming unit. It is possible to work out the hex code for each instruction from information in the instruction set in the data sheet (Table A.4). This could then be entered directly into the programming software on a host computer and downloaded to the PIC using a programming unit connected to the serial port. The 14-bit codes are loaded into memory starting at address 000. When the chip is powered-up, the program counter resets automatically to 000, and the first instruction is fetched from this address, copied to the instruction register in the control block, decoded and executed.

6.2.2 Program Counter: File Register 02

The program counter keeps track of the program execution by holding the address of the current instruction. It is automatically incremented to point to the next instruction during the execution cycle. If there is a jump in the program, the program counter is modified by the jump instruction so that it then points to the required jump destination.

6.2.3 Working Register: W

This is the main data register (8 bits), used for holding the data that is currently being worked on. It is referred to as W in the PIC program. Literals (values given in the program) must be loaded into W before being moved to another register, or used in a calculation. Most data movements have to be through W, in two stages, since direct moves between file registers are not available in the PIC instruction set.

6.2.4 Port B Data Register: File Register 06

The 8 bits stored in the Port B data register will appear on the LEDs connected to pins RB0–RB7, if the port bits are initialized as outputs. The data direction is set as output by placing a data direction code in the register TRISB. A '0' placed in a bit position sets that pin as an output (0 = output). A '1' sets it to input (1 = input). In this case, 00000000 (binary) will be placed in TRISB, but any combination of inputs and outputs can be used.

6.2.5 Port A Data Register: File Register 05

The least significant five bits of File Register 05 are connected to pins RA0–RA4, the other three being unused. This port will be used later to read the push buttons. If not initialized as outputs, the PIC I/O pins automatically become inputs. We will use this default setting for Port A. However, these inputs will have no effect unless the program uses them; the first program will not.

6.2.6 General Purpose Register 1: File Register 0C

The first general purpose register will be used later in a timing loop. It is the first of a block of 68 such registers, numbered 0C–4F. They may be allocated by the programmer as required for temporary data storage, counting and so on.

6.3 Program BIN1

The simple program called BIN1, introduced in Chapter 2, is listed as Program 6.1. The program consists of a list of 14-bit binary machine code instructions, represented as four hex-digit numbers. If bits 14 and 15 are assumed to be zero, the codes are represented by hex numbers in the range 0000–3FFF. The program is stored at addresses 000–0004 (five instructions) in program memory.

6.3.1 Program Analysis

The meaning of the program instructions must be related to the internal hardware of the PIC 16F84, as shown in Fig. 6.4.

Program 6.1 BIN1 machine code

Memory address	*Machine code instruction*	*Meaning*
000	3000	Load Working Register (W) with number 00
001	0066	Store W in Port B Direction Code Register
002	0186	Clear Port B Data Register
003	0A86	Increment Port B Data Register
004	2803	Jump back to address 0003 above

Address 0000: Machine Code Instruction = 3000

The code 3000 means move (copy) a literal (number given in the program) into the Working register (W). All literals must be placed initially in W before transfer to another register. The literal, which is zero in this case, can be seen in the code as the last two digits, 00.

Address 0001: Machine Code Instruction = 0066

This means copy the contents of W to the Port B Data Direction Register (DDRB). W now contains 00, which was loaded in the first instruction. This code will set all 8 bits of DDRB to zero, making all bits output. The file register address of Port B (6) is given as the last digit of the code.

Address 0002: Machine Code Instruction = 0186

This instruction will clear file register 6 (last digit), which means set all bits in the Port B data register to zero. Operations can be carried out directly on the port data register, and the result will appear immediately on the LEDs. On start-up, the register bits default to '1', switching the LEDs on. When the 'clear' instruction is executed, they will go out.

Address 0003: Machine Code Instruction = 0A86

Port B data is modified again. The binary value is increased by one, and this value will be seen on the LEDs.

Address 0004: Machine Code Instruction = 2803

This is a jump instruction, which causes the program to go back and repeat the previous instruction. This is achieved by the instruction overwriting the current program counter contents with 03, the destination address, which is given as the last two digits of the instruction code.

6.3.2 Program Execution

BIN1 is a complete working program, which initializes and clears Port B, and then keeps incrementing it. The last two instructions, that is, increment Port B and jump back, will repeat indefinitely, with the value being increased by one each time. In other words, Port B data register will act as an 8-bit binary counter. When it reaches FF, it will roll over to 00 on the next increment operation. If you study the binary count given in Fig. 2.5, you can see that the least significant bit is inverted each time the binary count is incremented. The least significant bit, RB0, will thus be toggled (inverted) every time the increment operation is repeated. The next bit, RB1, will toggle at half this rate, and so on, with each bit toggling at half the frequency of the previous bit. The MSB therefore toggles at 1/128 of the frequency of the LSB. The output pattern generated is illustrated in Fig. 6.5.

An instruction in the PIC takes four clock cycles to complete, unless it causes a jump, in which case, it will take eight clock cycles (or two instruction cycles). The repeated loop in BIN1 will therefore take $4 + 8 = 12$ clock cycles, and thus it will take 24 cycles for the RB0 to go low and high, giving the output period of the LSB. With the clock component values

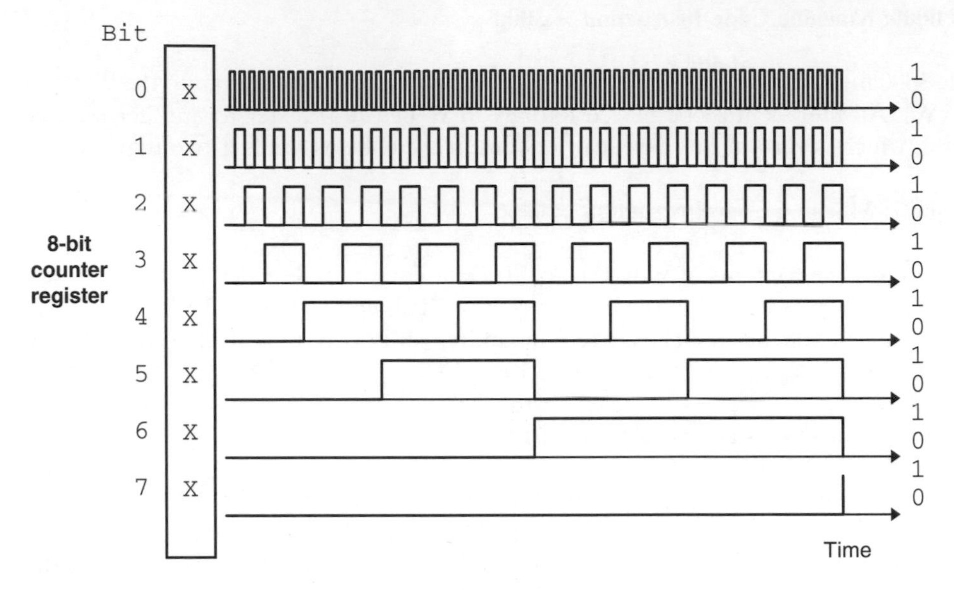

Figure 6.5 Waveforms produced by counting register.

as indicated on the circuit diagram in Fig. 6.3, the clock can be set to run at about 100 kHz, so RB0 will flash at 100 kHz/24 = 4.167 kHz, and RB7 will then flash at 4167/128 = 32.5 Hz. This is too fast to see unaided, but it is possible to reduce the clock speed by increasing the value of the capacitor C in the clock circuit. Alternatively, the outputs can be displayed on an oscilloscope. We will see later how to slow the outputs down without changing the clock.

The frequencies generated are actually in the audio range, and they can be heard through a small loudspeaker or piezo buzzer. This is a handy way of checking quickly that the program is working, and also immediately suggests a range of PIC applications – generating signals and tones at known frequencies by adjusting the clock rate or using a crystal oscillator. Again, we will come back to this idea later, and see how to generate a tune or/and audio outputs.

6.4 Assembly Language

It should be apparent that writing the machine code manually, as suggested above, for any but the most trivial applications is going to be a bit tedious. Not only do the actual instruction hexadecimal or binary codes have to be worked out, but so do jump destination addresses and so on. In addition, the codes are not easy to recognize or remember.

6.4.1 *Mnemonics*

For this reason, microcontroller programs are normally written in assembly language, not machine code. Each instruction has a corresponding mnemonic defined in the instruction set in the data sheet. The main task of the assembler program supplied with the chip is to convert a source code program written in mnemonic form into the required machine code. The mnemonic form of the program BIN1 is shown in Program 6.2.

Program 6.2 Mnemonic form of program BIN1

Top left of edit screen

Line number	*Column 0*	*Column 1*	*Column 2*	*Column 3*
0		MOVLW	00	
1		TRIS	06	
2		CLRF	06	
3		INCF	06	
4		GOTO	03	
5		END		
6				

The instructions are now written as recognizable (when you get used to them!) code words. These are then converted to the operation code part of the machine code instruction by the assembler program, which runs on the host computer. The program can now be typed into a text editor, spaced out as shown, using the tab key to place the code in the correct columns. Note that the first column (column 0) must be kept blank – we will see why later. The instruction mnemonics are placed in column 1, and the operands in column 2. The operand 00 is the data direction code for the port initialization, 06 is the file register number of the port data register, and 03 is the jump destination address, line 3 of the program. The PIC instructions are all 14-bits long, so each line of source code becomes the corresponding 14-bit code, which we have already seen. Other processors have variable length instructions, which makes this conversion more complicated.

In this program, then, the operands are still given in numeric form, which raises a problem when working out the jump back – the line numbers have to be counted up to obtain the jump destination address. The start address, if it is not 000, also has to be known and the line number must then be added to the start address to calculate the required address. We will soon see how to avoid this problem by using 'labels', but for the time being, we will stick with the numerical operands.

6.4.2 Assembly

The source code program can be created using a general purpose text editor, for example, DOS Editor, or within a dedicated software development environment such as MPLAB, the Windows package. The source code text is then converted into the corresponding machine code by the assembler program, which analyses the source code, character by character, and works out the binary code required for each instruction. The terminology is confusing here; the *assembly* language program (source code) is created in the text editor, while the software tool which does the conversion is the *assembler* program.

The source code for the PIC, in common with other processors, is saved on disk as a text file called PROGNAME.ASM, where 'progname' represents any suitable filename. This is then assembled by the assembler program MPASM.EXE (DOS software version), which creates the machine code file PROGNAME.HEX. This appears as hexadecimal code when listed. At the same time, the list file, PROGNAME.LST, is created which contains both the source and hex code, which is useful later on.

MPASM.EXE is started at the DOS command line with the name of the source code file to be assembled, as follows:

```
C: \PIC>MPASM /P16C84 A:\PICPROGS\BIN1.ASM
```

This command line assumes that the assembler is in the subdirectory 'PIC' of C: drive, the hard disk. The source code is on floppy disk in drive A: , in a directory called PICPROGS. The processor type must also be specified as a command line option, as there is some variation between PIC devices in the instruction set.

6.4.3 Labels

As we have seen, the mnemonic form of the program with numerical operands is still far from easy to work out. We need a method of representing the operands in a more easily recognizable form. The assembler is therefore designed to recognize labels. Operand labels are 'declared' at the beginning of the program, and the assembler will then substitute the numerical operand for the label. The process is similar to the replacement of the instruction mnemonic by the corresponding binary code, except that in this case the user defines the labels. The jump destinations can also be defined by label, by simply placing the label at the beginning of the destination line, and using a matching label as the jump instruction operand. The label is then replaced by the address, which has already been noted by the assembler.

The program BIN1 can thus be re-written using labels as shown in BIN2 source code, Program 6.3. The literal value 00 and the port register address 06 have been replaced with labels which are assigned at the beginning of the program. These are 'equate' statements, which allow the numbers that are to be replaced in the source code to be declared. In this case, the label 'allout' will represent the Port B data direction code, while the data register address itself, 06, will be represented by the label 'portb'. 'EQU' is an example of an assembler directive, which is an instruction to the assembler program and will not be translated into code in the executable program.

Note that lower case is used for the labels, while upper case is used for the instruction mnemonics and assembler directives. Although this is not obligatory, this convention

Program 6.3 BIN2 source code using labels

PC Screen

```
allout    EQU      00
portb     EQU      06

          MOVLW    allout
          TRIS     portb

          CLRF     portb
again     INCF     portb
          GOTO     again

          END
```

Program 6.4 BIN2 source code with comments

```
;       BIN2.ASM                M.Bates                     25/5/99
;
;       Outputs a binary count at Port B
;.................................................................

allout  EQU     00              ; Define Data Direction Code
portb   EQU     06              ; Declare Port File Register Address

        MOVLW   allout          ; Load W with Direction Code
        TRIS    portb           ; Send Code to Direction Register

        CLRF    portb           ; Start output at 00000000
again   INCF    portb           ; Increase output by 1
        GOTO    again           ; Repeat endlessly

        END                     ; Terminate Source Code
```

will be used because the instruction mnemonics are given in upper case in the instruction set. The labels can then be distinguished by using lower case. The jump destination label is simply defined by placing a label in column 1 of the line containing the destination instruction. The 'GOTO label' instruction then uses a matching label. Labels can be up to six characters long, starting with a letter.

The source code is terminated with the assembler directive END, which indicates to the assembler to stop assembling. The programs BIN1 and BIN2 are functionally identical, and the machine code will be the same.

6.4.4 Layout and Comments

A final version of BIN2 will now be created which includes comments in the program to explain the action of each line, and the overall program (Program 6.4). As much information as possible should be provided with the program, both when learning and developing real applications. Comments must be preceded with a semicolon (;), which tells the assembler to ignore the rest of that line. Comments and information can thus occupy a whole line, or can be added after each instruction in column 3. A minimal header has been added to BIN2, with the source code file name, author and date, and a comment added to each line. Blank lines can be used without a comment 'delimiter' (the semicolon). They are used to break up the source code into functional blocks, and thus make the operation of the program easier to understand. In BIN2.ASM, the first block contains the operand label equates, the second block the port initialization, and the third block the output sequence. The layout of the program is very important for understanding how it works.

We now have a program that can be entered into a text editor, assembled and downloaded to the PIC chip. The exact way of doing this will vary with the version of the PIC software and programming hardware that you use. The two main options are the DOS version, which will run on virtually any PC compatible, and the Windows software MPLAB.

Summary

- A block diagram can be used to specify the hardware, and the circuit designed from it.
- The PIC 16F84 program is stored in flash ROM, at addresses from 000. The instructions are decoded and executed by the processor control logic.
- The CPU registers are modified according to the program, and the sequence can be modified by the instructions.
- The PIC 16F84 has 14-bit instructions, containing both the operation code and operand.
- The program is written using assembler mnemonics and labels to represent the machine code instructions and operands.
- Layout and comments are used to document the program operation.

Questions

1. State the four hex-digit code for the instruction INCF 06.
2. State the two hex-digit code for the instruction MOVLW.
3. What is the meaning of the least significant two digits in the PIC machine code instruction 2803?
4. Why must the instruction mnemonic be in the second column of the source code?
5. Give two examples of a PIC assembler directive. Why are they not represented in the machine code?
6. What are the numerical values of the labels 'allout' and 'again' in BIN2?

Answers

1. 0A86.
2. 30.
3. Jump destination.
4. Labels only in first column.
5. EQU, END.
6. 00, 03.

Activities

1. Check the machine code for BIN1 against the information given in the PIC instruction set in the data sheet, particularly Table A.4, so that you could, if necessary, work out a program entirely in machine code. Modify the machine code program by deleting the 'Clear Port B' operation and changing the 'Increment Port B' to 'Decrement Port B'. What would be the effect at the output when the program was run?

2. Construct the hardware shown in Fig. 5.3 using a suitable hardware prototyping method. Refer to Chapter 12 if necessary. A socket must be used for the PIC chip. Enter the machine code for BIN1 directly into the programming software memory buffer and download to the chip. Run the program in a target system as specified in the circuit diagram in Fig. 5.3, or an equivalent circuit. Feed the outputs to a small loudspeaker with a 220R current limiting resistor in series. Use an oscilloscope to measure the clock and output frequencies. Confirm the relationship between the clock frequency and the output frequencies. Increase the capacitor value to 220 nF, which should make the MSB flash at a visible rate. Predict this a figure for the output frequency. (Hint: the rate is proportional to the RC clock components product.)

3. Enter the program BIN2, using labels, into the text editor, assemble and test as above. Check that the machine code and function is identical to BIN1.

Chapter 7
PIC Program Development

We have now seen how assembly language can be used to write programs for the PIC 16F84 microcontroller, and how these are converted to machine code for execution in the PIC chip. We will now take a look at the set of software tools available for developing further programs, and how each is used in the program development process. The program BIN2 will be developed further to illustrate the process, using the same hardware as described in Chapter 6. The intention at this stage is to outline how to assemble and test BIN3 and BIN4, but not to provide full details of each step. The reader can refer forward for more details if necessary.

The flowchart in Fig. 7.1 gives an overview of the development process. The software tools are provided in the form of utility programs that are used to create, test and download the application program to the PIC chip. The DOS file set is shown in Table 7.1. The PIC Windows development system, MPLAB, provides a set of equivalent tools in one integrated package. However, the DOS version allows us to identify each tool separately, is less complex than the Windows software, and may possibly be easier for the beginner to use. This chapter will outline the facilities of the development systems that are currently available, but relevant hardware and software are being continuously developed by the manufacturer and independent suppliers, designers and users. The Internet should be used to access the most recent information on the range of PIC chips and support software available at any given time; just search for 'PIC'!

7.1 Program Design

There are national standards for specifying engineering designs which should be applied, as necessary, in commercial work. The design rules for different types of products will vary;

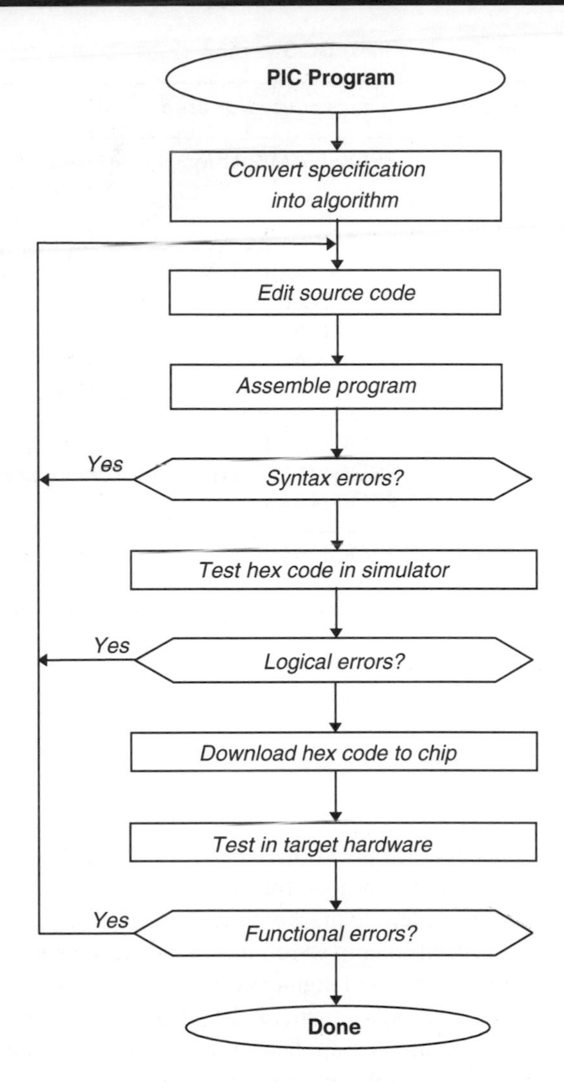

Figure 7.1 Flowchart of PIC program development process.

for instance, a military application will be more rigorously tested and documented than an industrial one. Our designs here are artificial in that they are intended to illustrate features of the PIC microcontroller, rather than meet a user's requirement. Nevertheless, we can follow the design process through the main steps.

7.1.1 *Hardware Design*

The first step in the application design process is to devise the hardware on which the program will run. We will not be studying hardware design in great detail here, and we will simply confirm, for the purpose of this design exercise, that the hardware configuration for the BINx applications already described in Chapter 6, is suitable. This includes the assumption that the instruction set and programming features of the microcontroller selected (PIC 16F84) are also suitable. If further features, such as an audio output, were

Table 7.1 PIC development system files, DOS version

Software tool	*Executable file*	*Files produced or used*	*Function*
Text Editor	EDIT.COM QBASIC.EXE	PROGNAME.ASM	Used to create and modify source code text file (uses QBasic Editor)
Assembler	MPASM.EXE	PROGNAME.HEX PROGNAME.ERR PROGNAME.LST PROGNAME.COD	Generates machine code from source code, reports syntax errors, generates list and symbol files
Simulator	MPSIM.EXE	PROGNAME.HEX PROGNAME.COD PROGNAME.INI PROGNAME.STI	Simulates program execution in software, optionally using Initialization and Stimulus files
Programmer	MPSTART.EXE	PROGNAME.HEX	Downloads machine code to chip

required, the existing hardware design could be modified. If more ports were required, another PIC chip, or other types of hardware, such as a conventional microprocessor system, must be considered.

7.1.2 Program Specification

The operational requirements of the application to be created must be clearly identified. In the commercial environment, a customer may do this, or if the application is a more speculative venture, the requirements of the potential market must be analysed. A specification must then be written in a way that lends itself to conversion into a software product using the language and tools available. Each programming language offers a different combination of features which must be matched to the user requirement as closely as possible. Similarly, the hardware system type must be selected to suit the application, before attempting the detailed circuit design. Choosing the most suitable microprocessor or microcontroller is clearly crucial. To make this choice, one needs a knowledge of the whole range of options. Chapter 14 provides a starting point for the development of this knowledge of the available solutions.

For so-called real-time or control applications of microprocessors, the main choice of language is between assembler and a high level language (HLL) such as 'C'. The HLL allows such features as screen graphics, file handling and network communication to be integrated more easily into a complex application, while assembly code is generally faster and requires less memory. Of course, ultimately all languages are converted into machine code to run on the selected processor.

HLLs are normally used to develop applications for conventional processor systems, especially if using the Windows/Intel PC as a standard hardware platform. However, for less complex applications, a suitable microcontroller may be used, programmed in assembler language. The PIC family contains an expanding range of devices that offer a combination of different features, for example, built-in serial communication ports or analogue to digital converters. The 16F84 is one of the less complex PICs, but it has the great advantage of flash ROM memory, making it easily reprogrammable, and thus ideal for prototyping simple applications.

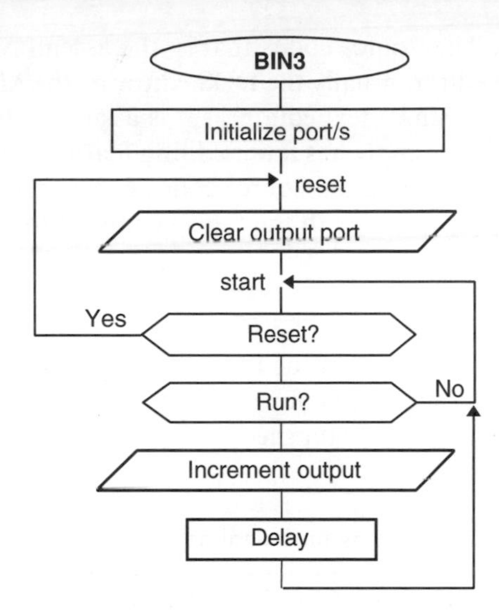

Figure 7.2 Flowchart for BIN3.

BIN3 specification

A system is required which outputs a visible 8-bit binary count to a set of eight LEDs. The output should run when a count enable button is pressed, and stop when it is released, holding the count. A reset button should reset the count to zero, with all LEDs off.

7.1.3 Program Algorithm

The program 'algorithm' is the process whereby the specification is implemented, taking into account the features of the programming language to be used. The BIN3 specification is much less rigorous than would normally be the case for real software product, and it is artificial in that it has been devised to suit the 16F84, rather than the other way round. The frequency of operation of the output could, for instance, be specified. As the specification is not very specific, it should be easy to meet!

A flowchart is useful for clarifying the algorithm. A flowchart for BIN3 is shown in Fig. 7.2. The program title is placed in the start symbol at the top of the flowchart, and the process required defined as a conditional sequence. Each flowchart box contains a description of the action at each stage, using shaped boxes for processes (rectangle), input and output (sloping) and decisions (pointed). The pointed box has *two* outputs, to represent a conditional branch in the program. This decision box should contain a question with the answer yes or no, and the active selection labelled Yes or No as appropriate; only one needs to be labelled. Software design techniques, including flowcharts, will be covered in more detail later.

7.2 Program Editing

The program is constructed using the instruction set of the processor selected. This is provided with the hardware data sheet, in this case, Section A.9 of Appendix A. A summary is also found in Table A.4. The 'source code', that is, the assembly code program, must be entered into a suitable text editor, usually the DOS editor or the MPLAB edit window. We will not go into to details of using a text editor, as it is assumed that the reader has used a wordprocessor. The text editor simply has fewer editing features, because it is used only for creating plain text files, such as program source code. A non-proportionally spaced typeface, such as 'Courier', must be used with the tab spacing set to eight characters.

7.2.1 Instruction Set

Table 7.2 is a more user-friendly form of the PIC 16F84 instruction set organized by function. The example given with each instruction should be easier to understand than the more detailed information provided in the data sheet.

Table 7.2 PIC 16F84 instruction set by functional groups

PIC 16F84 INSTRUCTION SET BY FUNCTIONAL GROUPS

F = Any file register (specified by number or label), example is 0C
W = Working register, W
L = Literal value (follows instruction), example is 0F9

Operation		*Example*	
Move			
	Move data from F to W	MOVF	0C,W
	Move data from W to F	MOVWF	0C
	Move literal into W	MOVLW	0F9
Register			
	Clear W (reset all bits and value to 0)	CLRW	
	Clear F (reset all bits and value to 0)	CLRF	0C
	Decrement F (reduce by 1)	DECF	0C
	Increment F (increase by 1)	INCF	0C
	Swap the upper and lower four bits in F	SWAPF	0C
	Complement F value (invert all bits)	COMF	0C
	Rotate bits left through Carry Flag	RLF	0C
	Rotate bits right through Carry Flag	RRF	0C
	Clear (reset to zero) the bit specified (e.g. bit 3)	BCF	0C,3
	Set (to 1) the bit specified (e.g. bit 3)	BSF	0C,3
Arithmetic			
	Add W to F	ADDWF	0C
	Add F to W	ADDWF	0C,W
	Add L to W	ADDLW	0F9
	Subtract W from F	SUBWF	0C
	Subtract W from F, placing result in W	SUBWF	0C,W
	Subtract W from L, placing result in W	SUBLW	0F9
Logic			
	AND the bits of W and F, result in F	ANDWF	0C
	AND the bits of W and F, result in W	ANDWF	0C,W
	AND the bits of L and W, result in W	ANDLW	0F9

Table 7.2 continued

Operation	Example	
Logic (continued)		
OR the bits of W and F, result in F	IORWF	0C
OR the bits of W and F, result in W	IORWF	0C,W
OR the bits of L and W, result in W	IORLW	0F9
Exclusive OR the bits of W and F, result in F	XORWF	0C
Exclusive OR the bits of W and F, result in W	XORWF	0C,W
Exclusive OR the bits of L and W	XORLW	0F9
Test and Skip		
Test a bit in F and Skip next instruction if it is Clear (=0)	BTFSC	0C,3
Test a bit in F and Skip next instruction if it is Set (=1)	BTFSS	0C,3
Decrement F and Skip next Instruction if it is now Zero	DECFSZ	0C
Increment F and Skip next Instruction if it is now Zero	INCFSZ	0C
Jump		
Go To a Labelled Line in the Program	GOTO	LOOP1
Jump to the Label at the start of a Subroutine	CALL	DELAY
Return at the end of a Subroutine to the next instruction	RETURN	
Return at the end of a Subroutine with L in W	RETLW	0F9
Return from Interrupt Service Routine to next instruction	RETFIE	
Control		
No Operation - delay for 1 cycle	NOP	
Go into Standby Mode to save power	SLEEP	
Clear Watchdog Timer to prevent automatic reset	CLRWDT	
Load Port Data Direction Register from W	TRIS	06
Load Option Control Register from W	OPTION	

The result of Arithmetic and Logic operations can generally be stored in W instead of the file register by adding ',W' to the instruction. General Purpose Register 1, address 0C, represents all file registers (00–2C). Literal value 0F9 represents all values 00–FF. Bit 3 is used to represent File Register Bits 0–7. For MOVE instructions data is copied to the destination but retained in the source register.

7.2.2 BIN3 Source Code

The same instructions are used as in the example BIN2 (Chapter 6), with additional statements to read the switches and control the output. The type of operations that we are selecting from the instruction set have been discussed in Chapter 5. Program 7.1 is the result.

First, note the general layout and punctuation required. The program header block contains as much information as necessary at this stage. These comments are preceded by a semicolon on each line to indicate to the assembler that this text is not part of the program. Assembler directives, such as EQU and END, are also not part of the program proper, but used to define labels and the end of the program source code. The labels 'porta', 'portb' and 'timer' refer to file registers 05, 06 and 0C, respectively; 'inres' and 'inrun' are input bit labels representing the push buttons. The program uses 'Bit Test and Skip' instructions followed by 'GOTO label' for conditional jumping.

At this stage, the reader can type the source code into the editor without full analysis in order to try out its features. The instructions are placed in the first three columns. The comments can

Program 7.1 BIN3 source code

```
;
;       BIN3.ASM                        M. Bates              1/6/99
;...............................................................
;
;       Slow output binary count is stopped, started
;       and reset with push buttons.
;
;
;***************************************************************

; Register Label Equates........................................

porta   EQU     05                      ; Port A Data Register
portb   EQU     06                      ; Port B Data Register
timer   EQU     0C                      ; Spare register for delay

; Input Bit Label Equates .....................................

inres   EQU     0                       ; 'Reset' input button = RA0
inrun   EQU     1                       ; 'Run' input button = RA1

;***************************************************************

; Initialize Port B (Port A defaults to inputs)............

        MOVLW   00                      ; Port B Data Direction Code
        TRIS    portb                   ; Load the DDR code into F86
        GOTO    reset


; Start main loop ..............................................

reset   CLRF    portb                   ; Clear Port B

start   BTFSS   porta,inres             ; Test RA0 input button
        GOTO    reset                   ; and reset Port B if pressed
        BTFSC   porta,inrun             ; Test RA1 input button
        GOTO    start                   ; and run count if pressed

        INCF    portb                   ; Increment count at Port B

        MOVLW   0FF                     ; Delay count literal
        MOVWF   timer                   ; Copy W to timer register
down    DECFSZ  timer                   ; Decrement timer register
        GOTO    down                    ; and repeat until zero

        GOTO    start                   ; Repeat main loop always
        END                             ; Terminate source code
```

Table 7.3 Layout of assembler source code in columns

Column 1	*Column 2*	*Column 3*	*Column 4*
LABEL Label EQUated to a value, or to indicate a program destination address for jumps.	COMMAND Mnemonic form of the instruction for the processor to carry out a specific operation. Only mnemonics specified in the instruction set may be used.	OPERAND/S The data or register contents to be used in the instruction. Registers are usually represented by a label. Some instructions do not need an operand.	COMMENT Explanatory text to the right of a semicolon on any line of code helps the programmer and user to understand the program. It has no effect on the operation of the program. Full line comments may also be used between program blocks.

be left out to save time. Labels go in the first column, instruction mnemonics in the second, and the instruction operands in the third. The source code text file should be saved as BIN3.ASM in a suitably named directory or folder on disk, PICPROGS for instance.

Syntax

'Syntax' refers to the way that words are put together to create meaningful statements, or a series of statements. In programming, the syntax rules are determined by the assembler which will be used to create the machine code, in our case, MPASM.EXE. The assembler must be provided with source code which it can convert into the required machine code without any ambiguity, that is, only one meaning is possible. This is why the assembler syntax rules are very strict.

Layout

The program layout should be in four columns, as described in Table 7.3. If using a word-processor to edit or print, a Courier font, as used on mechanical typewriters, should be selected. Each character then occupies the same space, and the columns are correctly aligned. The label, command and operand columns should be set to a width of eight characters, with the maximum label length of six characters, leaving a minimum of two clear spaces between columns. The tab key is normally used to place the text in columns, and tab spacing can usually be adjusted if necessary. Multiple spaces can be used instead, if necessary. This is sometimes a good idea if the source code is to be loaded into an editor with a different tab setting.

Comments

Consider the following example:

```
;       BIN3.ASM      F. Bloggs      26/2/98            Ver. 1.2
; ***********************************************************
```

Comments are not part of the actual program, but are there to help the programmer and user understand how the program works. Comments are preceded by a semicolon (;), which can be placed at the beginning of a line to indicate a comment which relates to a whole program block (functional set of statements), or at the start of column 4 for line comment. The comment and line are terminated with a line return ('Enter' key).

A standard header block is recommended. For simple programs, the first line should at least contain the source code filename (under which the text will be saved), the author and date, and/or version number. A program description should also be provided in the header, and for more complex programs, the processor type, hardware set-up and other relevant program information. Always keep, at least, the previous version of a program under a different file name, in case you lose the current file, and always keep back-up files (copies) on another disk. Use a numbered filename, so that as the program is developed the filename is updated to preserve the previous version, for example, PROG1, PROG2, PROG3 and so on. As these are small text files, backing-up to a floppy disk is strongly recommended, so that the disk can be removed for safe keeping, unless back-up to a network drive is available.

7.3 Program Structure

Structured programming means constructing the program, as far as possible, from discrete blocks. This makes the program easier to write and understand, more reliable and more easily modified, if required.

7.3.1 BIN4 Source Code

BIN3 is unstructured, in that the program instructions are essentially executed in the order given in the source code. An equivalent 'structured' program, BIN4, is given as Program 7.2.

Program 7.2 BIN4 source code

```
;
;        BIN4.ASM                      M. Bates                3/6/99
; ......................................................................
;
;        Slow output binary count is stopped, started
;        and reset with push buttons. This version uses a
;        subroutine for the delay....
;
;        Processor: PIC 16F84
;
;        Hardware: PIC Demo System
;        Clock: CR ~100kHz
;        Inputs: Push Buttons RA0, RA1 (active low)
;        Outputs: LEDs (active high)
;
```

```
;       WDTimer: Disabled
;       PUTimer: Enabled
;       Interrupts: Disabled
;       Code Protect: Disabled
;
; ***************************************************

; Register Label Equates.................................

porta   EQU     05                  ; Port A Data Register
portb   EQU     06                  ; Port B Data Register
timer   EQU     0C                  ; Spare register for delay

; Input Bit Label Equates ...............................

inres   EQU     0                   ; 'Reset' input button = RA0
inrun   EQU     1                   ; 'Run' input button = RA1

; ***************************************************

; Initialize Port B (Port A defaults to inputs)..........

        MOVLW   b'00000000'         ; Port B Data Direction Code
        TRIS    portb               ; Load the DDR code into F86
        GOTO    reset

; 'delay' subroutine ....................................

delay   MOVWF   timer               ; Copy W to timer register
down    DECFSZ  timer               ; Decrement timer register
        GOTO    down                ; and repeat until zero
        RETURN                      ; Jump back to main program


;       Start main loop
.........................................................

reset   CLRF    portb               ; Clear Port B Data

start   BTFSS   porta,inres         ; Test RA0 input button
        GOTO    reset               ; and reset Port B if pressed
        BTFSC   porta,inrun         ; Test RA1 input button
        GOTO    start               ; and run count if pressed

        INCF    portb               ; Increment count at Port B
        MOVLW   0FF                 ; Delay count literal
        CALL    delay               ; Jump to subroutine 'delay'

        GOTO    start               ; Repeat main loop always
        END                         ; Terminate source code
```

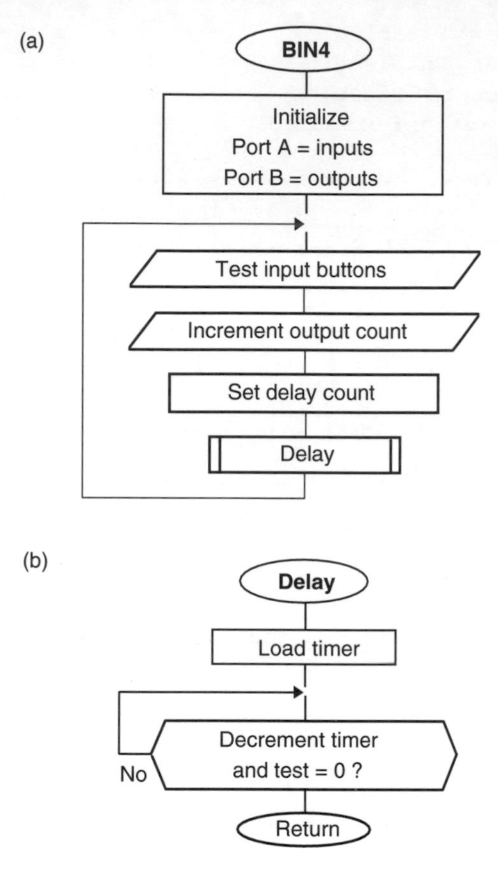

Figure 7.3 Flowcharts for BIN4. (a) Main sequence; (b) Delay subroutine.

The main difference between BIN3 and BIN4 is that the program now has the delay code as a 'subroutine'. The subroutine is inserted before the main program block, and assembled first. It is then 'called' from the main program by label. This arrangement is a first step towards structured programs constructed from reusable software components. The subroutine can be created as a standard program block, and re-used in the program if necessary. It can be called as many times as required, which means that the block of code only needs to be written once. It can also be saved as a separate file and re-used in another program, using the INCLUDE assembler directive.

A program flowchart has been given for BIN3 (Fig. 7.2). The same flowchart describes BIN4, but the delay routine should now be expanded as a separate subroutine flowchart, as in Fig. 7.3. The use of flowcharts in program design will be more fully examined in Chapter 10.

7.4 Program Analysis

The program BIN4 will now be analysed in some detail as it was designed to contain all the basic syntax.

7.4.1 Label Equates

Consider the example:

```
timer    EQU    0C
```

The use of labels in place of numbers makes programs easier to write and understand, but we have to 'declare' those labels at the beginning of the program. In assembly code, the assembler directive EQU is used to assign a label to a number, which can be a literal, file register number or individual register bit. In BIN4, 'porta' and 'portb' are the port data registers (05 and 06) and 'timer' is the first spare register (0C), which will be used as a software counter. The labels 'inres' and 'inrun' will refer to bit 0 and bit 1 of Port A; they are simply given the numerical value 0 and 1.

7.4.2 Port Initialization

Consider the example:

```
TRIS    portb
```

In BIN4, Port B will be used as the output for the 8-bit binary count. The data direction must be set up using the TRIS command, which loads the port data direction register with the data direction code. In this example, the code is given in binary, and defined as b'00000000'. This is useful, especially if the port bits are to be set as a mixture of inputs and outputs; the binary code identifies them more clearly than would the hexadecimal equivalent. This code is loaded into W using MOVLW, and the TRIS command follows.

7.4.3 Program Jumps

Consider the example:

```
GOTO    start
```

The 'GOTO label' command is used to make the program jump to a line other than the one following. In BIN4, 'GOTO reset' skips over the following DELAY routine, to start the main loop. We will come back to the reason for this in a moment. There is another unconditional jump at the end of the program, 'GOTO start', which makes the main loop repeat endlessly. Other 'GOTO label' instructions are used with 'Test and Skip' instructions to create conditional branches. In this program, the input buttons are checked using this type of instruction and the program branches, or not, depending on whether it has been pressed.

7.4.4 Bit Test and Skip if Set/Clear

Consider the example:

```
BTFSS    porta,inres
```

At the start of the main loop (address label 'reset'), port B is cleared to zero using 'CLRF portb'. The input button connected to Port A, bit 0 is then tested using 'BTFSS porta,inres', which means 'Bit Test File (register bit) and Skip if Set'. Without labels, the instruction 'BTFSS 05,0' would have the same effect. The buttons are connected 'active low', meaning that the input goes from '1' to '0' when the button is pressed. If the button connected to RA0 is not pressed, the input will be high, that is, set. The following instruction, 'GOTO reset' is therefore skipped, and the next executed. When the button is pressed, the 'GOTO reset' is executed, and the CLRF instruction repeated, clearing the previous count.

BTFSC means 'Bit Test and Skip if Clear'; it works in the same way as BTFSS, except that the logic is reversed. Thus, 'BTFSC porta,inrun' tests bit 1 of Port A register and skips the following 'GOTO start' if the 'run' button has been pressed. The program will then proceed to increment the output count. If the button is not pressed, the program waits by jumping back to the 'start' line. The combined effect is that the count runs when the 'run' button is pressed, and the count is reset to zero if the 'reset' button is pressed.

7.4.5 Decrement/Increment Register and Skip if Zero

Consider the example:

```
DECFSZ    timer
```

The other instructions for conditional branching allow a register to be incremented or decremented and then checked for a zero result. This is a common requirement for counting and timing applications, and in the delay routine in BIN3, a register 'timer' is loaded with the maximum value FF and decremented. If the result is not yet zero, the jump 'GOTO down' is executed. When the register reaches zero, the GOTO is skipped and the subroutine ends.

7.4.6 Subroutine Call and Return

Examples:

```
main   .....              ; start main program

       .....

       CALL delay         ; jump to subroutine

       ......             ; return to here

delay  ......             ; subroutine start

       ......

       ......

       RETURN             ; subroutine ends
```

In this program, the subroutine provides a delay by loading a register and counting down to zero. The delay is started using the 'CALL delay' instruction, when the program jumps to the label and runs from there. CALL means 'come back to the same place after the subroutine', so the return address has to be stored for later recall. The address of the instruction following (in this case 'GOTO start') is saved automatically in the special memory block called the 'stack' as part of the execution of the CALL instruction. The subroutine is terminated with the instruction 'RETURN', which does not require an operand because the return destination address is automatically pulled from the stack and replaced in the program counter. This takes the program back to the original place in the main program. The stack can store up to eight return addresses, so multiple levels of subroutine can be used. The return addresses must be pushed onto and pulled from the stack in the correct order, so if a CALL or RETURN is missed out of the program, a stack error message will be generated in the simulator.

7.4.7 End of Source Code

The source code must be terminated with assembler directive END so that the assembly process can be stopped in an orderly way, and control returned to the host operating system.

7.5 Program Assembly

The assembler program takes the source code text and analyses it character by character, line by line, starting at the top left, in the same sequence as it is entered. The DOS command is:

```
C:\PIC>mpasm↵
```

A dialogue screen allows the source code to be loaded, the processor type selected and other options to be set. When the assembler is then run, the corresponding 14-bit binary-machine code for each line in the source code is generated, until the END directive is detected. The binary code is saved as a file called BIN4.HEX.

7.5.1 Syntax Errors

If there are any syntax errors in the source code, such as spelling, layout, punctuation or failure to define labels properly, error messages will be generated by the assembler (Table 11.1). The error messages will be displayed, indicating the type of error, and line number. You must note the messages and line numbers, or print out the error file, BIN3.ERR, then go back and re-edit the source code and make the necessary changes. The error is sometimes on a previous line to the one indicated, and sometimes one error can generate more than one message. Warnings and information messages can usually be ignored. You may receive messages such as the following:

```
Warning[224] C:\MPLAB\BOOKPRGS\CL1.ASM 65 : Use of this
instruction is not recommended.

Message[305] C:\MPLAB\BOOKPRGS\CL1.ASM 81 : Using default
destination of 1 (file).
```

Warnings may be caused by using the instructions TRIS and OPTION, which the manufacturer warns may not be recognized in future. However, they are used here because they simplify port initialization and initialization of the Option register. The message is caused by not specifying the file register as the destination in some instructions, again in order to simplify the syntax. Assembler options can be set to suppress these messages if required.

In MPLAB, the assembler can be run from a selection button or the menus, at which point the current source code in the Edit window will be assembled. The error messages can be displayed on the same screen as the source code, and the erroneous lines are highlighted. When all errors have been cleared, the assembler will produce the machine code program BIN4.HEX. When loaded into the program downloading utility, it can be viewed as a list of four hex-digit numbers, representing the 14-bit binary instruction codes.

7.5.2 List File

A program 'list file' BIN3.LST is also produced by the assembler, which contains the source code, the machine code, error messages and other information all in one listing (Table 7.4). This is useful for analysing the program and assembler operations, and debugging (fault finding) the source code.

The list file header shows the assembler version used, and source file details. The column headings are then given:

LOC:	Memory location addresses at which the machine code will be stored
VALUE:	The numerical value with which equated labels will be replaced
OBJECT CODE:	Machine code produced for each instruction
LINE:	Line number of list file
SOURCE TEXT:	Source code including comments

At the end of the list file, additional information is provided:

SYMBOL TABLE:	Lists all the EQUate labels and address labels allocated
MEMORY USAGE MAP:	Shows the locations occupied by the object code

Note that there is no machine code produced by the lines that are occupied by a full line comment. The actual program starts to be produced at line 00040. The machine code for the first instruction is shown in column 2 (3000), and the address where it will be stored in the chip when downloaded is shown in column 1 (0000). The whole program will occupy locations 0000–000F (16 instructions).

If we study the machine code, we can see how the labelling works. For example, the last instruction 'GOTO start' is encoded as 2809, and the 09 refers to address 09 in column 1, the location with the label 'start'. The assembler program has replaced the label with the corresponding numerical address for the jump destination. Similarly, the label 'porta' is replaced with its file register number 05 in line 00060. The messages [305] inserted in the main body of the program refer to the fact that this program does not use the full syntax for register addressing.

The label values are listed again in the symbol table. These values will be used by the simulator to allow the user to display the simulated registers by label. The amount of program memory used (16 locations, 0000–000F) is shown in graphical format in the memory usage map, and finally a total of errors, warnings and messages given. If there are fatal errors, which prevent successful assembly of the program, the list file will not be produced.

Table 7.4 BIN4 list file

```
MPASM 01.21 Released            BIN4.ASM  1-5-1999  15:04:14              PAGE 1

LOC   OBJECT CODE LINE SOURCE TEXT
   VALUE

                  00001 ;
                  00002 ;       BIN4.ASM        M. Bates                    26/2/98
                  00003 ;       ...............................................
                  00004 ;
                  00005 ;       Output binary sequence is stopped, started
                  00006 ;       and reset with input buttons...
                  00007 ;
                  00008 ;       Processor:      PIC 16F84
                  00009 ;
                  00010 ;       Hardware:       PIC Demo System
                  00011 ;       Clock:          CR ~100kHz
                  00012 ;       Inputs:         Push Buttons RA0, RA1 (active low)
                  00013 ;       Outputs:        LEDs (active high)
                  00014 ;
                  00015 ;       WDTimer:        Disabled
                  00016 ;       PUTimer:        Enabled
                  00017 ;       Interrupts:     Disabled
                  00018 ;       Code Protect:   Disabled
                  00019 ;
                  00020 ;       Subroutines:    DELAY
                  00021 ;       Parameters:     None
                  00022 ;
                  00023 ;***********************************************
                  00024
                  00025 ;       Register Label Equates.........................
                  00026
 0005             00027 porta   EQU     05            ; Port A Data Register
 0006             00028 portb   EQU     06            ; Port B Data Register
 000C             00029 timer   EQU     0C            ; Spare register for delay
                  00030
                  00031 ; Input Bit Label Equates ..............................
                  00032
 0000             00033 inres   EQU     0             ; 'Reset' input button=RA0
 0001             00034 inrun   EQU     1             ; 'Run' input button = RA1
                  00035
                  00036 ; ***********************************************
                  00037
                  00038 ; Initialize Port B (Port A defaults to inputs)......
                  00039
0000  3000        00040         MOVLW   b'00000000'   ; Port B Data Direction Code
0001  0066        00041         TRIS    portb         ; Load the DDR code into F86
                  00042
0002  2808        00043         GOTO    reset         ; Jump to start of main
                  00044
                  00045 ;       Define DELAY subroutine.......................
                  00046
0003  30FF        00047 delay   MOVLW   0xFF          ; Delay count literal
                                                               continued . . .
```

Table 7.4 continued

```
0004 008C          00048         MOVWF   timer          ; is loaded into spare reg.
                   00049
Message[305]: Using default destination of 1 (file).
0005 0B8C          00050 down    DECFSZ  timer          ; Decrement timer register
0006 2805          00051         GOTO    down           ; and repeat until zero then
0007 0008          00052         RETURN                 ; return to main program
                   00053
                   00054
                   00055 ;       Start main loop ..............................
                   00056
                   00057
0008 0186          00058 reset   CLRF    portb          ; Clear Port B Data
                   00059
0009 1C05          00060 start   BTFSS   porta,inres    ; Test RA0 input button
000A 2808          00061         GOTO    reset          ; and reset Port B
                   00062
000B 1885          00063         BTFSC   porta,inrun    ; Test RA1 input button
000C 2809          00064         GOTO    start          ; and run count if pressed
                   00065
                   00066
Message[305]: Using default destination of 1 (file).
000D 0A86          00067         INCF    portb          ; Increment count at Port B
000E 2003          00068         CALL    delay          ; Execute delay subroutine
000F 2809          00069         GOTO    start          ; Repeat main loop
                   00070
                   00071
                   00072         END                    ; Terminate source code

SYMBOL TABLE
  LABEL                      VALUE

__16C84                      00000001
delay                  00000003
down                         00000005
inres                  00000000
inrun                  00000001
porta                  00000005
portb                  00000006
reset                  00000008
start                  00000009
timer                  0000000C

MEMORY USAGE MAP ('X' = Used, '-' = Unused)

0000 : XXXXXXXXXXXXXXXX ---------------- ---------------- ----------------
0040 : ---------------- ---------------- ---------------- ----------------

All other memory blocks unused.

Errors : 0
Warnings : 0
Messages : 2
```

7.6 Program Simulation

The BIN4.HEX file could now be downloaded to the PIC chip and tested, and it would run, because the program given here has already been tested. However, during actual program development, it is quite possible that, although the program has been assembled without any syntax errors, 'logical' errors in the program may still be present. Logical errors make the program work incorrectly, that is, not to the specification. If this is so, the source code must be analysed again to try to find the errors. In complex programs, this process might have to be repeated several times, making it time consuming and inefficient. This is where a software simulator is useful; it allows the program to be 'run' on the host PC, as if it were being executed in the chip, but without having to download to the actual hardware. It can then be checked for logical errors and the source code changed and re-tested much more easily.

In the simulator, the program can be executed continuously or in single step mode; in the latter case, each instruction is displayed after execution, so the sequence can be checked. The registers can also be displayed and monitored for correct operation, and inputs simulated. If an error is identified in the program logic, it can be temporarily changed and checked in the simulator program, and then the source code re-edited. Suppose that in developing BIN4, for example, we had failed to analyse the switching logic correctly, and instead of the instruction BTFSS (Bit Test and Skip if Set) at the line labelled 'start', BTFSC (Bit Test and Skip if Clear) had been entered. The program would assemble successfully, but when tested, would not start correctly. While the 'inres' switch was not pressed, the program would jump back to the 'reset' line instead of going to the second switch ('inrun') test to allow the output sequence to be started. This fault could be detected in the simulator, by running the program, stepping through the start sequence and simulating the switch inputs.

7.6.1 DOS Simulator

In the DOS system, assuming that the MPSIM.EXE file is in directory 'PIC', the simulator is started at the command line by typing:

```
C:\PIC>mpsim↵
```

A simulator screen is displayed with the simulator command prompt '%'. At this prompt, commands to run, single step, modify the display, modify the inputs and so on, can be entered. Selected registers, including program counter and ports, which change as the program executes, are displayed at the top of the screen. At the top right, the step count records the number of instructions executed so far, and the elapsed time clock displays the time since the start of program execution. This is calculated from a simulator clock rate which can be selected to match the target hardware clock.

An initialization file can be created that runs a predefined sequence of commands when the simulator is started. This is a text file that can be created using the same (DOS) editor as the program source code. The initialization file (MPSIM.INI) displayed in Fig. 7.4 can be studied and modified in order to set up the display for a particular program. A list of the some commonly used commands is shown in Table 7.5. The full list of commands can be seen by typing 'h' at the command prompt.

(a)

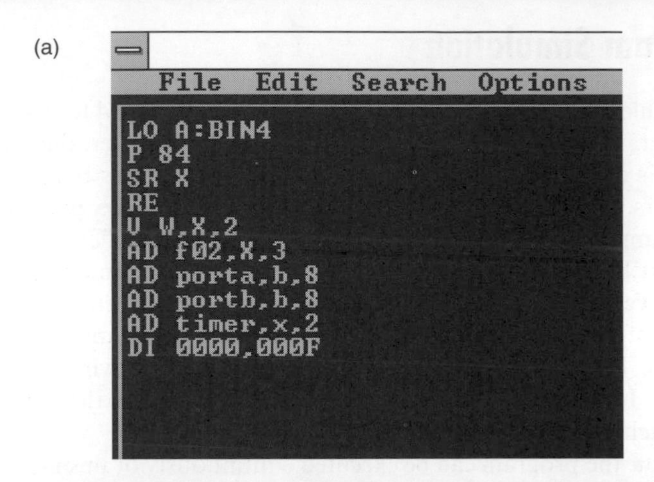

(b)

```
BIN4            RADIX=X   MPSIM 5.11.09   16c84   TIME=0.00µ 0
W: 00  f02: 000  porta: 00000000  portb: 00000000  timer: 00

001:  0066           TRIS    portb          ; Load the DDR code into F86
002:  2808           GOTO    reset          ; Jump to start of main program
003:  30FF  delay    MOVLW   0xFF           ; Delay count literal
004:  008C           MOVWF   timer          ; is loaded into spare register
005:  0B8C  down     DECFSZ  timer          ; Decrement timer register
006:  2805           GOTO    down           ; and repeat until zero then
007:  0008           RETURN                 ; return to main program
008:  0186  reset    CLRF    portb          ; Clear Port B Data and restart
009:  1C05  start    BTFSS   porta,inres    ; Test RA0 input button
00A:  2808           GOTO    reset          ; and reset Port B if pressed
00B:  1885           BTFSC   porta,inrun    ; Test RA1 input button
00C:  2809           GOTO    start          ; and run count if pressed
00D:  0A86           INCF    portb          ; Increment count at Port B
00E:  2003           CALL    delay          ; Execute delay subroutine
00F:  2809           GOTO    start          ; Repeat main loop unconditional
%
```

(c)

```
BIN4            RADIX=X   MPSIM 5.11.09   16c84   TIME=8458.00µ 2813   ?=Help
W: FF  f02: 005  porta: 00011101  portb: 00000110  timer: 9B

0006 2805           GOTO    down           ; and repeat until zero then
%
0005 0B8C  down     DECFSZ  timer          ; Decrement timer register
%
0006 2805           GOTO    down           ; and repeat until zero then
%
0005 0B8C  down     DECFSZ  timer          ; Decrement timer register
%
0006 2805           GOTO    down           ; and repeat until zero then
%
0005 0B8C  down     DECFSZ  timer          ; Decrement timer register
%
0006 2805           GOTO    down           ; and repeat until zero then
% e
User halted processor at address 5
0005 0B8C  down     DECFSZ  timer          ; Decrement timer register
%
```

Figure 7.4 PIC simulator screen, DOS version. (a) Initialization file; (b) Program listing; (c) Single step program execution.

A typical sequence of operations to test BIN4 is shown in Table 7.6. The execution sequence must be predicted from the program design stage, and the simulated sequence checked against it. A printout of the list file may be useful here, particularly if annotated with the predicted sequence.

7.6.2 Windows Simulator

The Windows version of the PIC development system is started from the MPLAB icon (Fig. 7.5). It allows the program source code to be loaded, edited and assembled within

Table 7.5 Selected MPSIM commands

Command	*Description*	*Example*
GENERAL		
h	Display full list of MPSIM commands	`% h`
q	Quit MPSIM	`% q`
ge	Get and run an initialization file (test.ini) for your program	`% ge b:test`
go	Reset and run loaded program	`% go`
e	Execute the loaded program from current address	`% e`
INITIALIZATION		
p	Select processor type	`% p 84`
lo	Load a machine code file (test.hex) for testing	`% lo b:test`
	Execute the loaded program from address given	`% e 0`
st	Load stimulus file	`% st b:test`
ra	Reset all registers	`% ra`
re	Reset program timer and step count	`% re`
rs	Reset processor	`% rs`
nv	Clear the view screen	`% nv`
ad	Add a file register by number to the view screen	`% ad f5`
	Add a file register by label, using number type and digits	`% ad porta,b,8`
dw	Disable watchdog timer	`% dw d`
sc	Modify the cycle time (microseconds)	`% sc 4.0`
DEBUGGING		
di	Display the loaded program from current address	`% di`
	Display the loaded program and addresses from start	`% di 0`
ss	Single Step – from current address	`% ss`
	Single Step – from given address	`% ss 3e`
dr	Display current state of all registers	`% dr`
b	Set a break point at an address	`% b 01c`
	Set a break point at a label	`% b moton`
bc	Clear break point at an address	`% bc 1c`
se	Modify an input pin at the next step	`% se ra4`
MODIFICATION		
ia	Modify the code at an address	`% ia 2f`
in	Insert new code at an address	`% in 2f`
de	Delete code between addresses	`% de 3e,45`
o	Save the modified machine code	`% o filename`

the same package. A 'Project' must be created to associate the program files with, which would in this case be called BIN4. The source code is created or modified in the edit window, and assembled by selecting 'Make Project'.

When the program execution is simulated, 'Step' and 'Animate' mode highlight the instruction currently being executed in the source code window. The special function registers and file registers can be displayed in a 'Watch' window, and the Stopwatch window allows the program timing to be studied. The inputs can be simulated using buttons

Table 7.6 Typical command sequence in MPSIM

Command	*Meaning*	*Effect*
`% h ↵`	Show help	Help shown
`% ↵`	More help	More help shown
`% [space]`	Quit help	Return to command prompt
`% lo a:bin3 ↵`	Load the hex file from floppy disk	Green message if OK
`% sc 4 ↵`	Set the simulator clock to 4 MHz	Clock step rescaled
`% ad timer ↵`	Add 'timer' register to display	Timer display in file view screen
`% ss ↵`	Single step through initialization	Instruction executed and displayed
`% ↵`	Next step (repeat)	Next instruction
`% se ra1 ↵`	Toggle the input Port A, Bit 1 ('Run')	Current bit value displayed for change
`% e ↵`	Execute (through the delay loop)	Count (F6) increments if working OK
`% [space]`	Stop execution	Registers, clock and step count stop
`% se ra1 ↵`	Simulates release of 'Run' button	F6 held
`% se ra0 ↵`	Toggle the input Port A, Bit 0 ('Reset')	Current bit value displayed for change
`% ss ↵`	Single step through reset sequence	F6 reset to zero
`% se ra0 ↵`	Release 'Reset' button	Waits for run
`% se ra1 ↵`	Simulates 'Run' button active again	Current bit value displayed for change
`% e ↵`	Execute	Count runs again
`% [space]`	Stop	Count stops
`% q ↵`	Quit MPSIM	DOS prompt displayed

which have to be allocated to the input pins, RA0 and RA1, by selecting 'Asynchronous Stimulus'.

MPLAB is best run with a reasonably high resolution screen, as several windows will be open at once. BIN4 would be tested using similar steps to those detailed for MPSIM, except that, when the simulator has been set up, most steps are run from a screen button. More details on program testing using MPLAB are provided in Chapter 13.

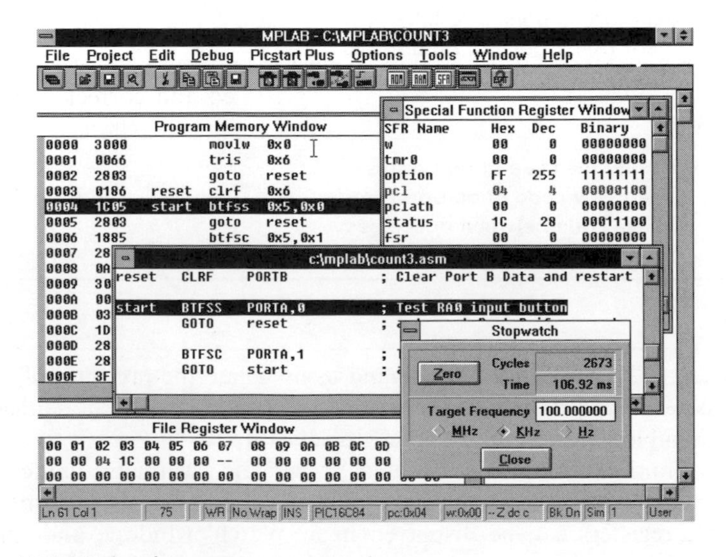

Figure 7.5 MPLAB development system windows.

7.7 Program Downloading

After testing in the simulator for correct operation, the machine code program must now be blown into the flash ROM on the chip. A DOS programmer utility provided by the manufacturer is called MPSTART, supplied with a serial programming unit. The chip is fitted into a socket in the hardware, which is connected to the serial port (COM1) of the host PC (see Fig. 1.11). MPSTART is run from the DOS command line:

```
C:\PIC>mpstart↲
```

If the programmer unit is connected to the PC serial port COM2, select this option from the MPSTART menus in the programming window. The type of processor (16C84) must now be selected; note that the 16C84 and 16F84 are practically identical, so selection of the 16C84 is OK if the 16F84 is not listed. The program hex file, BIN4.HEX, is loaded from disk into the PC memory buffer using the [File],[Load] menu selections. [Enter] on the default screen, and the chip fuse options can be selected, that is, Power-on Timer enabled, and Watchdog Timer and Code Protection disabled. The type of clock circuit used in the target system must also be selected: XT (crystal) or RC for the BINx hardware. The program can now be sent to the chip by hitting [F5]. Ensure that a chip is in the programmer (the right way round!) and the programmer unit is connected correctly to the selected COM port on the PC and is powered up.

Many third party software/hardware packages are also available for programming PIC chips, and designs for simple programmer units are regularly published in hobby magazines. These designs are continuously updated as new PIC chips become available. An upgraded serial programmer unit is available for use with MPLAB, which is controlled from the Windows Integrated Development Environment (IDE), but the programming options will essentially be as described above.

7.8 Program Testing

The program can now be run on the actual hardware, and checked for correct operation. In a commercial product, a test schedule must be devised and correct operation to that schedule confirmed and recorded. The test procedure should check all possible input sequences, not just the correct ones! It is in fact quite difficult to be sure that complex programs will always be 100% reliable. A test schedule for BIN4 is suggested in Table 7.7.

Table 7.7 Basic test schedule for BIN3

	Test	Correct operation	Checked ✓
1	Check PIC connections	Correct orientation and pins	
2	Power up	LEDs off	
3	Clock frequency	100 kHz	
4	Press RUN	Count on LEDs	
5	Release RUN	LED count halted	
6	Press and release RESET	LEDs off	
7	Press RUN	Count on LEDs from zero	

If the !MCLR input is enabled, the program will start immediately on power up. To ensure a 'clean' start, the Power-on Timer fuse should be enabled when programming the chip as indicated above.

Summary

- The development process consists of application specification, hardware selection and design, and program development and testing.
- The program is converted from an algorithm to assembler source code, FILENAME.ASM, using the instruction format specified for the assembler.
- The assembler converts the source code text into object code, FILENAME.HEX. Any syntax errors are detected and listed.
- A list file, FILENAME.LST, is created which lists the source code, object code, label and memory allocation.
- The simulator allows the machine code to be tested without downloading to the actual chip. Logical errors can be detected at this stage.
- The program can then downloaded and tested in the target hardware, using a test schedule developed from the specification.

Questions

1. Place the following program development steps in the correct order: test in hardware, simulate, assemble, edit source code, download.
2. Suggest some advantages of using 'C' as the programming language.
3. State two advantages of using subroutines.
4. State the instruction for incrementing the register 0F.
5. In which register must a port data direction code be placed prior to using the TRIS instruction?
6. How could you halve the delay time in BIN4?
7. State the MPSIM command to simulate an input connected to Port A, bit 0.
8. State the chip fuse settings which should be selected when downloading BIN4. Why is it generally desirable to enable the Power-on Timer?

Activities

1. Start up DOS editor or MPLAB, create a source code file for BIN3, and enter the assembler code program, leaving out the comments. Assemble, correct any errors and simulate. Check that the Port B (F6) file register operates as required.

2. Construct a prototype circuit and test the program to the test schedule given in Table 7.7. Refer forward to Chapter 12 if necessary.

3. Modify the program as BIN4, and confirm that its operation is essentially the same.

4. Modify the program to rotate a high bit (LED on) through each output bit position, starting with RB0, continuously, without the switch control.

Answers

4. INCF 0F.

5. W.

6. Delay count = 80 h.

7. sc ra0.

Chapter 8
PIC 16F84 Architecture

An overview of programming the PIC microcontroller has been provided in Chapter 7, and we now need to look at the PIC internal hardware arrangement. The key reference is Fig. A.1 in Appendix A, the 'PIC 16F8X block diagram' (the 'X' means any number, in this case 3 or 4). The data sheet contains all the details of the internal architecture discussed in this chapter.

8.1 Block Diagram

A simplified internal architecture (Fig. 8.1) has been derived from the block diagram given in the data sheet. Some features seen in the manufacturer's block diagram have been left out because they are not important at this stage. The functional parts of the chip are shown as blocks, with the main address paths identified as block arrows. The 8-bit data paths are shown as single arrows in this diagram, and the timing and control block has connections to all other blocks that set up the processor operations.

The file register set contains various control and status registers, as well as the port registers and the program counter. The most commonly used are the ports (A and B), status register (STATUS), real time clock counter (TMR0) and interrupt control (INTCON). There are also a number of spare general purpose registers (GPRx) which can be used as data registers, counters and so on. The file registers are numbered 00–4F, but are usually given suitable labels in the program source code. File registers also give access to a block of EEPROM, a non-volatile data memory.

The timing and control circuits contain start-up timers which means that the reset input !MCLR can simply be connected to Vdd, the positive supply. The processor can be clocked at any frequency between 0 and 10 MHz, giving a minimum 400 ns instruction execution period (four clock cycles), and a maximum instruction execution rate of 2.5 Mips (millions of instructions per second).

It can be seen in the block diagrams that the memory and file register address lines are separate from the data paths within the processor. This is referred to as Harvard

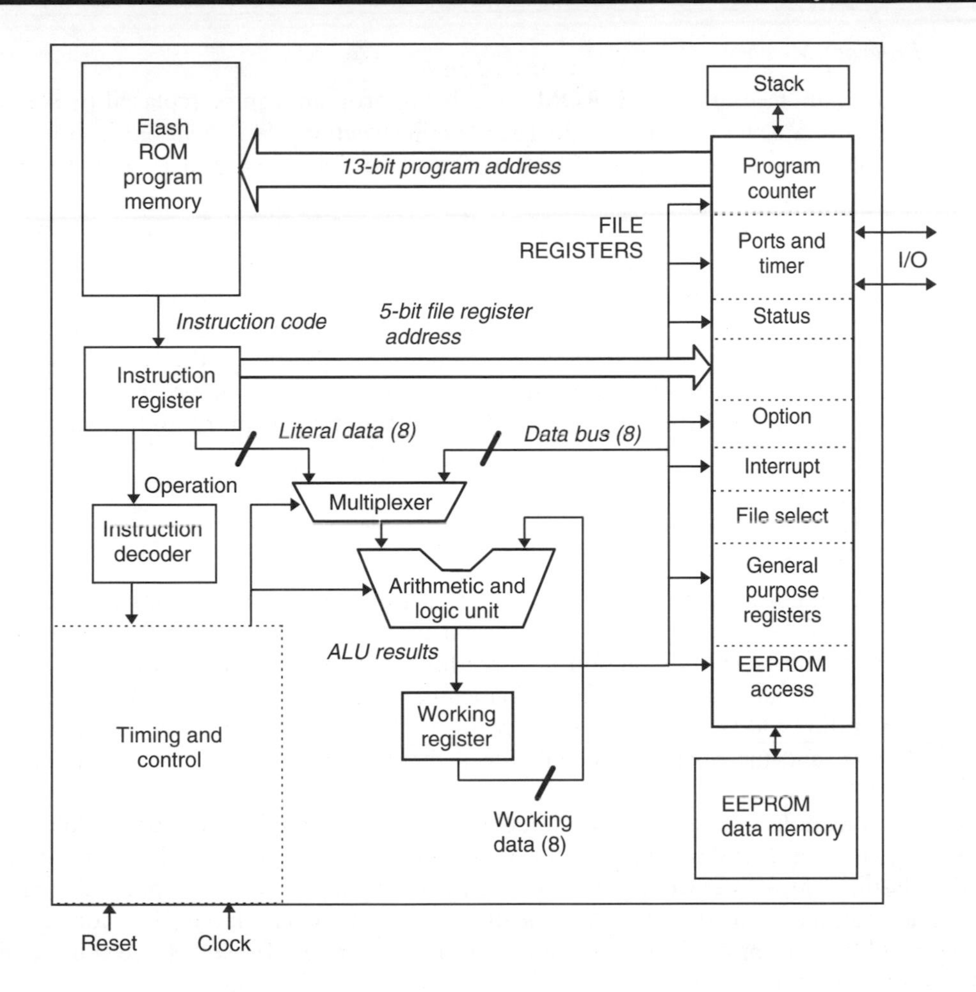

Figure 8.1 PIC 16F84 simplified block diagram.

architecture; it improves the speed of processor operation because data and addresses do not have to share the same bus lines. It also allows 'pipelining' of the instruction execution; as one instruction is executed, the next is being fetched from program memory. The reduced size of the instruction set also speeds up decoding, and the inherently short data path length in a single chip design reduces data transmission time. All these features contribute to a high data bandwidth and speed of operation, compared with CISC microcontrollers.

8.2 Program Execution

The program consists of a list of 14-bit codes which contain both operation code and operand in a fixed length instruction. This machine code program is normally created using a host PC, and downloaded, as outlined in Chapter 7. We are not too concerned here exactly how the downloading is carried out; suffice it to say that the program is received in serial form from the host PC via the I/O port pin, RB7, and written to the program memory while a programming voltage is applied to !MCLR. RB6 acts as a serial data clock.

8.2.1 *Program Memory*

Since the program memory is flash ROM, an existing program can be replaced by simply overwriting with a new program. Up to 1024 (1k) instructions can be stored in program memory. The program counter (PCL) holds the current address, and is reset to 0000 when the chip is powered up or reset.

The user program must therefore start at address 0000, but the first instruction can be GOTO the start of the program at some other labelled address. This is essential when using interrupts, as we shall see later, because the interrupt service routine must be placed at address 0004. The address is fed to the program memory via a 13-bit address bus. As the 16F84 memory only contains 1k locations, the actual address required is only 10 bits, but a standard 13 bit address is used in order to maintain compatibility with other PIC processors with more program memory and other future products.

8.2.2 *Instruction Execution*

This section contains the instruction register, instruction decoder, and timing and control logic. The 14-bit instructions stored in program memory are copied to the instruction register for decoding. The instruction decoder logic converts the instruction input into settings for all the internal control lines. Remember that the 14-bit instruction contains both the operation code and the operand. The instruction decoder will only use the operation code bits, and the operand will provide a literal, file register address or program address.

For example, if the instruction is MOVLW (move a literal into W), the control lines will be set up to feed the literal operand to W via the multiplexer and ALU. If the instruction is MOVWF, the control lines will be set up to copy the contents of W to the specified file register. The operand will be the address of the file register (00–4F) required. If we look at the binary codes for the 'move' instruction codes in the data sheet, Table A.4, we can see the difference in the code structure:

```
 MOVLW k = 11 00xx kkkk kkkk
 MOVWF f = 00 0000 1fff ffff
MOVF f,d = 00 1000 dfff ffff
```

In the MOVLW instruction, the operation code is the high 4 bits (1100), 'x' are 'don't care' bits, and 'k' represents the literal bits, the low byte of the instruction. In the MOVWF instruction, the operation code is 0000001 (7 bits) and 'f' indicates the file register address bits. Only 7 bits are used for the register address, allowing a maximum of 2^7 (= 128) registers to be addressed. In fact, the 16F84 has only 80 registers in all, but 7 bits are still needed to address this number.

In the MOVF instruction the operation code is 001000, and the file register address is needed as before to identify the data source register. However, there is one bit, d, which controls the data destination. This bit must be 0 to direct the data into W, which is the usual operation. To move an 8-bit data word from file register 0C to W, for example, requires the syntax MOVF 0C,W. W is always replaced by 0, the required value of 'd', by the assembler. The default value of 'd' is '1', so where the destination is the file register, we can leave out the second operand in the source code.

8.2.3 Data Processing

This is carried out by the ALU, in conjunction with the data multiplexer and Working Register. The multiplexer allows new data to be fed from the instruction (if a literal) or registers. This may be combined with data from W, or register data manipulated in a single register operation. W is used in register pair operations as the temporary store. Final results are stored back in the file registers, or W, depending on the setting of the data destination bit.

8.2.4 Jump Instructions

If a GOTO instruction is decoded, the program counter will be loaded with the program memory address of the jump destination given as the instruction operand. A program label in the source code will have been replaced by the destination address by the assembler. Any file register bit can be used in a 'Bit Test and Skip' instruction for conditional branching with GOTO.

If a CALL instruction is decoded, the destination address is loaded into the PC in the same way, and the subroutine executed until a RETURN instruction is encountered. The program then jumps back to the address following the CALL. This is achieved by storing the return address on the stack, as part of the subroutine call sequence, and putting it back into PC when the program returns from the subroutine. It works on a last in, first out (LIFO) basis, with the last address stored being the first to be recovered.

8.3 Register Set

All the file registers are 8 bits wide. They are divided into two main blocks, the special function registers, which are reserved for specific purposes, and general purpose registers, which can be used for temporary storage of any data byte. The file register set is shown in Fig. 8.2 in numerical order.

The registers in page 0 (file addresses 00–4F) can be directly addressed, and it is recommended that the register labels given in Fig. 8.2, which match the data sheet, are used as the register labels. Standard header files can be included in your programs that define all the register names. Special instructions are available to access the page 1 registers (80–CF). We have used the instruction TRIS to access the data direction registers TRISA and TRISB already. In a similar way, we will use the instruction OPTION to access this register; this will be needed later to set up the hardware timer, TMR0. Alternatively, a Register Bank select bit in the Status Register can be used to access page 1 file registers, and this method is recommended for more advanced programming.

8.3.1 Special Function Registers

The operation of the special function registers is summarized below, with the emphasis on those that are used most frequently. The functions of all the registers are detailed in the chip data sheet. The shaded registers either do not exist, or are repeated at addresses 80–CF (page 1).

PCL Program Counter Low byte
File register number = 02

Address	Page 0	Page1
0	INDO	
1	**TMRO**	**OPTION**
2	**PCL**	
3	**STATUS**	
4	FSR	
5	**PORTA**	**TRISA**
6	**PORTB**	**TRISB**
7		
8	EEDATA	EECON1
9	EEADR	EECON2
A	PCLATH	
B	**INTCON**	
C	**GPR1**	
D	**GPR2**	
E	**GPR3**	
F	**GPR4**	
10	**GPR5**	
. . . . .	General purpose registers	
4F	**GPR68**	

Figure 8.2 PIC 16F84 file register set.

The program counter contains (points to) the address of the instruction currently being executed (or the next), and counts from 000 to 3FF. The order of this count is modified when the program jumps. The PCL register contains only the eight low bits (00–FF) of the whole program counter, with the two high bits (00–03) stored in the PCLATH register (0A). We only need to worry about the high bits if the program is longer than 255 instructions in total, which is not the case for any of the demonstration programs.

PORTA Port A data register
File register number = 05

Port A has five I/O bits, RA0–RA4. Before use, the data direction for each pin must be set up by loading the TRISA register with a data direction code. If a bit is set to output, data moved to this register appears at the output pins of the chip. If set as input, data presented to the pins can be acted on immediately, or stored for later use by moving the data to a spare register. Examples of this have already been seen in earlier chapters. RA4 can alternatively be used as an input to the Counter Timer Register (TMR0) for counting applications. The use of the hardware timer will be covered in Chapter 9. The PORTA register bit allocation is shown in Table 8.1.

All registers are read and written in 8-bit words, so we sometimes need to know what will happen with unused bits. When the Port A data register is read within a program (MOVF), the three unused bits will be seen as ‘0’. When writing to the port, the three high bits are simply ignored. When used as outputs, the port lines are able to provide up to 20 mA of

Table 8.1 PIC 16F84 port bit functions

Register bit	*Chip pin label*	*Function*
PORT A		
0	RA0	Input or Output
1	RA1	Input or Output
2	RA2	Input or Output
3	RA3	Input or Output
4	RA4/TOCKI	Input or Output or Input to TMRO
5	–	None
6	–	None
7	–	None
PORT B		
0	RBO/INT	Output or Input or Interrupt Input
1	RB1	Output or Input
2	RB2	Output or Input
3	RB3	Output or Input
4	RB4	Output or Input + Interrupt on change
5	RB5	Output or Input + Interrupt on change
6	RB6	Output or Input + Interrupt on change
7	RB7	Output or Input + Interrupt on change

current (except RA4), in or out, which is enough to drive our LEDs in the demonstration circuit. An equivalent circuit for each port pin is given in the data sheet.

TRISA Port A data direction register
File register number = 85

The data direction of the port pins can be set, bit by bit, by loading this register with a suitable binary code, or the hex equivalent. A '1' sets the corresponding port bit to input, while a '0' sets it to output. Thus, to select all bits as inputs, the data direction code is 1111 1111 (FFh), and for all outputs is 0000 0000 (00h).

When the chip is powered up, these bits default to '1', so it is not necessary to initialize for input, only for output. This makes sense if you think about it, because if the pin is incorrectly wired up, it is more easily damaged if set to output. For instance, if the pin is accidentally grounded, and then driven to a high state, the short circuit current is likely to damage the output circuit.

The data direction register TRISA is loaded by placing the required code in W and then using the instruction TRIS 05 or TRIS 06 for Port A and Port B, respectively. Alternatively, all file registers with addresses 80–CF can be addressed directly, using the page selection bits in the Option register, and this may be seen in more advanced programs.

PORTB Port B data register
File register number = 06

Port B has the full set of eight I/O bits, RB0–RB7. If a bit is set to output, data moved to this register appears at the output pins of the chip. If set as input, data presented to the pins

Table 8.2 Status register (STATUS) bit functions

Bit	Label	Name	Function
0	C	Carry Flag	*Set if register operation causes a carry out of bit 8 of the result (8-bit operations)*
1	DC	Digit Carry Flag	*Set if register operation causes a carry out of bit 3 of the result (4-bit operations)*
2	Z	Zero Flag	*Set if the result of a register operation is zero*
3	PD	Power Down	*Cleared when the processor is in Sleep mode*
4	TO	Time Out	*Cleared when Watchdog Timer times out*
5	RP0	Register Bank Select Bits	*RP0 selects File Registers 00–7F or 80–FF*
6	RP1		*RP1 not used*
7	IRP		*IRP not used*

can be read at this address. The data direction is set in TRISB, as described above, and all bits default to input on power up. The PORTB register bit allocation is shown in Table 8.1.

Bit 0 of Port B has a dual function, and it can be initialized, using the Interrupt Control Register (INTCON), to enable the processor to respond to this input with an INTERRUPT sequence. In this case, the processor is forced to jump to a predefined Interrupt Service Routine upon completion of the current instruction. The processor can also be initialized to respond to a change on any of the bits RB4–RB7.

TRISB Port B data direction register
File register number = 86

As for Port A, the data direction can be set, bit by bit, by loading this register with a suitable binary code, or the hex equivalent, where '1' (default) sets an input, and '0' sets an output (must be initialized). The program instruction 'TRIS 06' moves the data direction code from W to the TRISB register.

STATUS Status (or flag) register
File register number = 03

Individual bits in the status register record information about the result of the previous instruction. Possibly the most commonly used is the zero flag, bit 2; when the result of any operation is zero, this zero flag bit is set to '1'. It is used by the Decrement/Increment and Skip if Zero instructions, and can be used by the Bit Test and Skip instructions, to implement conditional branching of the program flow. The status register bit functions are shown in Table 8.2. We can leave consideration of the rest of these for the moment; the data sheet and more advanced programming references will provide more information.

TMR0 Timer zero register
File register number = 01

A timer/counter register counts the number of pulses applied to a clock input. The binary count can be read from the register when the count is finished. TMR0, being an 8-bit register, can count up to 255 pulses. For external inputs, the pulses are applied at pin

Table 8.3 Option register (OPTION) bit functions

Bit	*Label*	*Name*	*Function*
0	PS0	Prescaler Rate Select Bit 0	*These 3 bits form a 3-bit code to select one of 8 prescale values for the counter/timer TMR0 or Watchdog Timer (WDT)*
1	PS1	Prescaler Rate Select Bit 1	
2	PS2	Prescaler Rate Select Bit 2	
3	PSA	Prescaler Assignment	*Assigns Prescaler to WTD or TMR0*
4	T0SE	Timer Zero Source Edge Select	*Select rising or falling edge trigger for T0CKI input at RA4*
5	T0CS	Timer Zero Clock Source Select	*Select Timer/Counter input as RA4 or internal clock*
6	INTEDG	Interrupt Edge Select	*Select rising or falling edge trigger for RB0 interrupt input*
7	RBPU	Port B Pull-up Enable	*Enable pull-ups on Port B pins so input data defaults to '1'.*

RA4. When used as a timer, the internal clock is used to supply the pulses. If the processor clock frequency is known, the time taken to reach a given count can be calculated. When the counter rolls over from FF to 00, an interrupt flag (see INTCON below) is set, if enabled. This allows the processor to check if the count is complete, or to be alerted when a set time interval has elapsed, even if it is doing something else at the time. The timer register can be read and written directly, so a count can be started at a preset value. The timer zero label refers to the fact that other PICs have more than one timer/counter register, but the 16F84 has only one. More details on using the TMR0 are given in Chapter 9.

OPTION Option register
File register number = 81

Table 8.3 details the Option register bit functions. The counter/timer operation is controlled by OPTION register bits 0–5. When used as a timer, the processor clock signal is used to increment the counter register. If a crystal clock is in use, the timing will be very accurate. Pre-scaling can be selected to increase the maximum time interval; this means dividing down the timer input frequency by a factor of 2, 4, 8, 16, 32, 64, 128 or 256. As is the case with the TRISA and TRISB registers, the OPTION register is accessed using a special instruction, namely 'OPTION'. The alternative method, which is recommended by the manufacturers, uses page selection.

INTCON Interrupt control register
File register number = 0B

The INTCON bit functions are given in Table 8.4. An interrupt is a signal which causes the current program execution to be suspended, and an interrupt service routine (ISR)

Table 8.4 Interrupt control register (INTCON) bit functions

Bit	*Label*	*Name*	*Function*
0	RBIF	Port B Change Interrupt Flag	*Set when any one of RB4 to RB7 changes state*
1	INTF	RB0 pin Interrupt Flag	*Set when RB0 detects interrupt input*
2	T0IF	Timer Overflow Interrupt Flag	*Set when Timer TMR0 rolls over from FF to 00*
3	RBIE	Port B Change Interrupt Enable	*Set to enable Port B change interrupt*
4	INTE	RB0 pin Interrupt Enable	*Set to enable RB0 interrupt*
5	T0IE	Timer Overflow Interrupt Enable	*Set to enable Timer Overflow interrupt*
6	EEIE	Data EEPROM Write Interrupt Enable	*Set to enable interrupt on completion of write operation to non-volatile data memory*
7	GIE	Global Interrupt Enable	*Enable all interrupts which have been selected*

carried out. An interrupt can be generated by an external device, via Port B, or from the timer. However, in all cases, the ISR must start at address 004 in the program memory. If interrupts are in use, an unconditional jump from address zero (i.e., the program start address) to a higher start address, is normally needed. The INTCON register contains three interrupt flags and five interrupt enable bits, and these must be set up as required during the program initialization by writing a suitable code to the INTCON register.

Other special function registers

Registers EEDATA, EEADR, EECON1 and EECON2 are used to access the non-volatile ROM data area. PCLATH acts as a holding register for the high bits (12:8) of the program counter. The File Select Register (FSR) acts as a pointer to the file registers. It can be used with IND0, which gives indirect access to the file register selected by FSR. This is useful for

Table 8.5 Other SFRs

Number	*Name*	*Function*
00 04	INDF FSR	*File Register Memory indirect addressing for block access*
0A	PCLATH	*Program Counter High Byte*
08 09	EEDATA EEADR	*Data EEPROM indirect addressing for block access*
88 89	EECON1 EECON2	*Data EEPROM Read and Write Control*

a block read or write to the GPRs, particularly for saving a set of data which has been read in at port. More information on this is given in Section 9.4.3.

8.3.2 General Purpose Registers (GPR1–GPR68)

The GPRs are numbered 0C–4F. They are also referred to as SRAM registers, because they can be used as a small block of static RAM for storing blocks of data, such as a data table of values needed temporarily in the program. We have already seen an example of using the first register (address 0C) as a counter register in a delay loop. The register was labelled 'timer', pre-loaded with a value, and decremented until it reached zero. This is a common type of operation, and not only used for timing loops. For example, a counting loop can be used for providing an output a certain number of times. We could have used any of the GPRs for this function because they are all operationally identical; however, we do need to declare a different name for each when using more than one.

Summary

- The following features of PIC chips enhance performance: Harvard/RISC architecture, instruction pipelining, high clock rate, single chip system.
- The 16F84 internal architecture can be represented as a block diagram showing the main functional blocks, which are: program ROM, execution logic, data processing, file registers and data EEPROM.
- The program memory stores up to 1024 14-bit instructions. The program execution starts at address 0000.
- The 14-bit instruction contains operation code and operands, which can vary in length.
- The ALU processes data from the instruction, registers or W.
- Jump instructions modify the program counter to go to another point in the program sequence.
- The file register set contains special function registers and general purpose registers.
- The most important of the SFRs (with their address/number) are the Timer (01), Program Counter (02), Status Register (03), Port A (05), Port B (06) and Interrupt Control (0B).
- The GPRs are a block of registers that can be used separately, or in blocks, to store temporary data, act as counters, and so on.

Questions

1. State the function of the following blocks within a PIC microcontroller: program memory, program counter, instruction decoder, ALU, W.
2. Why is it not necessary to initialize a PIC port for input?

3. List the main functions of the ALU, and state one single register operation and one register pair operation.

4. Why is the stack needed for subroutine execution?

5. State the function of the following PIC file registers: PORTA, TRISA, TMR0, PCLATH, GPRxx.

6. State the function of the register bits: STATUS,2; INTCON,1; OPTION,5.

7. Which port pin gives access to TMR0?

8. What is the default destination of a 'move' operation?

Activities

1. Compile a logic table which shows the binary code on the following internal connections, and contained in the registers, after each instruction cycle while the program BIN1 is executed. The first is given as a guide.

INSTRUCTION NUMBER	
Address:	0000
Instruction:	MOVLW 00
Machine Code:	3000
Program Address Bus (13):	0 0000 0000 0000
File Register Address (5):	X XXXX
Instruction Code Register:	0011 0000
Literal Bus:	0000 0000
8-Bit Data Bus:	XXXX XXXX
Working Register:	0000 0000
PORTB:	XXXX XXXX
TRISB:	XXXX XXXX

Create a table, and complete additional columns for each of the remaining four instructions.

2. Study the PIC 16F84 data sheet, and any other suitable references, and detail the meaning of the following terms, in the context of the PIC architecture: instruction pipelining, indirect addressing, watchdog timer, context saving, configuration bits, reset vector.

Chapter 9
Further Programming Techniques

Now that the basic programming methods have been introduced, we can look at some more advanced techniques. Sample programs demonstrating use of the timer, interrupts and data table are included in this chapter.

9.1 Program Timing

The PIC 16F84 data sheet, Fig. A.2, shows the clock signals in the processor. Each instruction takes four clock periods to execute. The diagram shows four internal clock signals (Q1–Q4) derived from the oscillator that provide a pulse during each of the four clock cycles. These are used to trigger sequential operations in the processor that fetch the instruction code from the program memory, and copy it to the instruction register. The instruction is then executed by the instruction decoder, and the processor control lines are set up to carry out the required process.

9.1.1 Pipelining

The data sheet diagram, Example A.1, shows how the instruction fetch and execute cycles are carried out simultaneously. This is achieved by overlapping the instruction execution operations, such that, while one instruction is being executed, the next is being fetched from the program memory into the instruction register. This overlapping of execution stages is called pipelining, and the PIC is described as having a two stage pipeline. CISC microprocessors such as the Pentium use more elaborate multistage pipelining.

Table 9.1 Sequence execution time

Label	*Instruction*	*Operand*	*Time (cycles)*
delay	MOVLW	0xFF	1
	MOVWF	timer	+ 1
down	DECFSZ	timer	+ (1x255)
	GOTO	down	+ (2x254) + 1
	RETURN		+ 2
		Total	768

If	Clock Frequency	=	4 MHz
then	Instruction Frequency	=	1 MHz
and	Instruction Period	=	1 μs
and	Total Delay Time	=	768 μs

9.1.2 Execution Time

The fact that all instructions take four clock cycles (except jumps) means that it is relatively easy to work out the overall execution time for, say, a software delay loop. The delay loop from program BIN3 is shown in Table 9.1. The PIC 16F84 clock can be driven at any speed from 0 up to 10 MHz; using a clock rate of 4 MHz makes the calculations straightforward, because the instruction cycle time is 1 μs. The sequence execution time has been worked out for this clock rate.

The move instructions take one cycle each, and the DECFSZ instruction is then repeated 254 times. The GOTO takes two cycles, because each time the GOTO is executed, the RETURN is pre-fetched into the pipeline, and then not executed, so a cycle is wasted. On the 255th loop, the register becomes zero and the GOTO is skipped, and the RETURN executed. This also takes two cycles, because of the wasted pre-fetch cycle, but is only executed once per delay sequence. The total loop time can then be calculated, by totalling the time taken for each instruction and the loop. As we can see, this comes to 768 μs, at 4 MHz. This figure can be confirmed if the program containing the loop is run in the simulator with the clock frequency set to 4 MHz.

The full instruction set in the data sheet gives the number of cycles taken, one or two, by each instruction. The block execution time for a section of code can therefore be predicted before testing in simulator or hardware. Alternatively, the timing can be checked and modified using the simulator. Incidentally, NOP (no operation) is useful here. For time critical sequences, NOP may be used to insert a delay of one instruction cycle, that is, four clock cycles; it has no other effect.

9.2 Hardware Counter/Timer

Accurate event timing and counting is frequently needed in microcontroller programs. For example, if we have a sensor on a motor shaft that gives one pulse per revolution of the shaft, the number of pulses per second will give the shaft speed. Alternatively, the interval between pulses can be measured, using a timer, to obtain the speed by calculation. A process for doing this would be:

1. wait for pulse;
2. read and reset the timer;
3. restart the timer;
4. process previous timer reading;
5. go to 1.

If an independent hardware timer is used, the program can carry on with other operations, such as processing the timing information, controlling the outputs and checking the sensor input, while the timer keeps an accurate record of the time elapsed.

9.2.1 Using TMR0

The special file register 01, Timer Zero (TMR0), which can be used as a counter or timer which, once started, runs independently of the program execution. This means it can count inputs or clock pulses concurrently with (at the same time as) the main program. The counter/timer can also be set up to generate an interrupt when it has reached its maximum value, so that the main program does not have to keep checking it to see if a particular count has been reached. A block diagram of TMR0 and its associated hardware and control registers is shown in Fig. 9.1.

As an 8-bit register, TMR0 can count from 00 to FF (255). The operation of the timer is set up by moving a suitable control code into the OPTION register. The counter is then clocked by an external pulse train, or from the chip oscillator. When it reaches its maximum value, FF, and is incremented again, it 'rolls over' to 00. This register 'overflow' is recorded by the INTCON (Interrupt Control) register, bit 2 (T0IF), going to '1' (assuming that it has been previously enabled and cleared). This condition can be checked by bit testing in the program, or can trigger an interrupt.

9.2.2 Counter Mode

The simplest mode of operation of TMR0 is counting pulses applied to RA4, which has the alternative name T0CKI, Timer Zero Clock Input. These pulses could be input manually from a push button, or, more likely, would be produced by some other signal source, such as the sensor on the motor shaft. If the sensor produces one pulse per revolution of the shaft, and one of the PIC outputs controls the motor, the microcontroller could be programmed to rotate the shaft by a set number of revolutions. If the motor were geared down, a positioning system could be designed to move the output through a set angle, in a robot, for example.

In order to increase the range of this kind of measurement, the prescaler allows the number of pulses received by the TMR0 register to be divided by a factor of 2, 4, 8, 16, 32, 64, 128 or 256. The ratio is selected by loading the least significant three bits in the OPTION register as follows: 000 selects divide by 2, 001 divide by 4, and so on up to 111 for divide by 256. TMR0 can also be pre-loaded with a value, and the overflow detected when it has been 'topped up' by a set number of pulses.

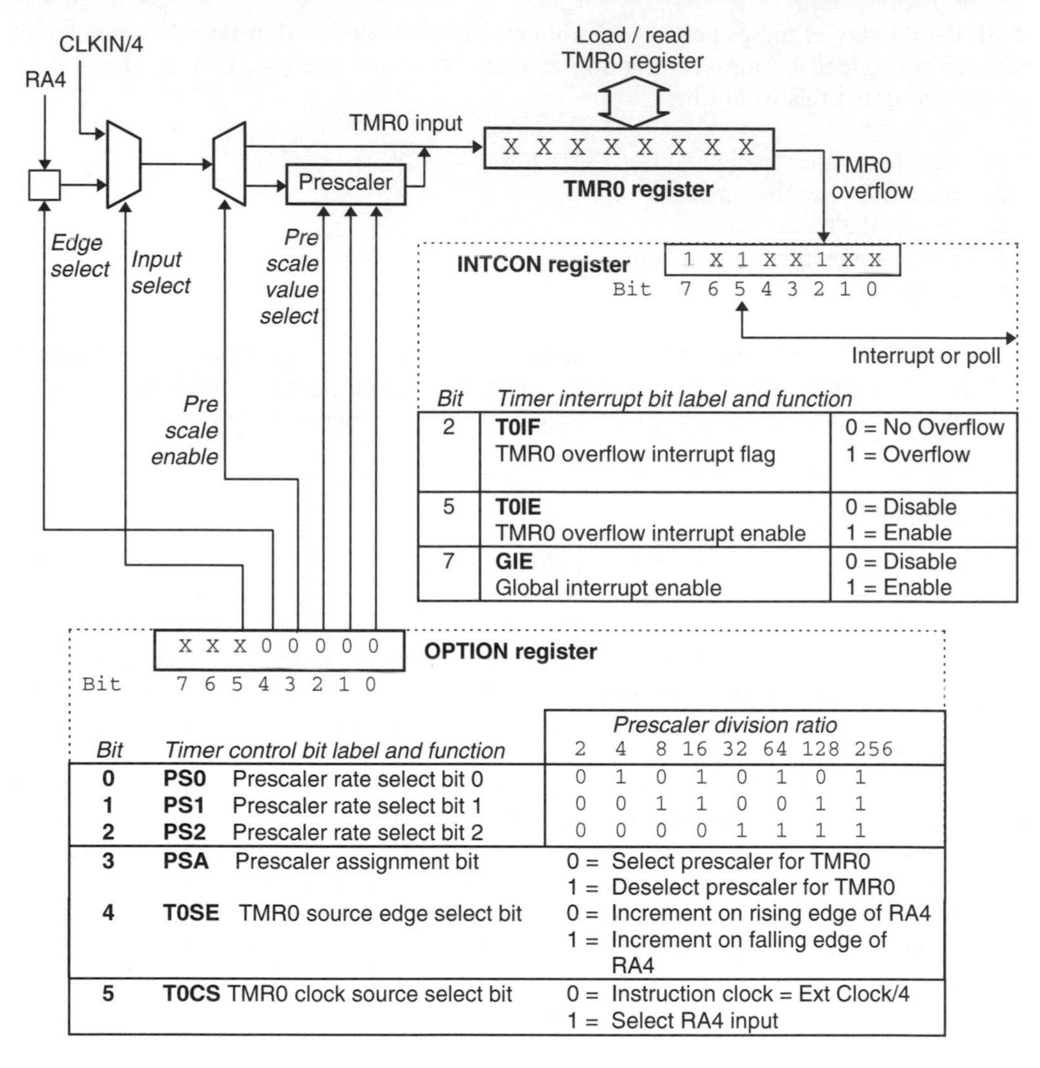

Figure 9.1 Hardware Counter/Timer set-up and operation.

9.2.3 Timer Mode

The internal clock is selected by setting the OPTION register, bit 5, to zero. To use TMR0 as an accurate hardware timer, a crystal oscillator must be used as the chip clock source. A convenient crystal frequency is 4 MHz, because it is divided by four before it is fed to the input of TMR0, giving a pulse frequency of 1 MHz. The counter would then be clocked every 1 µs exactly, and would take 256 µs to count from zero to zero again. Again, by pre-loading with a suitable value, a smaller time interval could be selected, with time out indicated by the timer interrupt flag. For example, by pre-loading with the value 156 (9C), the overflow would occur after 100 µs. Alternatively, the time period measured can be extended by selecting the prescaler. The maximum timer period would then be 512 µs, 1024 µs and so on to 65.536 ms. Crystals are also available in frequencies that are more conveniently divisible by 2. For example, a 32.768 kHz crystal frequency will produce a time out every 1.0000 seconds, if the prescale value of 32 is selected.

In Fig. 9.1, TMR0 is set up with xxx00000 in the OPTION register, selecting the internal clock source, with a prescale value of 2. The INTCON register has been set up with the timer interrupt enabled and the timer overflow interrupt flag has been set (overflow has occurred).

9.2.4 *Program TIM1*

Program TIM1 source code, which demonstrates the use of the timer, is listed as Program 9.1. It is designed to increment the output once per second. The program uses the same demonstration hardware as the previous programs, with eight LEDs displaying the contents of Port B. An adjustable CR clock is used, set to give a frequency of 65536 Hz (approximately). This frequency is divided by 4, and is then divided by 64 in the prescaler, giving an overall frequency division of $4 \times 64 = 256$. The timer register is therefore clocked at $65536/256 = 256$ Hz. The timer register counts from 0 to 256, and so overflows every second. The output is then incremented; it will take 256 s to complete the 8-bit binary output count.

9.2.5 *Timing Problems*

Each instruction in the program takes four clock cycles to complete, with jumps taking eight cycles. If the program sequence is studied carefully, extra time is taken in completing the program loop before the timer is restarted. In this application, it will cause only a small error, but in other applications it may be significant.

In TIM1, notice that the program has to keep checking to see if the time-out flag has been set by the timer overflowing. It is often more efficient to allow the processor to carry on with

Program 9.1 TIM1 source code

```
; ***********************************************************
;       TIM1.ASM          M. Bates 6/1/99              Ver 1.2
; ***********************************************************
;
; Minimal program to demonstrate the hardware timer operation.
;
; The counter/timer register (TMR0) is initialized to
; zero and driven from the instruction clock with a
; prescale value of 64.
; T0IF is polled while the program waits for time out.
; When the timer overflows, the Timer Interrupt Flag (T0IF) is
; set. The output LED binary display is then incremented.
; With the clock adjusted to 65536 Hz, the LSB LED flashes at
; 1 Hz.
;
;       Processor:        PIC 16F84
;
                                                   continued...
```

```
;       Hardware:          PIC BIN Demo Hardware
;       Clock:             CR = 65536 Hz
;       Outputs:           RB0 - RB7: LEDs (active high)
;       WDTimer:           Disabled
;       PUTimer:           Enabled
;       Interrupts:        Disabled
;       Timer:             Internal clock source
;                          Prescale = 1:64
;       Code Protect:      Disabled
;
;       Subroutines:       None
;       Parameters:        None
;
; ************************************************************

; Register Label Equates.....................................

TMR0    EQU     01                  ; Counter/Timer Register
PORTB   EQU     06                  ; Port B Data Register (LEDs)
INTCON  EQU     0B                  ; Interrupt Control Register

T0IF    EQU     2                   ; Timer Interrupt Flag

; ************************************************************

; Initialize Port B (Port A defaults to inputs)...............

        MOVLW   b'00000000'         ; Set Port B Data Direction
        TRIS    PORTB

        MOVLW   b'00000101'         ; Set up Option register
        OPTION                      ; for internal timer/64

        CLRF    PORTB               ; Clear Port B (LEDs Off)


; Main output loop ..........................................

next    CLRF    TMR0                ; clear timer register
        BCF     INTCON,T0IF         ; clear time-out flag

check   BTFSS   INTCON,T0IF         ; wait for next time-out
        GOTO    check               ; by polling time-out flag

        INCF    PORTB               ; Increment LED Count
        GOTO    next                ; repeat forever...

        END                         ; Terminate source code

;.............................................................
```

some other process while the timer runs, and allow the time-out condition to interrupt the main program when it has finished. This would then be an example of a timer interrupt.

9.3 Interrupts

Interrupts can be generated by an internal or external event, and the interrupt signal can be received at any time during the execution of the main process. For example, when you hit the keyboard or move the mouse on a PC, an interrupt signal is sent to the processor from the keyboard interface to request that the key be read in, or the mouse movement transferred to the screen. This action is called the 'interrupt service routine' (ISR). When the ISR has finished its task, the process that had been interrupted must be resumed as though nothing has happened. This means that any information being processed at the time of the interrupt may have to be stored temporarily. The program counter is saved automatically on the stack, as is the case when a subroutine is called, so that the program can return to the original execution point after the ISR has been completed. This system allows the CPU to get on with other tasks, such as printing or disk access, without having to keep checking all the possible input sources.

9.3.1 Interrupt Setup

A block diagram detailing the PIC 16F84 interrupt system is given in Fig. 9.2. The PIC has four possible interrupt sources.

1. RB0 can be selected as an edge-triggered interrupt input by setting INTCON,4 (INTE), with the active edge selected by OPTION,6 (INTEDG).
2. RB7–RB4 can be selected to trigger an interrupt if any of them changes state, by setting INTCON,3 (RBIE).
3. TMR0 overflow can be selected by setting INTCON,5 (T0IE).
4. Completion of an EEPROM write operation.

If interrupts are required, the interrupt source must be enabled in the INTCON register. First, the Global Interrupt Enable bit, which enables all interrupts, must be set (INTCON, 7) and then the specific interrupt bit must be set.

9.3.2 Interrupt Execution

Interrupt execution is also illustrated in Fig. 9.2. Each interrupt source has a corresponding flag, which is set if the interrupt event has occurred. For example, if the timer overflows, T0IF (INTCON,2) is set. When this happens, and the interrupt is enabled, the current instruction is completed and the next program address is saved on the stack. The program counter is then loaded with 004, and the routine found at this address is executed. Alternatively, location 004 can contain a 'GOTO addlab' (address label) if the ISR is to be placed elsewhere in program memory. If interrupts are to be used, a GOTO must also be used at the reset vector address, 000, to redirect the program counter to the start of the main program at a higher memory address, because the ISR (or GOTO addlab) will occupy address 004. The ISR must be created and allocated to address 004 (ORG 004) as part of the program source code.

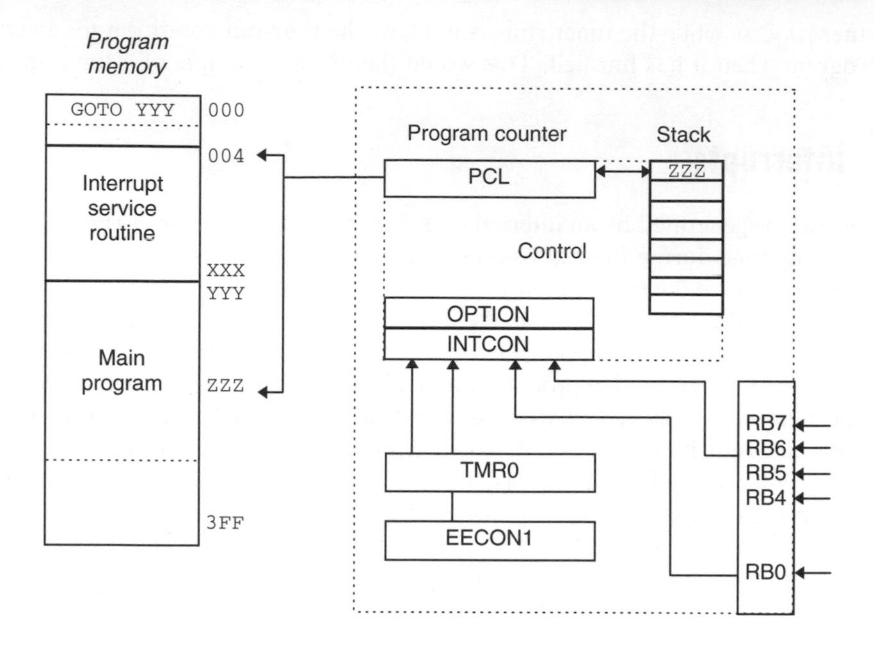

Interrupt Control Bit Functions

	Bit	*Label*	*Function*	*Settings*
INTCON	0	**RBIF**	Port B (4:7) interrupt flag	0 = No change 1 = Bit change detected
	1	**INTF**	RB0 interrupt flag	0 = No interrupt 1 = Interrupt detected
	2	**T0IF**	TMR0 overflow interrupt flag	0 = No overflow 1 = Overflow detected
	3	**RBIE**	Port B (4:7) interrupt enable	0 = Disabled 1 = Enabled
	4	**INTE**	RB0 interrupt enable	0 = Disabled 1 = Enabled
	5	**T0IE**	TMR0 overflow interrupt enable	0 = Disabled 1 = Enabled
	6	**EEIE**	EEPROM write complete interrupt enable flag	0 = Disabled 1 = Enabled
	7	**GIE**	Global Interrupt enable	0 = Disabled 1 = Enabled

OPTION	6	**INTEDG**	RB0 interrupt active edge select	0 = Falling edge 1 = Rising edge

Figure 9.2 Interrupt set-up and operation.

The ISR must be terminated with the instruction RETFIE (return from interrupt). This causes the original program address to be pulled from the stack, and program execution resumes at the instruction following the one which was interrupted. It may be necessary to save other registers as part of the ISR, so that they can be restored after the interrupt. This is called 'context saving'. This is illustrated below by saving and restoring the contents of Port B data register as part of the ISR.

9.3.3 *Interrupt Demo Program INT1*

A demonstration program, Program 9.2, illustrates the use of interrupts in a PIC program. The program outputs the same binary count to Port B, as seen in the BINx programs, to represent its normal activity. This process will be interrupted by RB0 being pulsed

Program 9.2 INT1 source code

```
;
*************************************************************
; INT1.ASM M.    Bates 12/6/99                      Ver 2.1
;
*************************************************************
;
;       Minimal program to demonstrate interrupts.
;
;       An output binary count to LEDs on PortB, bits 1-7
;       is interrupted by an active low input at RB0/INT.
;       The Interrupt Service Routine sets all outputs high,
;       and waits for RA4 to go low before returning to
;       the main program.
;       Connect push button inputs to RB0 and RA4
;
;       Processor:          PIC 16F84
;       Hardware:           PIC Modular Demo System
;                           (reset switch connected to RB0)
;       Clock:              CR ~100 kHz
;       Inputs:             Push Buttons
;                           RB0 = 1 = Interrupt
;                           RA4 = 0 = Return from Interrupt
;       Outputs:            RB1-RB7: LEDs (active high)
;
;       WDTimer:            Disabled
;       PUTimer:            Enabled
;       Interrupts:         RB0 interrupt enabled
;       Code Protect:       Disabled
;
;       Subroutines:        DELAY
;       Parameters:         None
;
*************************************************************

; Register Label Equates.................................

PORTA   EQU     05                  ; Port A Data Register
PORTB   EQU     06                  ; Port B Data Register
INTCON  EQU     0B                  ; Interrupt Control Register
timer   EQU     0C                  ; GPR1 = delay counter
tempb   EQU     0D                  ; GPR2 = Output temp. store

; Input Bit Label Equates ...............................

intin   EQU     0                   ; Interrupt input = RB0
resin   EQU     4                   ; Restart input = RA4
                                               continued...
```

```
INTF    EQU     1                    ; RB0 Interrupt Flag

;
*************************************************************

; Set program origin for Power On Reset......................

        org     000                  ; Program start address
        GOTO    setup                ; Jump to main program start

; Interrupt Service Routine at address 004...................

        org     004                  ; ISR start address

        MOVF    PORTB,W              ; Save current output value
        MOVWF   tempb                ; in temporary register

        MOVLW   b'11111111'          ; Switch LEDs 1-7 on
        MOVWF   PORTB

wait    BTFSC   PORTA,resin          ; Wait for restart input
        GOTO    wait                 ; to go low

        MOVF    tempb,w              ; Restore previous output
        MOVWF   PORTB                ; at the LEDs
        BCF     INTCON,INTF          ; Clear RB0 interrupt flag

        RETFIE                       ; Return from interrupt

; DELAY subroutine...........................................

delay   MOVLW   0xFF                 ; Delay count literal is
        MOVWF   timer                ; loaded into spare register
down    DECFSZ  timer                ; Decrement timer register
        GOTO    down                 ; and repeat until zero then
        RETURN                       ; return to main program

; Main Program ***********************************************

; Initialize Port B (Port A defaults to inputs)..............

setup   MOVLW   b'00000001'          ; Set data direction bits
        TRIS    PORTB                ; and load TRISB

        MOVLW   b'10010000'          ; Enable RB0 interrupt in
        MOVWF   INTCON               ; Interrupt Control Register

; Main output loop...........................................

count   INCF    PORTB                ; Increment LED display
        CALL    delay                ; Execute delay subroutine
        GOTO    count                ; Repeat main loop always

END                                  ; Terminate source code

; ***********************************************************
```

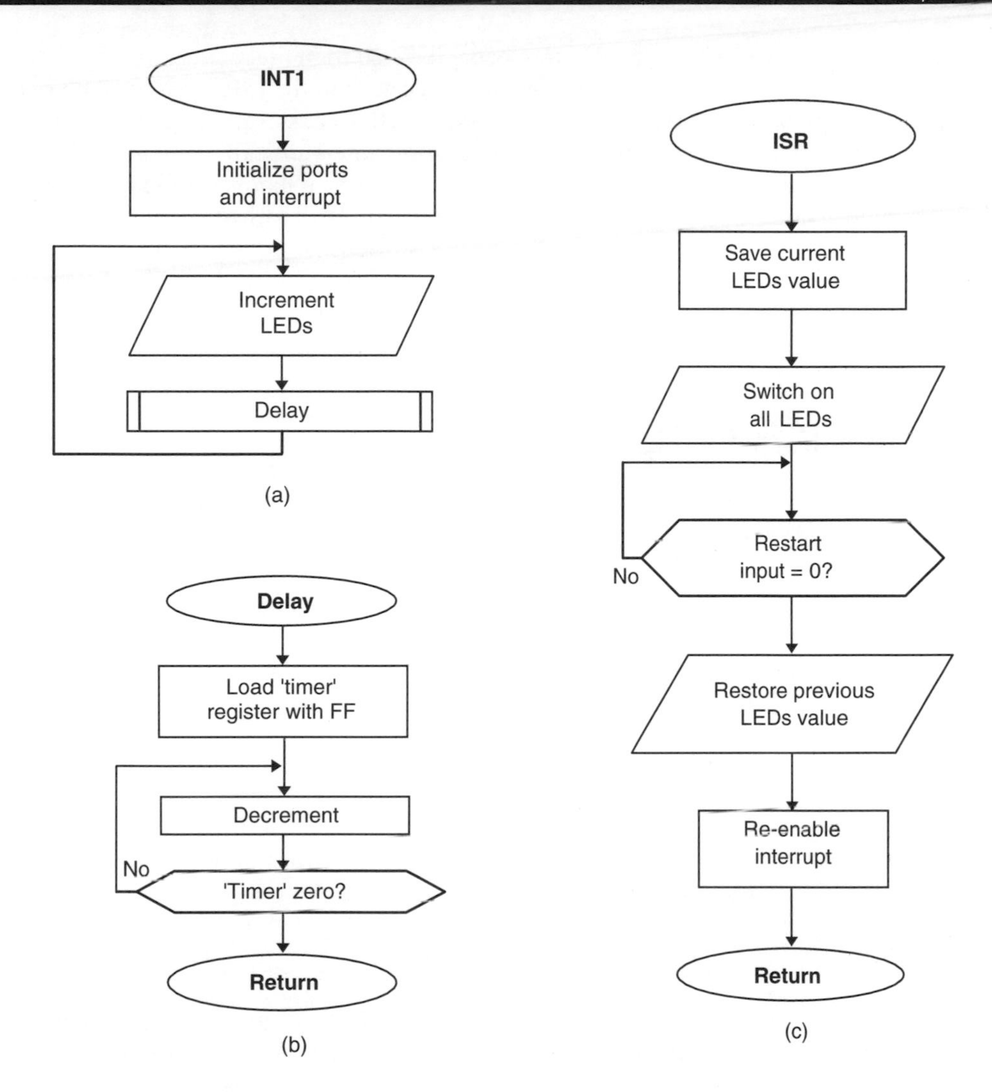

Figure 9.3 INT1 flowcharts. (a) Main sequence; (b) Delay routine; (c) Interrupt service routine.

manually. The interrupt service routine causes all the outputs to be switched on, and then waits for the button on RA4 to be pressed. The routine then terminates, restores the value in Port B data register and returns to the main program at the original point. The program structure and sequence can be represented by the flowcharts in Fig. 9.3.

The program is in three parts: the main program which runs the output count; the delay subroutine which controls the speed of the output count; and the interrupt service routine. The delay process in the main program is identified as a subroutine, and expanded in a separate flowchart. The ISR must be shown as a separate chart because it can run at any time within the program sequence. In this particular program, most of the time is spent executing the software delay, so this is the process that is most likely to be interrupted.

The interrupt routine is placed at address 004. The instruction 'GOTO setup' jumps over it at run time to the initialization process at the start of the main program. The interrupt and delay routines must be assembled before being referred to in the main program, so they must be listed *before* the main program. The last instruction in the ISR must be 'RETFIE'. This instruction pulls the interrupt return address from the stack, and places it back in the program counter, where it was stored at the time of the interrupt call.

To illustrate context saving, the state of the LEDs is saved in 'tempb' at the beginning of the interrupt, because port B is going to be overwritten with 'FF' to switch on all the LEDs. Port B is then restored after the program has been restarted. Note that writing a '1' to the input bit has no effect. During the ISR execution, the stack will hold both the ISR return address *and* the subroutine return address. The processor keeps a count of the addresses on the stack; the stack in the 16F84 can store up to eight return addresses.

9.4 More Register Operations

The functions of the most commonly used registers are described in Chapter 8, and further operations using the PIC registers are outlined in this section.

9.4.1 Data Destination W

The default destination for operations that generate a result is the file register specified in the instruction. For example

```
INCF    spare
```

increments the register labelled 'spare', with the result being left in the register. The above syntax generates a message when the program is assembled to remind the user that the 'default' destination is being used. This is because the full syntax is

```
INCF    spare,1
```

where '1' indicates the file register itself as the destination. If the result of the operation were required in the working register W, it could be moved using a second instruction

```
MOVF    spare,W
```

However, the whole operation can be done in one instruction by specifying the destination as W as follows

```
INCF    spare,0
```

or

```
INCF    spare,W
```

The label W is automatically given the value 0 by the assembler. The result of the operation is stored in W, while the original value is left unchanged in the file register. All the register arithmetic and logical byte operations have this option, except CLRF (clear file register) and CLRW (clear working register) which are by definition register specific, MOVWF and NOP (no operation). This option offers significant savings in execution time and memory requirements, which in PIC applications may be quite significant, and compensates for the lack of instructions to make direct moves between file registers.

9.4.2 *Register Bank Select*

The PIC 16F84 file register set (Appendix A, Fig. A.4) is organized in two banks, with the most commonly used registers in the default bank 0. Some of the control registers, such as the port data direction registers, TRISA and TRISB, and the OPTION register, mapped into bank 1. Many of the special function registers can be accessed in both banks. Others have special access instructions, namely TRIS to write the Port A and B data direction registers, and OPTION which is used to set up the Real Time Clock Counter.

The manufacturer recommends using bank selection to access all these registers, and the instruction set warns that the instructions TRIS and OPTION may not be supported by future assemblers. Bank 0 is enabled by default, and bank 1 registers OPTION, TRISA, TRISB, EECON1 and EECON2 can be selected by setting bit 5, RP0, in the STATUS register, prior to accessing the corresponding register number. Refer to the data sheet for more details.

9.4.3 *File Register Indirect Addressing*

File Register 04 is the File Select Register (FSR). It is used for indirect or indexed addressing of the other file registers, particularly the GPRs. If a file register address (00–4F) is loaded into FSR, the contents of that file register can be read or written through File Register 00, the Indirect File Register (INDF). This method can be used for accessing a set of data RAM locations, by reading or writing the data via INDF, and selecting the next file register by incrementing FSR. This indexed, indirect file register addressing is particularly useful for storing a set of data which has been read in at a port, in, for example, a data logging application. An output data table of predefined values, such as seven-segment display codes, can use the Program Data Table method described below in Section 9.6.

The demonstration Program 9.3 loads a set of file registers, 20–2F, with dummy data (AA), using FSR as the index register. FSR operates as a pointer to a block of locations, and is incremented between each read or write operation. Notice that the data actually has to be moved into INDF each time.

9.4.4 *EEPROM Memory*

The PIC 16F84 has a block of electrically erasable read only memory (EEPROM) which can store 64 bytes of data, and which is retained when the power is off. This is useful for security applications such as an electronic lock, where the correct combination can be stored and changed as required.

The four registers used are EEDATA, EEADR, EECON1 and EECON2 (Fig. 9.5). The data to be stored is placed in EEDATA, and the address (00–3F) in EEADR. Bank 1 must then be selected, and a read or write sequence included in the program as specified in the data sheet, Section 7. The complex write sequence is designed to reduce the possibility of an accidental write to EEPROM, whereby valuable data is lost. Reading the EEPROM is more straightforward.

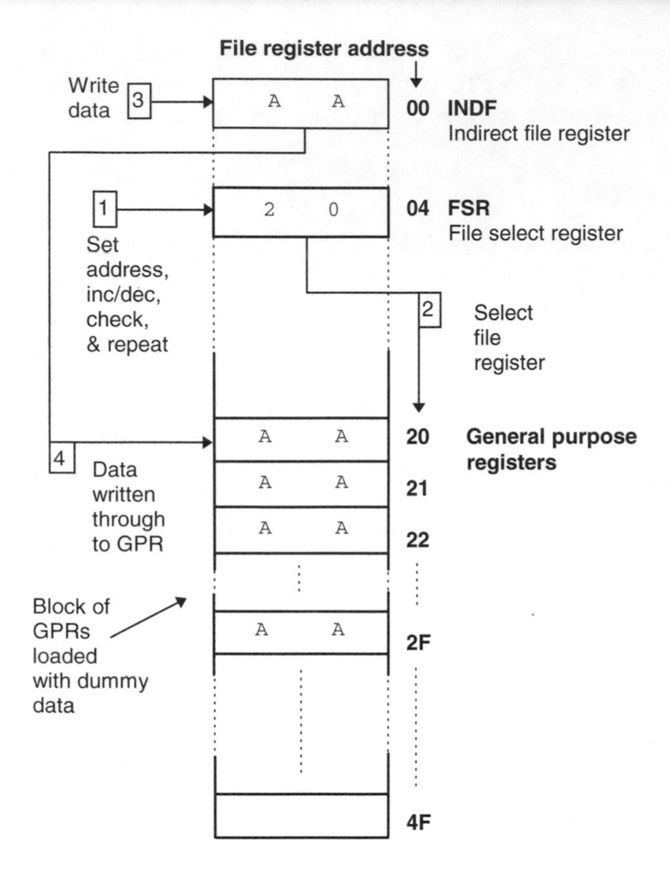

Figure 9.4 Indirect file register addressing.

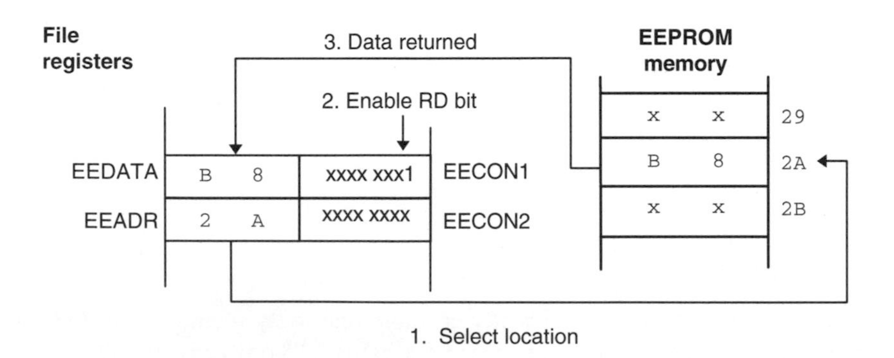

Figure 9.5 EEPROM read operation.

9.4.5 Program Counter High Register, PCLATH

The PIC 16F84 has 1k of program memory (000–3FF), requiring a 10-bit address; the 8-bit PCL (program counter low byte) can only select one of 256 addresses. The 1k of program memory is therefore divided into four 256-word blocks (pages), one of which is selected

Program 9.3 Indexed file register addressing

```
; index.asm

; Demonstrates indexed indirect addressing by
; writing a dummy data table to GPRs 20-2F

   PROCESSOR 16F84                  ; select processor

FSR     EQU     04                  ; File Select Register
INDF    EQU     00                  ; Indirect File Register


        MOVLW   020                 ; First GPR = 20h
        MOVWF   FSR                 ; to FSR

        MOVLW   0AA                 ; Dummy data
next    MOVWF   INDF                ; to INDF and GPRxx

        INCF    FSR                 ; Increment GPR Pointer
        BTFSS   FSR,4               ; Test for GPR = 30h
        GOTO    next                ; Write next GPR

        SLEEP                       ; Stop when GPR = 30h

        END                         ; of source code
```

with two extra bits in the PCLATH (program counter latch high) register. PCL provides the address within each page of memory and is fully readable and writable. PCL and PCLATH are modified automatically when a program jump is executed, that is, CALL and GOTO use a full 10-bit operand for jumps, so do not require any special manipulation of the address for jumping across page boundaries. However, if PCL is modified by a direct write under program control, PCLATH bits 0 and 1 may need to be manipulated to cross page boundaries successfully.

9.5 Special Features

The PIC 16F84 has a number of special features that enhance its flexibility and range of applications. Different oscillator types can be used, timers enabled to ensure reliable program start-up and recovery, and in-circuit programming and code protection are available.

9.5.1 Oscillator Types

The PIC chip can be operated with a simple RC network, a crystal oscillator or externally generated clock signal. Typical oscillator circuits are illustrated in Fig. 9.6. For applications

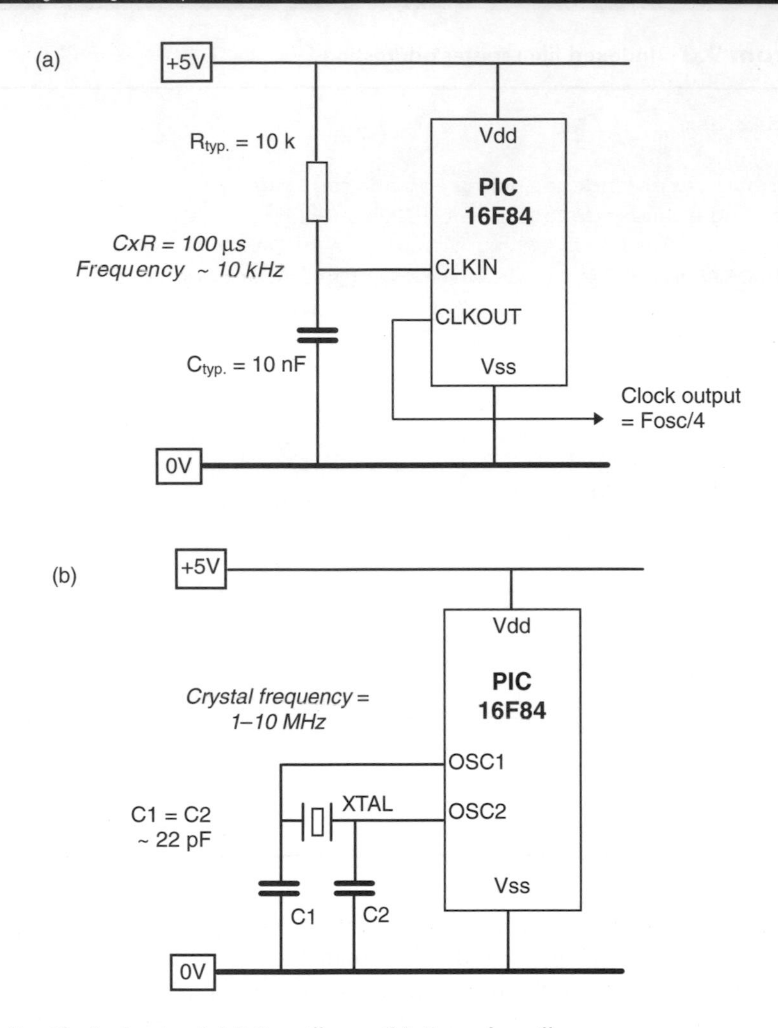

Figure 9.6 Clock circuits: (a) RC oscillator; (b) Crystal oscillator.

where the timing of the program is not particularly important, an inexpensive RC clock circuit (Fig. 9.6(a)) can be used. This requires only a resistor and capacitor connected as shown to the CLKIN pin of the chip. If a variable resistor is used, as in the BIN hardware, the clock rate can be adjusted, within limits, and therefore all output signal frequencies can be changed simultaneously (for example, the outputs from the program BIN1). The clock and output frequency can thus be 'trimmed' to a required value. On the other hand, the clock signal will not be very accurate or stable using an RC clock.

The crystal is slightly more expensive, but is far more precise, than the RC clock. In the oscillator circuit (Fig. 9.6(b)) the crystal resonates at a precise fixed frequency, with an accuracy of typically 30 ppm (parts per million), or 0.003%. This will allow the hardware timer to measure exact intervals. If the PIC chip is part of a larger system, or one with more than one processor, a system clock signal, generated by a master oscillator, can be input at CLKIN. One of the crystal options must then be selected. The clock type must be selected when programming the chip, to match the target system hardware design. There are three types of crystal which can be used: standard (XT), low power (LS) or high speed (HS). Assume that the standard type will be used, unless there is a particular reason to use one of the others.

9.5.2 Power-on Timers

When a power supply is switched on, the voltage and current initially rise in an unpredictable way, depending on the design of the supply and the circuits connected to it. If the processor program tries to start immediately, before the supply has settled down, it may not start correctly. In a conventional microprocessor, an external circuit is typically connected to the CPU reset input, which provides a delay between the power being switched on, and the processor starting.

The PIC has the required power-on timers built in to the chip. The reset input can therefore simply be connected to the positive supply (+5 V) for many applications, as is the case in the examples in this book. When the PIC is powered up, a Power-on Reset pulse is generated when the supply voltage detected at Vdd rises to about 1.5 V. This starts a Power-up Timer which times out after 72 ms, which in turn triggers an Oscillator Start-up Timer, which delays for another 1024 clock cycles, to allow the internal clock to stabilize. An internal reset is then generated, and the program is started. Enable the Power-up Timer when programming the chip, unless there is a reason to disable it.

9.5.3 Watchdog Timer (WDT)

This is an internal independent timer which, by default, forces the PIC to restart automatically after a fixed period (about 18 ms). The idea is to allow the processor to escape from an endless loop or other error condition, without having to be reset manually. This facility would be used by more advanced programs, so our main concern here is to prevent watchdog time out occurring when not required, because it will disrupt the sequence and timing of our programs. The Watchdog Timer can be disabled by selecting the appropriate fuse setting during program downloading, and this is the usual option for simple programs. Alternatively, the WDT can be regularly reset within the program loop using the instruction CLRWDT. If this happens at least every, say, 10 ms, the WDT auto-reset can be prevented. You may find that the WDT reset occurs when using the simulator. A command or selection is available to disable WDT.

9.5.4 Sleep Mode

The instruction SLEEP causes normal operation to be suspended and the clock oscillator to be switched off. Power consumption is minimized in this state, which is useful for battery powered applications. The PIC is woken up by a reset or interrupt; for example, when a key connected to port B is pressed.

The SLEEP instruction is used in Program 9.3, for the following reason. The empty memory locations after the program generally default to the code '3F', all high. This is in fact a valid PIC instruction, ADDLW FF, which means add literal 'FF' to W, so this instruction will be repeated throughout the unused locations. The program counter will roll over to zero after executing these meaningless instructions up to address 3FF, and the program at 000 will be restarted, so the program will loop by default. The SLEEP instruction prevents this occurring. If SLEEP is used to stop the program at the end, a power-on reset, !MCLR or an interrupt can restart the processor.

9.5.5 In-Circuit Programming

Normally, the PIC chip is placed in a programmer unit to download a program from a PC. However, it is possible for the PIC to be reprogrammed whilst remaining plugged into the application circuit. For instance, if controlling a remote system, it can be reprogrammed via the same serial communications link that is used for data transmission. This is a great advantage in applications where equipment is widely distributed in the field, as software fixes and upgrades can be carried out without the software engineer having to physically access the hardware.

9.5.6 Code Protection

In commercial applications, the PIC program designer does not want the product software to be copied by a rival manufacturer. The 'Code Protect' fuse, selected during programming, is designed to prevent such unauthorized copying. The chip can also be given a unique identification code during programming, if required.

9.5.7 Configuration Word

The Oscillator Selection Bits (2), Watchdog Timer, Power-up Timer and Code Protection are all selected by setting the bits of a Configuration Word, located at a special address that is only accessible when the chip is being programmed.

9.6 Program Data Table

A program may be required to output a set of pre-defined data bytes. This can be implemented by modifying the program counter register by adding an indexed pointer value, which is incremented or decremented between each output operation within a data table routine. Program 9.4, TAB1, shows how such a table may be used to generate a sequence at the LEDs in our BIN demonstration hardware. In this case, it is a bar graph display, which lights the LEDs from one end, using the binary sequence 0, 1, 3, 7, 15, 31, 63, 127, 255.

Spare registers labelled 'timer' and 'point' are used. Port B is set as outputs, and subroutines are defined for a delay and to provide a table of output codes. In the main loop, the table pointer register 'point' is initially cleared, and will then be incremented from 0 to 9 as each code is output. The value of the pointer is checked each time round the loop to see if it is 9 yet. When 9 is reached, the program jumps back to 'newbar', and the pointer reset to 0.

For each output, the pointer value (0–8) is placed in W and the 'table' subroutine called. The first instruction 'ADDWF PCL' adds the pointer value to the program counter. At the first call, this value is 0, so the next instruction 'RETLW 000' is executed. The program returns to the main loop with the value 00 in W. This is output to the LEDs, the delay run, and the pointer value is incremented. The new value is tested to see if it is 9 yet, and if not, the call is made to the table with the next value, 1, and so on to 8. Each time the pointer value is added to PCL, so that the program jumps to the second, then third, then fourth code and so on, until finally the ninth code, which is 0FF, is returned to the main output loop for display. After this, the test of the pointer, being equal to 9, succeeds, the

Program 9.4 TAB1 source code

```
;******************************************************************
; TAB1.ASM          M. Bates          13/6/99          Ver 1.3
;******************************************************************
;
;       Output binary sequence gives a demonstration of a
;       bar graph display, using a program data table...
;
;       Processor:        PIC 16F84
;
;       Hardware:         PIC Demo System
;       Clock:            CR ~10 kHz (Cycle time ~0.7 s)
;       Inputs:           none
;       Outputs:          LEDs (active high)
;
;       WDTimer:          Disable
;       PUTimer:          Enable
;       Code Protect:     Disable
;
;       Interrupts:       Disabled
;       Subroutines:      'delay' (no arguments)
;                         'table' (argument 'point')
;
; ******************************************************************

; Register Label Equates.....................................

PCL     EQU     02              ; Program Counter Low Register
PORTB   EQU     06              ; Port B Data Register
timer   EQU     0C              ; GPR1 used as delay counter
point   EQU     0D              ; GPR2 used as table pointer

; ******************************************************************

        ORG     000
        GOTO    start           ; Jump to start of main prog

; Define DELAY subroutine.....................................

delay   MOVLW   0xFF            ; Delay count literal
        MOVWF   timer           ; loaded into spare register

down    DECFSZ  timer           ; Decrement timer register
        GOTO    down            ; and repeat until zero
        RETURN                  ; then return to main program

; Define Table of Output Codes ...............................

table   ADDWF   PCL             ; Add pointer to PCL
        RETLW   000             ; 0 LEDS on
        RETLW   001             ; 1 LEDS on
                                                 continued...
```

```
        RETLW   003             ; 2 LEDS on
        RETLW   007             ; 3 LEDS on
        RETLW   00F             ; 4 LEDS on
        RETLW   01F             ; 5 LEDS on
        RETLW   03F             ; 6 LEDS on
        RETLW   07F             ; 7 LEDS on
        RETLW   0FF             ; 8 LEDS on

;       Initialize Port B (Port A defaults to inputs)..........

start   MOVLW   b'00000000'     ; Set Port B Data Direction Code
        TRIS    PORTB           ; and load into TRISB

; Main loop .............................................

newbar  CLRF    point           ; Reset pointer to start of table

nexton  MOVLW   009             ; Check if all outputs done yet
        SUBWF   point,W         ; (note: destination W)
        BTFSC   3,2             ; and start a new bar
        GOTO    newbar          ; if true...

        MOVF    point,W         ; Set pointer to
        CALL    table           ; access table...
        MOVWF   PORTB           ; and output to LEDs

        CALL    delay           ; wait a while...

        INCF    point           ; Point to next table value
        GOTO    nexton          ; and repeat...

; End of main loop ......................................

        END                     ; Terminate source code
```

jump back to 'newbar' is taken, and the process repeats. Note the use of 'W' as the destination for the result of the subtract (SUBWF) instruction. This is necessary to avoid the pointer value being overwritten with the result of the subtraction.

9.7 Assembler Directives

Assembler directives are commands inserted in PIC source code which control the operation of the assembler. They are not part of the program itself and are not converted into machine code. Many assembler directives will only be used when a good knowledge of the programming language has been achieved, so we will refer to a small number of selected examples at this stage. The use of some of these is illustrated in Program 9.5, ASD1. The assembler directives are placed in the second column, with the instruction mnemonics. We have already met some of the most commonly used directives, but END is the only one which is essential, all the others are simply available to make the programming process

Program 9.5 ASD1 source code

```
; ***********************************************************
; ASD1.ASM      M. Bates                    13/6/99     Ver 1.0
; *******************************************************
; Assembler directives, a macro and a pseudo-
; operation are illustrated in this counting
; program ...
; ***********************************************************

; Directive sets processor type:
  PROCESSOR 16C84

; SFR equates are inserted from disk file:
  INCLUDE ''C:\PIC\REG84.EQU''

; Constant values can be predefined by directive:
  CONSTANT    maxdel = 0xFF, dircb = b'00000000'

timer   EQU     0C      ; delay counter register


; Define DELAY macro ****************************************

DELAY   MACRO

        MOVLW   maxdel              ; Delay count literal
        MOVWF   timer               ; loaded into spare register

down    DECF    timer               ; Decrement spare register
        BNZ     down                ; Pseudo-operation:
                                    ; Branch If Not Zero
        ENDM

;***********************************************************

; Initialize Port B (Port A defaults to inputs)

        MOVLW   dircb               ; Port B Data Direction Code
        TRIS    PORTB               ; Load the DDR code into F86

;       Start main loop.....................................

        CLRF    PORTB               ; Clear Port B Data and restart
again   INCF    PORTB               ; Increment count at Port B
        DELAY                       ; Insert DELAY macro
        GOTO    again               ; Repeat main loop always

        END                         ; Terminate source code
```

more efficient. For definitive information refer to the documentation and help files supplied with your current assembler version.

9.7.1 Control Directives

PROCESSOR (Example: `PROCESSOR 16C84`)

This specifies the PIC processor for which the program has been designed, and allows the assembler to check that the syntax is correct for that processor. The simulator also uses this to automatically select the right processor. The processor can be selected in the assembler command line; if so, this supersedes the source code directive.

ORG (Example: `ORG 004`)

This sets the code 'origin', meaning the address to which the first instruction following this directive will be allocated. We have already seen (Program 9.2) how it is necessary to set the origin of the interrupt service routine as 004. The default origin is 000, so if not specified, the program will be placed at the beginning of the program memory space. This is the reset address where the processor always starts on power-up or reset. If using interrupts, an unconditional jump 'GOTO addlab' must be used at the reset address 000, to jump over the ISR.

END

The assembler is informed that the end of the source code has been reached. This is the one directive that *must* be present.

9.7.2 Conditional Directives

These directives allow selective assembly of source code blocks. That is, sections of code can be omitted during assembly, or repeated, by using high-level language type statements, such as IF...ELSE...ENDIF. Assembler 'variables' are used to define the conditions for assembly.

9.7.3 Listing Directives

LIST

This directive has a number of options that allow the format and content of the list file to be modified, for example, number of lines and columns per page, error levels reported, processor type, and so on.

PAGE

This directive forces a page break when printing

TITLE

This directive defines the program name printed in the list file header line, if you want it to be different from the source code file name (see also SUBTITL).

9.7.4 *Data Directives*

EQU (Example: `PORTA EQU 05`)

This is probably the second most commonly used directive, because it allows literal and register labels to be defined, and we have already used it routinely. It assigns a label to any numerical value (hex, binary, decimal or ASCII), and the assembler then replaces the label with the number. This allows recognizable labels to be used instead of numbers.

INCLUDE

This directs the assembler to include a block of source code from a named file on disk. If necessary, the full DOS file path must be given. The text file is included as though it had been typed into the source code editor, so it must conform to the usual assembler syntax, but any program block, subroutine or macro could be included in the same way. This allows separate source code files to be included, and opens the way for the user to create libraries of reusable program modules. In the example ASD1, it is used to include a standard header file (REG84.EQU) which defines labels for all the special function registers in the PIC. Use of this option is strongly recommended when the basics have been mastered.

DATA, ZERO, SET, RES

These allow program constants and data blocks to be defined and memory to be allocated for specified purposes.

9.7.5 *Macro Directives*

MACRO, ENDM

A macro is a block of source code which is inserted into the program when its name is used as an instruction. In ASD1, for example, DELAY is the name of the macro, and its insertion in the main program can be seen in the list file. Thus using a macro is equivalent to creating a new instruction from standard instructions, or an automatic copy-and-paste operation. The directive MACRO defines the start of the block (with a label), ENDM terminates it. It effectively allows you to create your own instruction mnemonics. See also LOCAL and EXITM.

9.8 Special Instructions

Special instructions are essentially macros which are predefined in the assembler. A typical example is shown in the program ASD1, 'BNZ down', which stands for 'Branch If Not Zero to label'. It is replaced by the assembler with the instruction sequence Bit Test and Skip and GOTO

```
BNZ down = BTFSS 3,2
           GOTO down
```

These two instructions are inserted into the program in place of the special instruction. The Zero Flag (bit 2) in the Status Register (register 3) is tested, and the GOTO skipped if it is set as a result of the previous operation being zero. If the result was not zero, the GOTO is executed, and the program jumps to the address label specified. Special instructions are designed to simplify operations using the carry or zero flag, and are equivalent to conditional branch instructions in complex instruction set processors.

Summary

- Each PIC instruction takes four clock periods to execute (instruction cycle time). Jumps take two instruction cycles. Block execution times can therefore be calculated.

- The hardware counter/timer TMR0 can be used to count inputs or instruction cycles. An 8-bit programmable prescaler is available. Timer overflow sets the T0IF flag, which can be used to trigger a timer interrupt.

- Interrupts allow an internal or external event to change to program sequence, and force the execution of an Interrupt Service Routine. The interrupt sources are port B, TMR0 and EEPROM write completion.

- Register and memory bank selection are sometimes necessary. Sixty-four bytes of data EEPROM are available for non-volatile storage.

- The clock signal which drives the chip can be obtained from an RC or crystal circuit, or master system clock. Power-on timers, Watchdog Timer, Sleep mode, In-circuit programming and Code Protection are available.

- Program data tables can be implemented using CALL and RETLW, with an incrementing PLC offset.

- Assembler directives are instructions to the assembler which are not converted into machine code.

- Macros are user-defined instructions. Special instructions are pre-defined macros.

Questions

1. State the number of clock cycles in a PIC instruction cycle, and the number of instruction cycles taken to execute the instructions: (a) CLRW, (b) RETURN.

2. If the PIC clock input is 100 kHz, what is the value of the instruction cycle time?

3. Calculate the pre-load value required in TMR0 to obtain a delay of 1 ms between the load operation and the T0IF going high, if the clock rate is 4 MHz and the prescale ratio selected is 4:1.

4. List the bits in the SFRs which have to be initialized to enable an RB7:RB3 interrupt.

5. Sketch the circuits for an RC and crystal clock, showing typical component values and the connections to the PIC 16F84. State one advantage of each type.

6. State the assembler directive that must be used in all PIC programs.

7. Explain the difference between a subroutine and a macro.

Answers

2. 40 μs.

3. 6.

4. TRISB, 3, 4, 5, 6 and INTCON bits, 0, 3, 7.

Activities

1. Calculate the time taken to execute one complete cycle of the output obtained from TAB1 with a clock rate of 10 kHz.

2. Modify the program TIM1 to use a timer interrupt rather than polling to control the delay.

3. Devise a program to measure the period of an input pulse waveform at RB0, which has a frequency range of 10–100 kHz. The input period should be stored in a GPR, called a 'period', as a value where $0A_{16} = 10\,\mu s$ and $64_{16} = 100\,\mu s$, with a resolution of 1 bit per microsecond. The clock uses a 4 MHz crystal. Estimate the accuracy of the frequency measurement at each end of the range.

Part C
Applications

Chapter 10
Application Design

Before designing hardware or writing a program, we have to describe as clearly as possible what an application is required to do. That means a specification is needed which defines the user's requirement. There are national and international standards which should be observed when designing commercial products, but here we will simply establish some basic 'common sense' rules.

Once the specification has been written, a useful starting point for hardware design is a block diagram. It should represent the main parts of a system and the information flow between them, in a simplified form. This can later be converted to circuit diagrams and the hardware wiring laid out and constructed on a PCB. In a similar way, software can be designed using techniques which allow the application program to be outlined, and then the details progressively filled in. Flowcharts have been used already, and this chapter will explain in more detail the basic principles of using flowcharts to help with program design.

Pseudocode is another useful method for designing software. The program outline is entered directly into the source code text editor as a set of general statements that describe each major block, which would normally be defined as functions and procedures in a high-level language, and subroutines and macros in a low-level language. Detail is then added under each heading until the pseudocode is suitable for conversion into source code statements for the assembler or compiler for the target processor or programming language. Pseudocode is most useful for so-called 'batch processing' programs found in business applications and large information processing systems. Here we will concentrate on flowcharts, because they are (arguably) more suitable for real-time applications, and their graphical nature makes them a better learning tool.

The first step in the software design process is to establish a suitable algorithm for the program; that is, a processing method which will achieve the specification using the features of an available programming language. This obviously requires some knowledge of the range of languages that might be suitable, and experience in the selected language. Formal software design techniques cannot be properly applied until the software developer is fairly familiar with the relevant language syntax. However, when learning programming we have to develop both skills together, so some trial and error is unavoidable. Here we will assume that it is acceptable for design techniques to be applied retrospectively as part of the learning process. For instance, a final version of a flowchart might be drawn after the program has been written and tested, when the suitability of the design algorithm has been proven.

In this chapter, a simple example application will be used to illustrate the development process. Real software products, of course, will generally be far more complex, but the same basic design principles may be applied. If the design brief is not specific about the hardware, considerable experience and detailed knowledge of the options available is required to select the most appropriate hardware and software combination. The relative costs in the planning, development, implementation, testing, commissioning and support of the product should also be estimated to obtain the most cost-effective solution. Naturally, the example used here to illustrate the software development process has been deliberately chosen to be suitable for PIC implementation.

10.1 MOT1 Design Requirements

A system is required to provide a pulse width modulated (PWM) drive signal for a small dc motor. Under PWM control, the motor runs at a speed that is determined by the average level of the signal, which in turn is dependent on the ratio of the on (mark) to off (space) time. This provides an efficient method of using a single digital output to control output power from a motor, heater, lamp or similar power output transducer. Pulse width modulation is used to control small digital position servo units, as used in radiocontrolled models, for example. The basic waveform is shown in Fig. 10.1.

A variable mark/space ratio (MSR) of 0–100%, with a resolution of 1%, is required. The frequency is not critical, but should be high enough to allow the motor to run smoothly. It is desirable to operate at a frequency above the audible range (>15 kHz) because some of the signal energy can radiate as sound from the windings of the motor, which can be quite irritating! However, a special PWM interface is needed to achieve this. We will aim only for low frequency operation to demonstrate the principles involved, despite the fact that in practice it would be of limited effectiveness. The hardware is also simplified for this example. Instead of a single FET, a full-bridge driver IC would normally be used to provide bidirectional motor control.

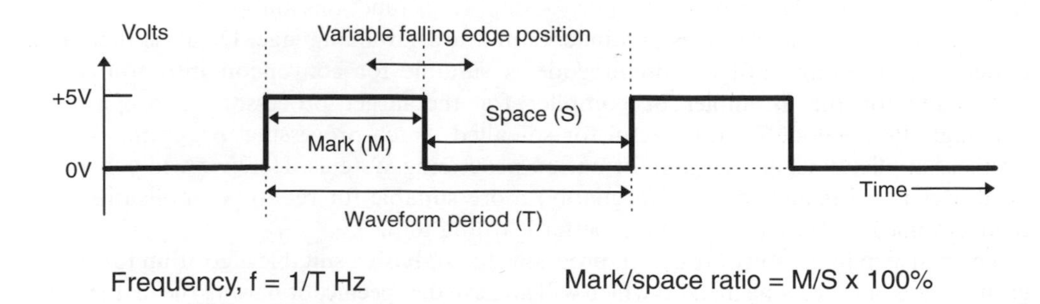

Figure 10.1 Pulse width modulation signal.

Table 10.1 MOT1 control logic

Inputs				*Output (PWM)*	*Motor (dc brush)*
!MCLR	*!RUN*	*!UP*	*!DOWN*		
Pulse	×	×	×	Restart	Off or default speed
1	1	×	×	0	Off
1	0	1	1	Initially: run with MSR = 50% or: run at current speed	Default speed or speed constant
1	0	0	1	Increment MSR (hold at max)	Speed increasing
1	0	1	0	Decrement MSR (hold at min)	Speed decreasing
1	0	0	0	Run at current speed	Speed constant

The motor speed will be controlled by two active low inputs that will increment or decrement the MSR output. An active low enable signal is also required to switch the drive on and off, while preserving the existing setting of the MSR. The system should start on reset or power-up at 50% MSR, that is, with equal mark and space, and a reset input should be provided to return the output to the default 50% MSR at any time. The increment and decrement operations must stop at the maximum and minimum values; in particular 0% must not roll over to 100%, causing a zero to maximum motor speed transition in a single step. The inputs and outputs must be TTL compatible for interfacing purposes, allowing PWM control from another master controller for multiple motor control. A programmed device will be used so that it is possible to modify the control algorithm to suit different motors and to enable future enhancement of the controller options and performance. A logic table (Table 10.1) is used to specify the operation required.

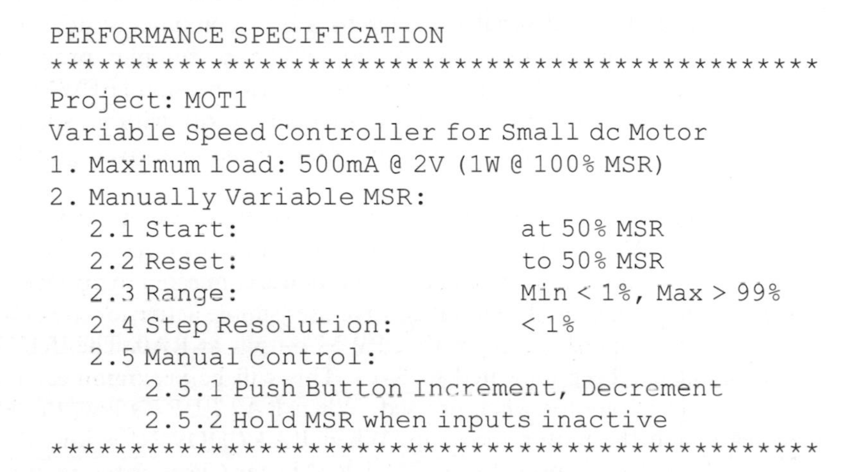
```
PERFORMANCE SPECIFICATION
***************************************************
Project: MOT1
Variable Speed Controller for Small dc Motor
1. Maximum load: 500mA @ 2V (1W @ 100% MSR)
2. Manually Variable MSR:
   2.1 Start:                    at 50% MSR
   2.2 Reset:                    to 50% MSR
   2.3 Range:                    Min < 1%, Max > 99%
   2.4 Step Resolution:          < 1%
   2.5 Manual Control:
       2.5.1 Push Button Increment, Decrement
       2.5.2 Hold MSR when inputs inactive
***************************************************
```

10.2 Block Diagram

In the block diagram, the system inputs and outputs must be identified, and if necessary, a provisional arrangement of subsystems worked out (Fig. 10.2). The direction and type of

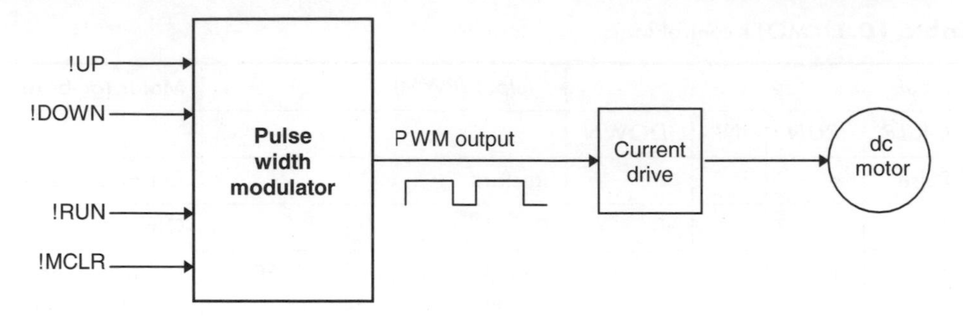

Figure 10.2 MOT1 block diagram.

information flow between the blocks should be identified clearly, using directed line segments (arrows). Small diagrams can be used to illustrate the nature of the signal, if appropriate. Parallel data paths should be shown as broad arrows, or with suitable signal labelling.

10.3 Hardware Design

Unless the program is being written for an existing hardware system, the general hardware configuration must be worked out as part of the design exercise. The nature and complexity of the software is an important consideration in the selection of a microprocessor or microcontroller, as is the number and type of inputs and outputs, data storage and interfacing.

The design requirements of MOT1 could be satisfied using a relatively complex controller system, based on a conventional CISC processor, such as the 68000, and additional features could then easily be included. More inputs and outputs could allow control of several motors simultaneously, a standard serial interface to a host computer system would be available, and the larger memory could accommodate a more complex program, or a program written in a higher-level language, such as 'C'. In the simple PWM example proposed here, however, the requirement is of minimal complexity with no special interfacing specified. Therefore a small microcontroller solution may be investigated, with a view to using the PIC 16F84.

A circuit derived from the block diagram is shown in Fig. 10.3. The motor is controlled by a field effect transistor (FET), which acts as a current switch operated by the PIC TTL level output. The motor forms an inductive load, so a diode is connected to protect the FET from any back emf from the motor. The input control uses simple active low push buttons.

The PIC provides motor speed control with a PWM output at RA0. The !RUN ('Not Run', active low) input has been allocated to RA4. This will be programmed to enable the PWM output to run the motor when pressed. When RA2 (!UP) is low, the MSR at RB0 should increase, and the motor speed up. When RA3 (!DOWN) is low, the MSR should be reduced, slowing the motor down. !MCLR (Master Clear) is the reset input to the PIC, which will restart the program when pulsed low, and hence reset the speed to the default value of 50% MSR.

We can now start work on the software using a flowchart to outline the program. A few simple symbols and rules will be used to help devise a working assembly code program; these are explained below.

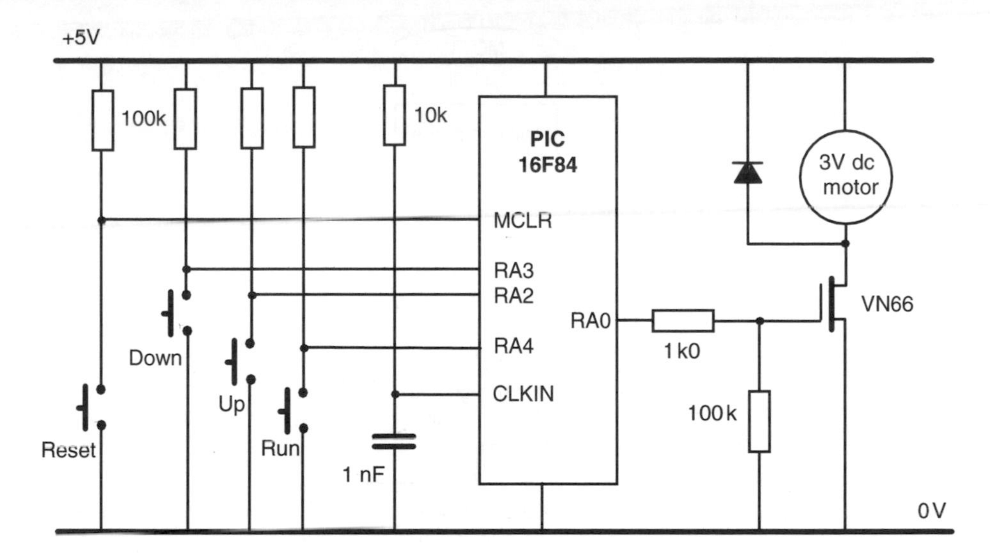

Figure 10.3 MOT1 circuit diagram.

10.4 Software Design

Computer programs incorporate three main types of operation:

1. Sequence of operations; program counter not modified.
2. Selection (conditional jump); test a condition for result true or false and select jump destination, or select from multiple options (HLLs only).
3. Iteration (repeating loop); loop using conditional jump back or loop endlessly.

A program consists of a sequence of instructions in a low-level language (LLL) such as PIC assembler, or statements in a high-level languages (HLL) such as Basic, Pascal or 'C'. These instructions are executed in the order that they appear in the source code, unless there is an instruction or statement that causes a jump, or branch. Usually jumps are 'conditional', which means that some input or variable condition is tested and the jump made, or not, depends on the result. In PIC assembler, 'Bit Test and Skip if Set/Clear' and 'Decrement/Increment File Register and Skip if Zero' provide conditional branching when used with a 'GOTO label' or a 'CALL label' immediately following.

A loop can be created by jumping back at least once to a previous instruction. In our standard delay loop, for instance, the program keeps jumping back until a register which is decremented within the loop reaches zero. In high level languages, conditional operations are created using the IF (condition) THEN (sequence), and loops created using the statements such as DO (sequence) WHILE (condition).

10.4.1 MOT1 Outline Flowchart

Flowcharts illustrate the program sequence, selections and iterations in a pictorial way, using a simple set of symbols. Some basic recommendations for laying out flowcharts

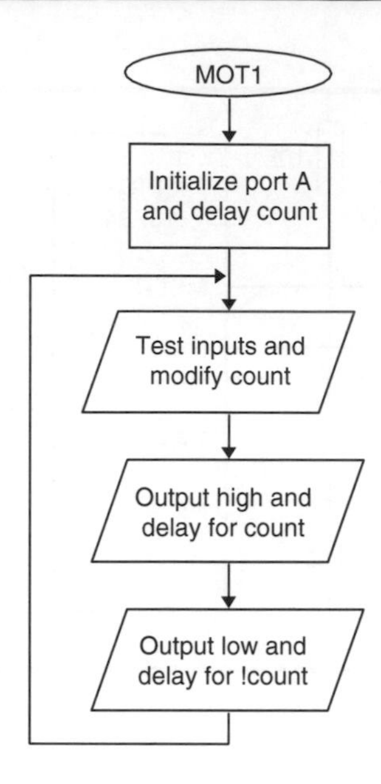

Figure 10.4 MOT1 outline flowchart.

will be made here which will help to ensure consistency in their use and will allow flowcharts to be used to create well structured programs. An outline flowchart for the motor speed control program MOT1 is shown in Fig. 10.4.

The outline flowchart shows a sequence where the inputs (Run, Speed Up and Speed Down) are checked and the delay count modified if either of the speed control inputs are active. The output is then set high and low for that cycle, using the calculated delays to give the mark/space ratio. The loop repeats endlessly, unless the Reset is operated. The reset operation is not represented in the flowchart because it is an interrupt, and therefore may occur at any time within the loop. The program name, MOT1, is placed in the start terminal symbol. Most programs need some form of initialization process, such as setting up the ports at the beginning of the main program loop. This will normally only need to be executed once. Any assembler directives, such as label equates, should not be represented, as they are not part of the executable program itself.

In common with most so-called 'real-time' applications, the program loops continuously until reset or switched off. Therefore, there is an unconditional jump at the end of the program back to the start, but missing out the initialization sequence. Since no decision is made here, the jump back is simply represented by the arrow, and no process symbol is needed. It is suggested here that the loop back should be drawn on the left side of the chart, and any loop forward on the right, unless it spoils the symmetry of the chart or causes line segment crossovers (see below).

10.4.2 MOT1 Detail Flowchart

The outline flowchart given in Fig. 10.4 may show enough information for an experienced programmer. If more detail is needed, boxes in the main program can be elaborated until there is enough detail for the less experienced programmer to translate the sequence into assembly code. A detail flowchart is shown in Fig. 10.5.

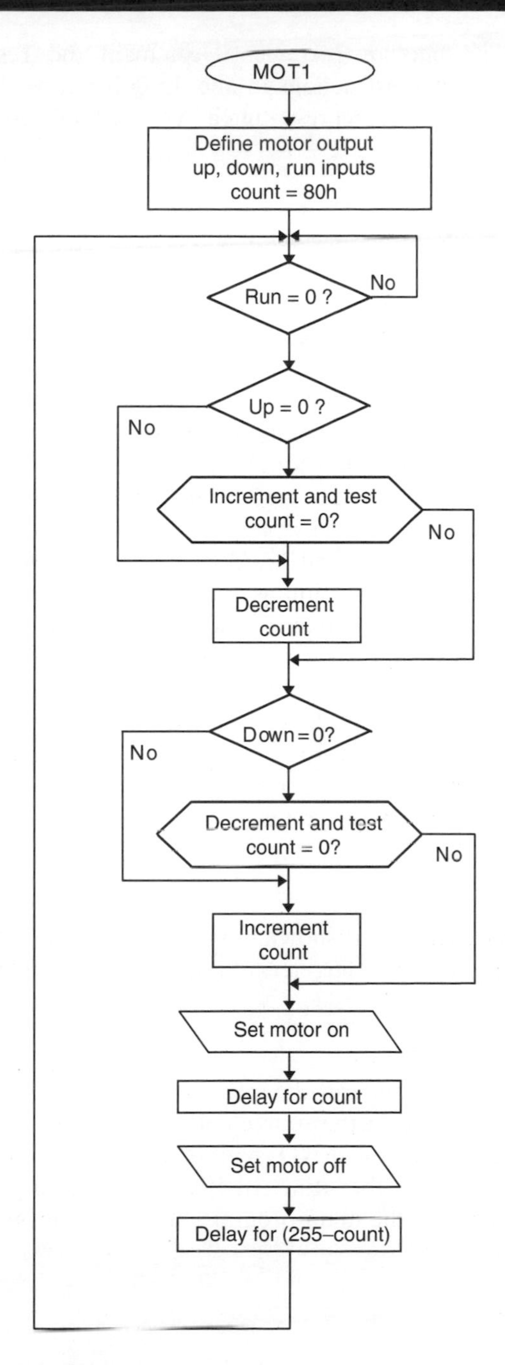

Figure 10.5 MOT1 detail flowchart.

After the initialization sequence, a set of conditional jumps is required to enable the motor, to check the 'up' and 'down' inputs, and to test for the maximum and minimum values of the value of 'Count' (FF and 01). Two different forms of the decision box have been used in this example, both of which may be seen in other references. The diamond-shaped decision symbol is used here to represent a 'Bit Test and Skip If Zero/Not Zero' operation, while the elongated symbol represents an 'Increment/Decrement and Test for Zero' operation, which essentially combines two instructions in one. In either case, the decision box should contain a question, with its outputs representing a 'Yes' or 'No' result of the test.

10.4.3 Program Structure

In the previous example program BIN4, a delay subroutine is used. You will recall that this is a process defined as a separate block of code which can be used more than once. It is indicated in the main program flowchart (Fig. 7.3(a)) with the subroutine box with double sides. The delay routine sequence is detailed in Fig. 7.3(b). It starts with a terminal symbol which contains the subroutine start label used in the program source code, and ends with 'Return'. It is essential that all subroutines which are invoked with the CALL instruction are terminated with a RETURN instruction. The CALL automatically pushes the return address onto the stack, and the RETURN pulls it back into the program counter.

'Decrement and Skip if Zero' is used to create the standard software delay loop. The two delays required in BIN4 could be written separately, but the function is the same, with only the delay count differing; using the same delay block twice is more code efficient. Sometimes separate blocks might be used if the timing is critical. Alternatively, a macro could be defined for the delay, which the assembler would insert twice (see Chapter 9) that would effectively create a 'delay' special instruction. A subroutine will often use a value set-up in the calling routine. In BIN4, the value for the delay time is placed in W for use by the delay routine; this is an example of 'parameter passing'. Other blocks in the main program can also be created as subroutines, but they are most useful when the routine is to be used more than once. The subroutine can then be copied for use in another program, or saved as a separate file for inclusion in new programs.

10.4.4 Flowchart Symbols

A minimal set of flowchart symbols is shown in Fig. 10.6, most of which have already been used. For commercial applications, the relevant international standards should be applied.

Terminals

These symbols are used to start or end the main program or a subroutine. The program name or routine start label used in the source code should also be used in the start box. If the program loops endlessly the END symbol is not needed, but RETURN must always be used to terminate a subroutine. In PIC programming, use the project name (MOT1) in the start symbol of the main program, and the subroutine start address label in subroutine start symbols.

Processes

The process box is a general purpose symbol which represents a sequence of instructions, possibly including loops inside it. The top level flowchart of a complex program can be simple, with a lot of detail concealed in each box. A subroutine is a process which will be implemented in the source code as a separate block, and which may be used more than once within a program. It should be expanded into a separate subroutine flowchart, using the same name in the start symbol as that shown in the calling process. Subroutines can be created at several levels in a complex program.

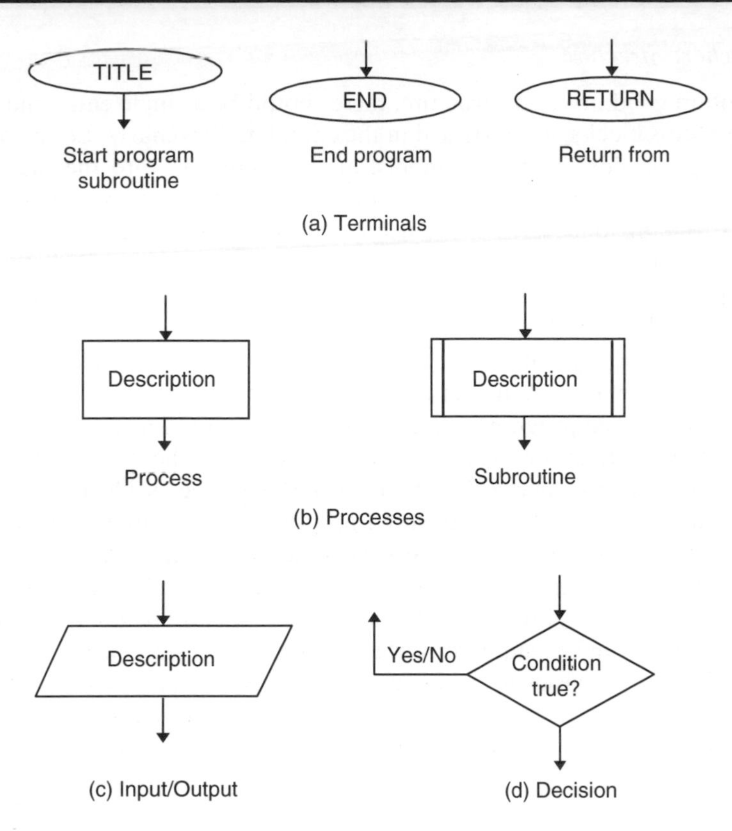

Figure 10.6 Flowchart symbols.

Input/Output

This represents processes whose main function is input or output using a port data register in the microcontroller or microprocessor system. In the PIC 16F84, this will refer to operations using file register 05 (port A) or 06 (port B). Use a statement in the box which describes the general effect of the I/O operation, for example, 'Switch Motor On' rather than 'Set RA0'.

Decisions

The decision symbol contains a description of the selection as a question. There will be two alternate exit paths, for the answer 'yes' and 'no'. Only the arrow looping back or forward needs to be labelled 'yes' or 'no'; the default option, which continues the program flow down the centre of the chart, need not be labelled. In PIC assembly language, this symbol would refer to the 'Test and Skip' instructions. In the MOT1 detailed flowchart, an enlarged decision box is used to represent the 'Decrement/Increment and Skip if Zero' operation. This symbol allows more text inside, so is a useful alternative to the standard diamond shape. If the 'Branch' special instructions, such as 'Branch if not Zero' are used in the source code, this same enlarged decision symbol may be used to represent them.

10.4.5 Flowchart Structure

In order to obtain good program structure, there should be a single entry and exit point to and from all process blocks, as illustrated in the complete flowcharts. Loops should, therefore, rejoin the main flow between symbols, and not connect into the side of a process symbol, as is sometimes seen. Terminal symbols have a single entry or exit point. Decisions in assembler programs only have two outcomes, branch or not, giving two exits. Loops back should be drawn on the left of the main flow, and loops forward on the right of the main flow, if possible. For the main flow down the page, the arrowheads may be omitted as forward flow is clearly implied.

Connections between pages are sometimes used in flowcharts, shown by a circular labelled symbol. It is recommended here that such connections be avoided; it should be possible to represent a well structured program with a set of separate flowcharts, each of which should fit on one page. An outline flowchart should be devised for the main sequence, and then each process detailed with a separate flowchart, so that each process, ideally, can be implemented as a subroutine or macro. More advanced programmers will make up their own minds whether to ignore these remarks, particularly when devising timing critical applications!

You should start designing the program with an outline flowchart on a single page, and expand each process using subroutines or functions on separate pages. Keep expanding the detail until each block can be readily converted to source code statements. A well structured program like this will be easier to debug and modify. Subroutines can be 'nested' to any required depth, depending on the stack size of the system. The PIC 16F84 has an eight-level deep hardware stack, which means that eight levels of subroutine are allowed.

10.5 Program Implementation

When the program logic has been worked out using flowcharts, or otherwise, the source code can be written using a text editor. This may be a general purpose editor, such as MSDOS editor, or part of an integrated development package such as MPLAB. Most high-level languages are supplied as an integrated edit and debug package.

10.5.1 Flowchart Conversion

A program design method should be applied so as to make the program as easy as possible to translate into source code. The PIC has a 'reduced' instruction set, meaning that the number of available instructions has deliberately been kept to a minimum to increase the speed of execution and reduce the cost of the chip. While this also means that there are fewer instructions to learn, the assembler syntax (the way the instructions are put together) can be a little more tricky to work out. For example, the program branch is achieved using the 'Bit Test and Skip' instruction. In most other assembly code languages, branching and subroutine calls are implemented using single instructions. The PIC assembler requires two instructions. However, recall that 'Special Instructions' (essentially pre-defined macros) are available which combine 'test', 'skip' and 'goto' instructions to provide equivalents to conventional conditional branching instructions.

The representation of the program with different levels of detail is illustrated in Fig. 10.7. Figure 10.7(a), shows the process in detail so that each process box converts into only one or two lines of code. This may be necessary when learning the programming syntax. Later,

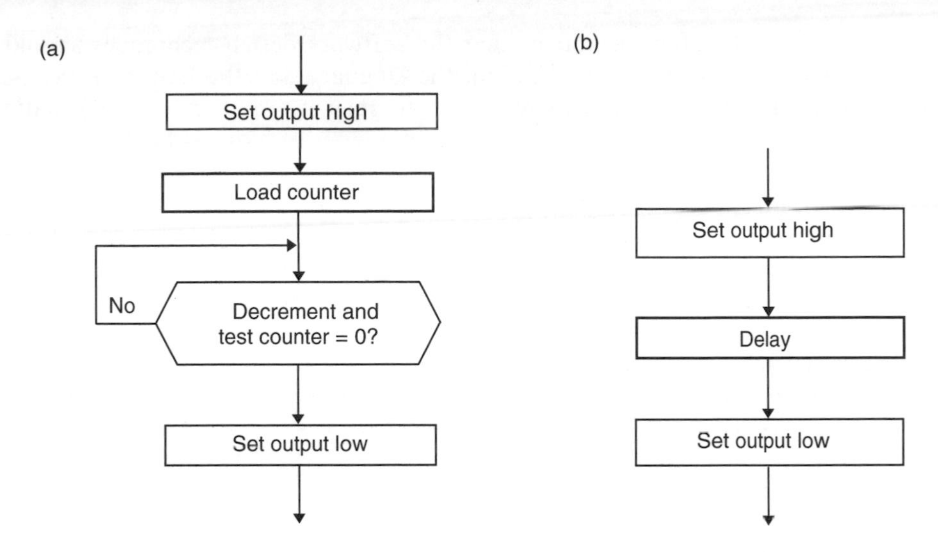

Figure 10.7 PIC program branch flowchart fragments: (a) detail flowchart; (b) outline flowchart.

when the programmer is more familiar with the language and the standard processes which tend to recur, such as simple loops, then a more condensed flowchart may be used, such as Fig. 10.7(b), where the loop is concealed within the 'delay' process. As we have seen above, this process can also be written as a separate, reusable, software component, that is, a subroutine. The corresponding source code fragment is shown in Table 10.2.

Another limitation in PIC assembler is found when moving data between registers. It is not possible to copy data directly between file registers, it has to be moved into the Working register, W, first, and then into the file register. This requires two instructions instead of the single instruction available in CISC processors. This problem is overcome, to some extent, by the availability of the destination register option with the byte processing operations. Nevertheless, the advantage of simplicity when learning PIC programming outweighs these limitations, especially when learning assembler programming for the first time!

Table 10.2 PIC program branch code fragment

```
;       Branch Program Fragment

        .
        .
        BSF     PortA,0     ; Set Output

        MOVLW   0FF         ; Set Count Value
        MOVWF   count1      ; Load Count
back1   DECFSZ  count1      ; Dec. Count and Skip if 0
        GOTO    back1       ; Jump Back

        BCF     PortA,0     ; Reset Output
        .
        .
```

We can see from the above examples that the software design techniques should be applied in a way which suits the application, the language and the level of expertise of the programmer. In a commercial environment, the relevant in-house and international standards should be applied to software design and documentation, as necessary.

10.5.2 MOT1 Source Code

The program source code for the MOT1 program is given in Program 10.1. The program does not use a subroutine for the delay, because the 'timer' value has to be modified for the

Program 10.1 MOT1 source code

```
; **********************************************************
; MOT1.ASM       M. Bates                        14/6/99
; **********************************************************
;
;        DC Motor Control using Pulse Width Modulation
;        Motor (RA0) starts with 50% MSR when enabled with
;        RA4. Speed controlled with RA2, RA3.
;
;        Hardware: Simple Motor Circuit
;        Clock: CR ~100 kHz
;        Inputs: Push Buttons (active low):
;        RA2 = Speed Up
;        RA3 = Slow Down
;        RA4 = Run
;        Outputs: RA0 (active high) = Motor
;
;        Chip Fuse Settings:
;        WDTimer: Disable
;        PUTimer: Enable
;        Interrupts: Disable
;        Code Protect: Disable
;
; **********************************************************

  PROCESSOR 16F84

; Register Label Equates..................................

PORTA   EQU     05              ; Port A
Timer   EQU     0C              ; Delay Counter
Count   EQU     0D              ; Delay Count Pre-load

; Input Bit Label Equates ................................

motor   EQU     0               ; Motor Output = RA0
up      EQU     2               ; Speed Up Input = RA2
down    EQU     3               ; Slow Down Input = RA3
run     EQU     4               ; Motor Enable Input = RA4

; **********************************************************
                                                continued...
```

```
; Initialize .............................................

        MOVLW   b'11111110'      ; Port A bit direction code
        TRIS    PORTA            ; Set the bit direction
        MOVLW   080              ; Initial value
        MOVWF   Count            ; ...for delay

; Next Page ..............................................

; Input Test .............................................

start   BTFSC   PORTA,run        ; Test Run input
        GOTO    start            ; and wait if HIGH

        BTFSS   PORTA,up         ; Test Up input, if high
        INCFSZ  Count            ; ...increment Count, test
        GOTO    test             ; and check down button
        DECF    Count            ; or decrement Count again if 00

test    BTFSS   PORTA,down       ; Test Down input, if high
        DECFSZ  Count            ; ...decrement Count, test
        GOTO    cycle            ; and do an output cycle
        INCF    Count            ; or increment Count again if 00

; Output High and Delay ..................................

cycle   BSF     PORTA,motor      ; Switch on motor

        MOVF    Count,W          ; Get delay count
        MOVWF   Timer            ; Load timer register
again1  DECFSZ  Timer            ; Decrement timer register
        GOTO    again1           ; and repeat until zero then

; Output Low and Delay ...................................

        BCF     PORTA,motor      ; Switch off motor

        MOVF    Count,W          ; Get delay count again
        MOVWF   Timer            ; Reload timer register
        COMF    Timer            ; Complement timer value
        INCF    Timer            ; Increment to avoid 00 value
again2  DECFSZ  Timer            ; Decrement timer register
        GOTO    again2           ; and repeat until zero then

; Repeat Endlessly .......................................

        GOTO    start            ; Restart main loop
        END                      ; Terminate source code
```

'low' delay. The 'timer' value also has to be checked to prevent it rolling over to 00 from FF when being incremented, and from FF to 00 when being decremented. A labelling convention has been adopted throughout this text whereby instruction mnemonics have been given in upper case to match the instruction set in the data sheet. Upper case characters for the special function register names (PORTA) have also been used to match the register names used in the data sheet, and lower case characters with the first letter capitalized used for general purpose registers (Timer, Count). The bit labels are lower case (motor, up, down, run), as are the address labels. This is not obligatory, and often assembler source code is all entered in lower case, or all in upper case. The PIC assembler can be set to be case sensitive, and in this case, the labels used must match exactly in respect of upper and lower case characters, because the ASCII codes are different.

10.6 Source Code Documentation

Most programming languages allow comments to be included in the source code as a debugging aid for the programmer, and information for other software engineers who may need to fix the code at a later date. Comments in PIC source code are preceded by a semicolon; the assembler ignores any source code text following, until a line return is detected.

A header should always be created for the main program and the associated routines. It should contain relevant information for program downloading, debugging and maintenance. Examples have already been given. The layout should be standardized, especially in commercial products. The asterisk symbol (*) is often used to separate and decorate comments; rows of dots are also useful, and there is some scope for individual touches!

The author's name, organization, date, and a program description is essential. Hardware information on the processor or system type is often required; for example, when a PIC program is assembled, the processor type must be specified, because there is some variation in the syntax required for each PIC chip type. The processor type may be specified in the header block as an assembler directive; this is essential in MPLAB. Target hardware details, such as input and output pin allocation, are useful, and the design clock speed needs to be specified in programs where code execution time is significant. Programmer settings which enable or disable hardware features such as the Watchdog Timer, Power-up Timer and Code Protection can be specified.

The general layout of the source code should be designed to make the structure clear, with subroutines headed with their own brief functional description. Blank lines should separate the functional program blocks; that is, instructions which together carry out an identifiable operation. In this way, the source code can be presented in a way that makes it as easy to interpret as possible.

Summary

- The application requirements and performance specification should be clearly stated as the first step in software design.
- A block diagram should be used to outline the hardware, hardware selected and a circuit designed.

- Programs consist of statements which allow sequence, selection and iteration.
- The software algorithm should be represented with a suitable software design aid (flowchart), and elaborated until sufficiently detailed to translate into source code.
- Flowcharts should be structured, using separate charts to expand the processes in the higher level chart.
- Program flowcharts can be constructed from symbols representing terminals, processes, input/output and decisions.
- Source code should include full comments for future reference, maintenance and modification.

Questions

1. Explain how pulse width modulation offers an effective way of controlling dc loads from a single digital output.
2. Explain briefly the role of the block diagram, circuit diagram and flowchart in application design.
3. State the three basic operations that make up a microcontroller program, state how they are represented in a flowchart, and give an example of how each is implemented in PIC assembler code.
4. Explain briefly the role of the subroutine in structured programming, and why it is generally desirable to use them.

Activities

1. Compare the source code for MOT1 with the flowchart in Fig. 10.5, and check that they correspond.
2. (a) Devise a block diagram for a motor control system which has a bi-directional drive, and inputs which select the motor on/off and direction of rotation. Separate active high outputs will be used to enable the motor in each direction; ensure that both outputs are not high at the same time.

   (b) Construct a flowchart for a PIC program which will allow the user to turn the motor on and off from a single active low input, but only allow the direction to be selected when the motor is off. Produce a logic table, outline and detail the flowcharts, and write the code using all the recommendations for source code documentation.
3. (a) Devise a set of structured flowcharts for making a cup of tea (manually!).

   (b) Draw a block diagram of a coffee machine, and devise a set of flowcharts for a control program. You may assume a PIC microcontroller will be used with suitable interfacing, sensors and actuators.

Chapter 11
Program Debugging

The design of a simple PIC motor control application MOT1 has been discussed in Chapter 10, and an assembler code program has been developed. In practice, it is unlikely that a program will be written without any errors, especially when learning the language, so we need now to look further at the techniques and tools available for debugging (removing the errors from) PIC programs. We are going to consider two main types of error: syntax errors and logical errors. Syntax errors are mistakes in the source code that are detected by the assembler, and error messages are generated. Logical errors are mistakes in the logic of the program and are detected using a simulator; they occur when program function does not meet design requirements. If either type of error is detected, the program must be re-edited to remove the errors.

11.1 Syntax Errors

When the program source code for a PIC program has been created in the editor, it must be converted into machine code for downloading to the chip. This is carried out by the assembler program, which analyses the source code text line by line, and converts the instruction mnemonics into the corresponding binary codes for loading into the chip program memory. Only valid statements, as defined in the PIC instruction set, will be recognized and successfully converted.

In the DOS version of the software, the assembler utility program is called MPASM.EXE. It is run from the DOS command line, and provides a dialogue screen where the source code filename and filepath are specified, and saves the machine code as PROGNAME.HEX (HEXadecimal code). In the Windows package, MPLAB, the menu option 'Compile Single File' will assemble PROGNAME.ASM from the edit window, while 'Build All' will assemble all files where other source code files have been 'included' in the main program. 'Make Project' first checks that the current source code has a more recent date than the previously assembled version. 'C' source code programs can also be converted by MPLAB into PIC machine code, so the menu options refer to 'compile' rather than 'assemble'.

The program must be written using the set of instruction mnemonics as defined in the assembler reference documentation. Labels, numerical formatting, assembler directives and so on must all be used correctly. If they are not, error messages will be generated when the source code is assembled. These describe the 'syntax errors' that have been found. The syntax of a language refers to the way that the words are put together; any language, for programming or not, must by definition follow certain rules so that the meaning is clear. The error messages, if any, are saved in a text file PROGNAME.ERR, which can be listed at the DOS prompt with the command 'type progname.err', or viewed in the source code editor. In MPLAB, the error messages will be displayed automatically.

For demonstration purposes, deliberate errors were introduced into the example program MOT1.ASM, and the error file MOT1.ERR generated by the assembler is listed in Table 11.1. There are three levels of error shown: 'Message', 'Warning' and 'Error'. The source code line number where the problem was found is indicated, and the type of problem that the assembler thinks is present. However, a word of warning – due to the presence of the error itself, the assembler may be misled as to the actual error. Consequently, the message generated is not always accurate. For example, the register label 'count' was left out of the register equate list, causing the assembler to misinterpret it as an illegal op-code in line 63. On the other hand, the same problem is correctly identified in line 51 as 'Symbol not previously defined (Count)'.

Table 11.1 Error messages from MOT1

```
Warning[205] C:\MPLAB\MOT1.ASM 27 : Found directive in column 1 (PROCESSOR).
Warning[224] C:\MPLAB\MOT1.ASM 49 : Use of this instruction is not recommended.
Error[113]   C:\MPLAB\MOT1.ASM 49 : Symbol not previously defined (PORA).
Error[126]   C:\MPLAB\MOT1.ASM 49 : Argument out of range (0000 not between
0005 and 0007).
Error[113]   C:\MPLAB\MOT1.ASM 51 : Symbol not previously defined (Count).
Error[113]   C:\MPLAB\MOT1.ASM 61 : Symbol not previously defined (Count).
Message[305] C:\MPLAB\MOT1.ASM 61 : Using default destination of 1 (file).
Warning[207] C:\MPLAB\MOT1.ASM 63 : Found label after column 1 (DEC).
Error[122]   C:\MPLAB\MOT1.ASM 63 : Illegal opcode (Count).
Error[113]   C:\MPLAB\MOT1.ASM 66 : Symbol not previously defined (Count).
Message[305] C:\MPLAB\MOT1.ASM 66 : Using default destination of 1 (file).
Error[113]   C:\MPLAB\MOT1.ASM 68 : Symbol not previously defined (Count).
Message[305] C:\MPLAB\MOT1.ASM 68 : Using default destination of 1 (file).
Error[113]   C:\MPLAB\MOT1.ASM 75 : Symbol not previously defined (Count).
Message[305] C:\MPLAB\MOT1.ASM 77 : Using default destination of 1 (file).
Error[113]   C:\MPLAB\MOT1.ASM 78 : Symbol not previously defined (again).
Error[113]   C:\MPLAB\MOT1.ASM 85 : Symbol not previously defined (Count).
Message[305] C:\MPLAB\MOT1.ASM 87 : Using default destination of 1 (file).
Message[305] C:\MPLAB\MOT1.ASM 88 : Using default destination of 1 (file).
Message[305] C:\MPLAB\MOT1.ASM 89 : Using default destination of 1 (file).
Error[125]   C:\MPLAB\MOT1.ASM 101 : Illegal condition (EOF encountered
before END or conditional end directive).
```

The PROCESSOR directive was misplaced, causing a non-fatal warning, which would not itself prevent successful assembly of the program. The TRIS instruction also caused a warning in the MPLAB assembler, as Microchip have warned that this instruction may not be supported in future. It is used in our examples because the alternative method of port initialization, using page selection, is more complicated to use. The instruction mnemonic DEC was used instead of DECF, causing the error at line 63. This contributed to the register label count being misinterpreted. The jump destination 'again1' has been incorrectly called 'again', causing the error at line 78. Finally, the END directive had been omitted at the end of the program, causing the message 'Illegal Condition'.

The message 'Using default destination of 1 (file)', which appears several times, refers to the fact that the full syntax for MOVWF instruction has not been used. The destination for the result of the operation should be specified as the file register or the working register, by placing a 1 or 0 after the destination register number or label. In the examples throughout this text, we take advantage of the assumption by the assembler that the destination is the file register if not specified in the instruction. More advanced programmers will use this feature of the instruction set to save instructions when the data destination is the Working register. These messages can be suppressed if required. When the error messages have been studied carefully, and printed out if necessary, the source code must be re-edited and re-assembled until it is correct.

11.2 Logical Errors

When all syntax errors have been eliminated, the program will assemble successfully, and the hex file created. However, this does not necessarily mean that it will function correctly when downloaded to the chip; in fact, it probably won't! Usually there will be logical errors, particularly when learning the programming method. Mistakes in the program functional sequence or syntax will prevent it operating as required by the specification. For instance, the wrong register may be operated on, or a loop may execute correctly, but the wrong number of times. There may also be 'run-time' errors, that is, mistakes in the program logic which only show up when the program is actually executed. A typical run-time error is 'Stack Overflow', which is caused by CALLing a subroutine, but failing to use RETURN at the end of the process.

11.2.1 Simulation of Microprocessors

Conventional microprocessor systems vary between each application because they are built from discrete chips. The configuration of the CPU, memory and I/O chips is designed to suit the application; therefore only the CPU itself can normally be simulated. In these systems, some logical errors can only be detected by running the program on the actual hardware, and testing for the correct outputs in response to the specified inputs. In some systems, it is possible to download the program to RAM for testing. In others the program can only be installed by programming an EPROM memory chip and actually fitting it in the hardware. This could mean erasing and reprogramming the EPROM repeatedly, in order to get the program right. This can obviously be quite time consuming.

If the program is being tested in the hardware, it may also be difficult to work out exactly what the problem is. The program will 'crash' or 'hang', giving no indication of the reason. This is where a simulator is useful – it allows the program to be tested in software, in the host

PC, before it is tried on the actual target hardware. The simulator will produce messages such as 'Stack Overflow' for run time errors, but, more importantly, allows the program sequence and operation of the key registers to be checked as the program progresses.

11.2.2 Program Testing

Complete simulation of the PIC, and microcontrollers in general, is possible because the internal chip design is fixed. The PIC simulator, in particular, represents a major advance in low cost microcontroller development systems for the hobbyist and the small application developer. The simulator incorporates a software model of each of the PIC range of chips that runs entirely within the host computer. The program can be tested for logical errors using this software, and only transferred to the chip when working correctly in the simulator.

When using the PIC simulator, the inputs to the chip that would be generated by the circuits around it in the real hardware also have to be simulated. There are two main options – asynchronous and programmed inputs. The simplest method is the asynchronous stimulus. Single bit inputs are generated at the required time while the program is executed in single-step mode. In MPSIM, the command 'se ra4' would be entered at the command prompt to change the state of Port A, bit 4. In MPLAB, an input button can be defined and displayed from the 'Debug' menu and operated at the required time.

Inputs and other register changes can also be defined in a 'stimulus file', which allows the register contents to be changed at any predetermined step in the program. This allows the same test sequence to be repeated each time the simulator is restarted; in more complex applications, this will obviously save time.

11.2.3 Logical Errors in MOT1

The flowchart for MOT1 shows the overall program sequence required. At this stage, it will also be useful to refer to the list file, MOT1.LST, which shows the program addresses and the hex machine code, as well as the source code. This can be printed out in the same way as the source code.

Most debugging toolsets allow the processor registers to be displayed and checked as the program executes, so, for example, the program address sequence can be followed in the program counter register. The first test can simply be to run the program in the simulator. Even if no registers or outputs are displayed, some run-time errors may be detected, such as the program running through into empty program memory locations, stack operation failures, and structural errors not detected by the assembler.

When a program is run in a continuous execution mode in a simulator, it runs much more slowly than it would in the real hardware. Therefore, a display is required of the time that would have elapsed after each instruction cycle (step) in the real system. To obtain the correct reading, the clock speed has to be set in the simulator to match the target hardware system in which the chip will later be fitted. It is unlikely, however, that correct operation can be proved by simply running the program. If there is a problem with the program sequence, you will want to find out where it starts to go wrong.

11.2.4 Single Stepping

A simple but powerful technique for finding logical errors is single stepping. The program is executed one instruction at a time so that the execution sequence can be studied at leisure.

Obviously, the programmer has to know what the sequence ought to be, and the flowchart and list file are useful here. It may be useful to sketch the expected program jumps on a printout of the source code.

In a conventional microprocessor system, single stepping is possible when running a program in the target hardware, if such debugging tools are included in the control (monitor) program. Single stepping a PIC program, however, can be carried out in the simulator, unless an in-circuit emulator is available to replace the chip in the actual application hardware.

In the DOS simulator, each program instruction is displayed after execution. In the Windows simulator, the current instruction is highlighted in the edit window. The sequence can thus be checked, and, if required, the registers displayed. To test MOT1 in MPLAB, for instance, a 'watch' window can be set up showing selected labelled registers, the stopwatch displayed (set to 100 kHz) and the asynchronous input buttons allocated to the inputs RA2 (speed up), RA3 (slow down) and RA4 (run). The program can then be run in 'animate' mode, giving continuous execution while allowing the inputs to be operated during program execution. In this way, the execution sequence and variation in the Count register, which determines the output (RA0) frequency, can be checked.

11.2.5 Break Points

Single stepping is useful for programs with no repetitive loops, but single stepping through a simple delay loop is pointless, because once the loop has been successfully entered, the same instructions are executed many times. To avoid this, the process can be shortened by changing the Count register value to a low value, either in the source code or the simulator. Alternatively, the routine can be 'stepped over' using the relevant command or selection.

Another way of skipping loops and getting to a point of interest further down the program is to use a 'break point'. The program can be forced to stop at a selected position by specifying a program label or address at that point. Alternatively, a register can be tested, and the program stopped when it reaches a specified value. The state of all the relevant registers can then be studied, and single stepping restarted from that point, or program execution restarted to run to the next break point.

11.2.6 Tracing

Tracing is another way of checking the program sequence; the instructions are executed and the changes in the relevant registers are logged to a file. In MPLAB, 'Debug', 'Trace Settings' and a dialogue box allows you to set the start and stop point of the recording process, using program labels or addresses. After running the program, the instruction sequence can be displayed by selecting 'Window', 'Trace Memory Window'. Note that the trace is *not* updated while the window is open.

11.3 DOS Simulator

It is not possible to include a full tutorial on the use of the PIC simulator, since there are different versions available, principally the DOS toolset and integrated MPLAB package which has superseded it. The DOS version, introduced in Chapter 7, does have some advantages for the beginner, in that each part of the development system (assembler, simulator and downloader) is separate and can be studied more easily. MPASM Version 1.21, MPSIM Version 5.11.9 and MPSTART Version 3.4 have been used to prepare the earlier examples in this text.

11.4 Windows Simulator

The Windows development software is integrated into a software package which allows the user to switch easily between the different utilities. MPLAB Version 3.12 has been used to assemble and test MOT1. To create the program, a new file was created, MOT1.ASM, and the source code typed into the edit window. The processor type must be specified as an assembler directive, in this case as PROCESSOR 16F84, in all programs. Programs must also be terminated with the END directive.

When the source code file has been created and saved in a suitable user subdirectory of MPLAB, a project must be created called MOT1. MOT1.ASM can then be assigned to this project using the 'Edit Project' dialogue. The project files created when the source code is assembled will then be placed in this directory.

MPLAB has four sets of buttons, or toolbars, which are selected by the leftmost button: Edit, Debug, Project and User Defined. As in most Windows applications, the buttons simply provide quick access to the most frequently used menu options. The function of each button is displayed at the bottom of the screen, and this gives more information than the equivalent menu item.

Edit toolbar

Edit provides the most useful functions for creating the program, such as file save and load. Debug gives access to tools for detecting logical errors in the simulation mode, such as single step and set a break point. The Project options are used for project management, such as assemble and save all the project files and set-up. If the set-up is saved, when the project is reopened the screen windows should be arranged as before. The User toolbar can be customized as required; the default selection provides a mixture of the most commonly used buttons.

Project toolbar

When the program source code has been entered, and saved, a new project should be created, and the source code file selected for the project, and assembled by selecting the 'Build All' button. If there are errors, these are listed, and highlighted in the source code.

Debug toolbar

When the syntax errors have been eliminated, the code can be tested by selecting 'Run'. This may not tell you much, if no other diagnostic windows have been opened. The following are recommended, and a typical arrangement of the MPLAB widows for assembler program debugging is shown in Fig. 11.1.

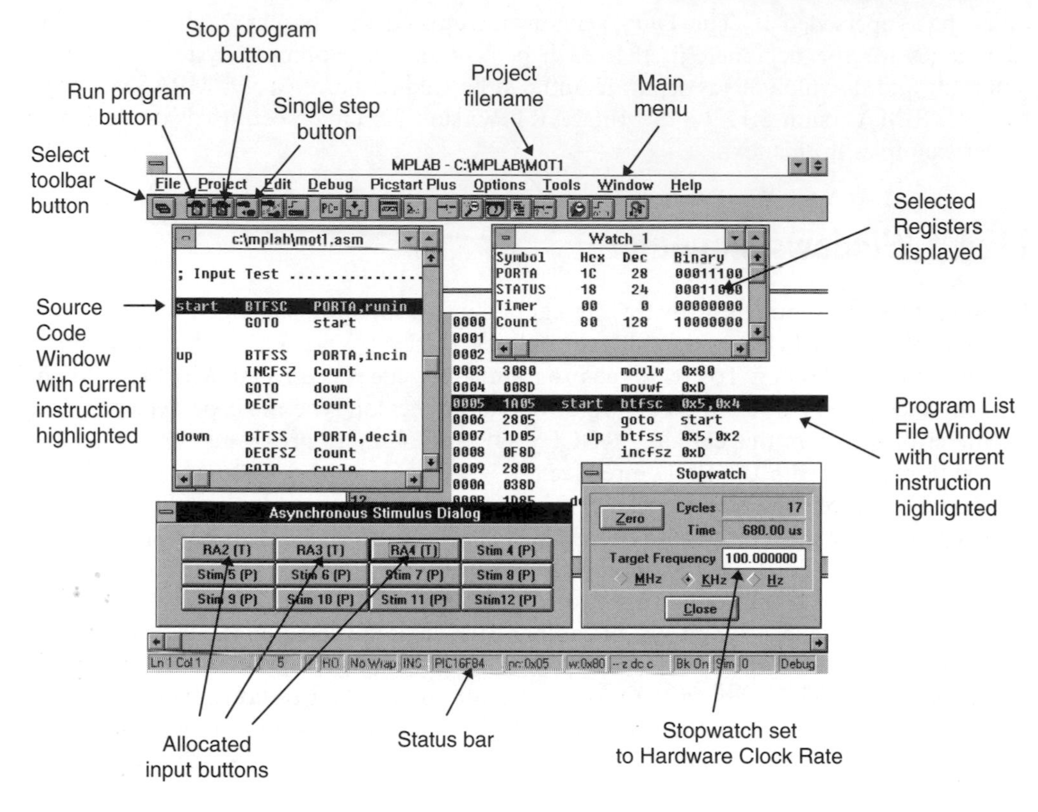

Figure 11.1 MPLAB debug screen.

11.4.1 Debug Windows

Source code edit window (displays source code)

The edit window is used for creating and editing the program. When one of the assembly options is selected, the source code in the current edit window is converted to machine code (PROGNAME.HEX). If syntax errors are then detected, the error file is displayed automatically, and the number of the incorrect line in the source code is given. The current line number is displayed in the status bar at the bottom of the screen. When 'Single Step' or 'Animate' mode is selected, the instruction being executed is highlighted.

Watch window (displays selected file registers)

A watch window allows you to select registers for display, so that the changes in their contents can be monitored and checked against the expected sequence. The button to create a new watch window is found in the debug toolbar. The list of pre-defined and user labelled file registers is displayed, and the relevant ones added to the watch window. The window should also be given a suitable title. The registers selected for display when testing the MOT1 project were PORTA, STATUS, Timer and Count.

Asynchronous stimulus (displays programmable input buttons)

To test the MOT1 program, the inputs which enable the motor and change the speed must be simulated. The menus must be used here to select [Debug],[Simulator Stimulus], [Asynchronous Stimulus]. The dialogue box offers 12 screen buttons; click with the right mouse button on the 'Stim 1(P)', and select [Toggle] mode with the left mouse button. Repeat to select [Assign Pin...] and double click the left mouse button to assign the input pin RA2 to the first screen button. Repeat for the other two inputs.

Stopwatch (displays selected clock rate and step count)

The stopwatch window allows the simulated chip clock to be set to the frequency that will be used in the actual hardware. For simulating MOT1, the value 100 Hz is entered. The total number of instructions executed, and the corresponding elapsed time are then correctly displayed. These can be reset to zero at any time. When testing MOT1, the time taken for the delays and the overall output period, can be checked. This kind of check can also be carried out using a break point.

Program memory window (displays list file)

If the 'Break Settings' toolbar button is selected, a dialogue box is displayed which allows the program to be run and then stopped at a specific point. For example, to check the period of the PWM 'on' pulse when MOT1 is run, a break point can be set where the output goes high, and another where it goes low. The time interval between the break points can then be measured with the stopwatch. The break points can be specified by address label or the actual address. In the example above, the instruction address where the output is set is labelled 'start'. However, the address where it is reset (0014) is not labelled, so the actual program memory address must be entered. In order to see this address, the Program Memory Window must be opened in the debug toolbar. The break points are displayed when set and the code turns red when the break point has been used.

11.4.2 Test Schedule for MOT1

Using a simulator assumes that the programmer is testing the program function against the specification. The specification therefore has to be converted into a test procedure that will test all its functions, and all possible incorrect input sequences, especially where these are generated by an operator. A test procedure for MOT1 is specified in Table 11.2.

The test procedure looks excessively detailed when written down, but in practice it does no more than test all the features of MOT1. The software product needs to meet the specification only once, in a prototype, but in subsequent production units the hardware must still be tested. Only the basic features of the simulator have been used, but these are sufficient to validate the design.

11.4.3 Further Debugging

The simulator thus allows the program logic to be tested before running the program in the actual hardware, to ensure that it is functionally correct. The following is given as an

Table 11.2 Simulation test schedule for MOT1

MPLAB simulator Set-up:	*Source Code Window: MOT1* *Watch 1 Window shows registers: select PORTA, Timer, Count* *Asynchronous Stimulus Dialogue: select RA2, RA3, RA4 (toggle mode)* *Stopwatch: enter target frequency 100.00 kHz* *Optional: Program Memory Window*

	Test	*Action*	*Required performance*	✓/×	*Fault/comment*
1	Initialize	RA2, RA3, RA4 = 1	Check Watch Window, PORTA		*All inputs inactive*
2	Select	Debug, Run, Animate	Count = 80 Waits in Start Loop		*Waiting for Run Enable*
3	Enable Run	Input: RA4 = 0	RA0 = 1 Runs into high delay Timer decrements to 0 RA0 = 0 Runs into low delay Timer decrements to 0 Repeats		*One cycle of output at default MSR 50%* *Stopwatch: output period ≈ 33 ms*
4	Disable Run	Input: RA4 = 1	Returns to Start Loop		*Waiting for Run Enable*
5	Test Increment	Input: RA2 = 0 RA4 = 0	Count increments to 81 Increments to 82 after next cycle		*Wait for Count = 82 then next test*
6	Test Decrement	Input: RA3 = 0 RA4 = 0	Count decrements to 81 Decrements to 80 after next cycle		*Decrement Count back down to 80*
7	Test for Count Roll-over	ModifyWatch 1 Address = 'Count' Data = FE (hex)	Run for 2 cycles Maximum Count = FF Count does NOT go to 00		*Roll-over prevented*
8	Test for Count Roll-under	Modify Watch 1 Address = 'Count' Data = 02 (hex)	Run for 2 cycles Minimum Count = 01 Count does NOT go to 00		*Roll-under prevented*
9	Test for Output Disable	Set all inputs = 1	Returns to start loop		*Output disabled*
10	Program Reset	Hit Stop and Reset Buttons	Simulation reset to first instruction		*Reset*
11	Output = 50%	RA4 = 0 and Animate	Timer Runs Count = 80		*Output Toggles*

example of the kind of error that might easily be missed, but seriously affects the operation of the program, and would compromise safe operation of the real system.

The simulation showed that after initializing Port A, with bit 0 (RA0) set as output, this data bit goes high by default (all port bits default to logic '1'). The motor will therefore come on, even if the enable input has not yet been operated. This is obviously a major flaw, which can be fixed by following the port initialization instructions with an instruction to clear the motor output bit. This would still result in a very short pulse to the motor at the start of the program; if this caused a problem to the motor drive, an alternative fix would be needed; for example, an instruction to operators that the motor drive should only be switched on after the controller has been started. Alternatively, the motor drive bit could be re-defined as active low.

11.5 Hardware Testing

When the program has been fully debugged in the simulator, it can be downloaded to the chip, which is then placed in the target system. The target hardware layout and connections should be checked and tested separately before inserting the chip. The board should be carefully inspected for correct assembly. Solder bridges between tracks and dry joints are common faults. The connections can be buzzed out with a continuity tester and checked against the circuit diagram.

Before fitting the chip, apply the power and check the rest of the circuit. The supply current should not be excessive, and components should be checked for overheating. The voltages at the chip power supply pins (Vdd and Vss) of the chip should also be checked, as incorrect connection of the supply is likely to damage any chip. In some applications, make sure the components connected to the chip outputs can be powered-up with an open circuit input. For instance, the FET gate input should not be allowed to float – there is a pull-down resistor fitted.

If all is well, switch off the power and fit the chip using a suitable tool. Anti-static precautions should be observed, but the 16F84 has been found in practice not to be particularly sensitive. Make sure it is the right way round! Pin 1 should be marked on the board. Switch on and check that the chip is not overheating or drawing excessive current. If left to overheat for more than a few seconds, the chip will probably be destroyed.

11.5.1 In-circuit Program Testing

Connect an oscilloscope to the output. On power-up, there should be no output. When the 'Run' button is pressed, the default output waveform with a 50% mark/space ratio should be observed, running at a frequency of about 30 Hz. The speed 'Up' and 'Down' buttons should be operated to ensure that the speed control stops at the minimum and maximum value, and does not 'roll over' from zero to full speed in one step. Note that the program algorithm does not give an MSR of 100% or 0%, but stops one step short of the maximum and minimum. Since there are 255 steps altogether, the step size should be less than 1%.

The circuit should also be tested for 'fail-safe' operation, that is, no unplanned or potentially dangerous output is caused by an incorrect input operating sequence. In this case, operating both the 'Up' and 'Down' buttons together would be an erroneous input combination, that should result in no change in speed, because the increment and decrement operations cancel out.

Table 11.3 Test schedule for MOT1 system

TEST SCHEDULE	Project name: MOT1		
Project:		Specification number:	
Test	*Required result*	*Tick*	*Comment*
Inspected	*No faults observed*		
Power Up	*Motor Off*		*Output low*
!RUN operated	*Motor On, 50% MSR*		*Frequency ~30 Hz*
!UP operated	*MSR increases >99%*		*Step Resolution <1%*
!UP released	*MSR constant*		*Speed held*
!DOWN operated	*MSR decreases <1%*		*Step Resolution <1%*
!DOWN released	*MSR constant*		*Speed held*
!MCLR pulsed	*Speed reset 50% MSR*		*Restarts program*
!RUN released	*Motor Off*		*Output low*
Tested by:			
Signed:			Date:

Other examples of potential problems that would need to be considered are input switch bounce, variation in component performance (check specifications), dynamic operation of the motor, minimum MSR required to make the motor run, and so on. More complex applications will probably have more incorrect input conditions and component-related problems, so the test schedule should be devised to anticipate all possible fault modes. If the circuit is being produced on a commercial basis, a formal test schedule would be needed, and the performance certificated as correct to the product specification. A basic test schedule for the PWM program running in a PIC 16C84 is given in Table 11.3. Additional documentation should be prepared according to circumstances (education, commercial, research) to provide the application user or product customer with the relevant information on using the system.

11.5.2 In-Circuit Emulator

The simulator allows the program to be tested without the target hardware. However, the execution speed will be much lower than in the real hardware, and input conditions due to the real components such as non-ideal voltage levels and transient signals are not simulated. A more complete test can be carried out using an 'In-Circuit Emulator' which allows the host computer to pretend to be the PIC chip via a header connector with the same pin-out as the chip. A pod containing an interface circuit which provides the specific processor emulation hardware, is usually connected between the PC and the target hardware (Fig. 11.2).

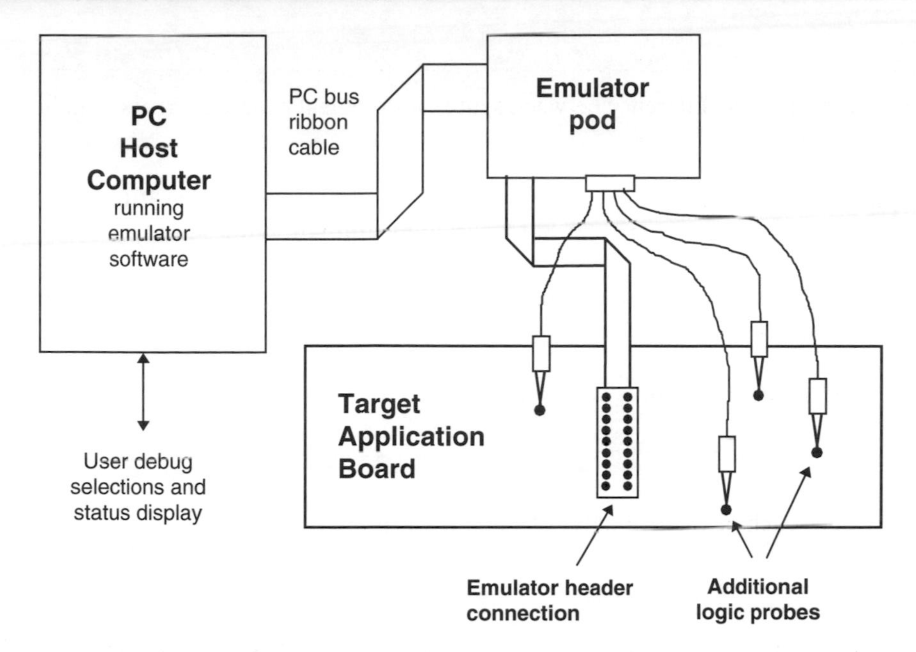

Figure 11.2 In-circuit emulator.

The emulator system provides execution at full speed on the actual hardware. Professional development systems use this technique as it allows the complete application hardware system to be tested as it interacts with a virtual processor. A low cost emulator is available, the amusingly named 'ICEPIC'. This type of equipment is usually regarded as essential in the commercial development environment.

Summary

- There are two main types of error that can occur in source code, syntax and logical errors.
- Syntax errors are invalid statements which are detected by the assembler; error messages are generated to assist debugging.
- Logical errors are mistakes in the program design or implementation, which can be detected by using the simulator to test the program operation.
- MPLAB is an Integrated Development Environment, featuring an editor, assembler, simulator, emulator and downloading software.
- A typical MPLAB simulation set-up would use the Edit, Program, File Register, Watch, Stopwatch and Asynchronous Stimulus windows.
- Logical errors can be identified using single stepping, tracing, break points, and register and bit modification
- The In-Circuit Emulator allows the software to be tested at full speed on the target hardware.

Questions

1. Explain briefly the difference between syntax and logical program errors, and how they are detected.

2. Explain the difference between 'simulation' and 'emulation' in the PIC system. Why does emulation need some extra hardware, and simulation does not?

3. How are the following used in program debugging: single stepping, tracing, break point, asynchronous stimulus, watch window?

4. The highlighted instruction in Fig. 11.1 is:

```
0005    1A05    start    btfsc    0x5,0x4
```

   What does each of the five parts of the line mean?

Activities

1. (a) Start your program editor, enter the source code for MOT1 and assemble it. Note any error messages generated. If the program assembles correctly first time, put some deliberate errors in the source code and inspect the error messages.
   (b) If using MPLAB, create a project MOT1, assign MOT1.ASM to the project and test the program using the set-up suggested in Section 11.4. Use the test schedule in Table 11.2 to check the program operation and confirm that it meets the specification.

2. Modify the application by eliminating the inputs and generating a PWM output whose MSR increments automatically, once per output cycle. Allow the incremented Count register to roll over from 100% MSR to restart at 0% MSR. Devise a test schedule to confirm correct operation. Download to a PIC chip and run the program in the hardware, monitoring the output with an oscilloscope. What output should you see? Predict the time taken (to the nearest half second) for a complete cycle from 0 to 100% MSR.

Answer

2. 8.5 s.

Chapter 12
Prototype Hardware

We now come to the stage where we need to look at the techniques available for building our PIC circuits. The design, layout and construction of printed circuit boards (PCBs) will not be discussed in any detail, because it involves learning to use specific PCB design software, and so needs to be studied separately. Suffice it to say that circuit design, simulation and layout software has developed to the point where it is generally available at a reasonable cost, and PIC circuits can be designed using the software of your choice. The software will typically allow the circuit to be drawn, simulated, netlisted and laid out in one integrated package.

12.1 Construction Methods

In this chapter we will look briefly at some techniques which are suitable for building one-off boards and prototypes. A general purpose test board will be designed and laid out, and some programs provided to demonstrate its features and the related programming principles. The DIZI board (DIsplay and buZzer with Interrupt) has a seven-segment display and an audio output for some simple display and sound applications. In the next chapter a motor board design, MOTA, will be described, for which programs, based on MOT1, involving open and closed loop control of a small dc motor, will be developed.

12.1.1 Printed Circuit Board

The PCB is the standard method for making electronic circuits. In its basic form, it starts life as a sheet of insulating board with a layer of copper on one, or both, sides. The circuit connections are made by photographically transferring a pattern of conducting tracks and pads for the component connections onto the copper, and dissolving (etching) the copper between the tracks. The components are then soldered in place.

Typically, the circuit diagram can be drawn and converted into a PCB layout within a single suite of programs. These packages will also test the circuit design for correct function by computer simulation using mathematical models for the behaviour of each component. At present, full models for microcontrollers are not generally available in low cost circuit simulators, so the PIC has to be simulated separately, using MPSIM. In due course, microcontroller models will, we can assume, be fully integrated into low cost circuit design packages. However, even with the current software, the PCB layout can take some time to create, and of course, time must initially be invested in learning to use the software. Therefore, we will build our prototype hardware using methods which do not require access to special software or PCB fabrication equipment.

12.1.2 Breadboard

A common method of constructing prototype circuits uses breadboard (plugblock). The connecting wires are simply pushed into an array of interconnected sockets, and the connections can be changed at will. The breadboard module normally has sets of terminals laid out on a 0.1 inch grid that will accept the manual insertion of component leads and insulated tinned copper wire (TCW) links. It has rows of contacts interconnected in groups placed either side of the centre line of the board, where the ICs (integrated circuits) are normally inserted, giving four contacts on each IC pin. At each side of the board, there are longitudinal rows of common contacts that are often used for the power supplies. Some types of breadboard can be supplied in blocks that plug together to accommodate larger circuits, or are mounted on a base with 4 mm power sockets and mountings for control components such as rotary potentiometers (pots).

The layout for a simple circuit is shown in Fig. 12.1, with a PIC 16F84 driving an LED at RB0 via a current limiting resistor. The only other components required are a capacitor and

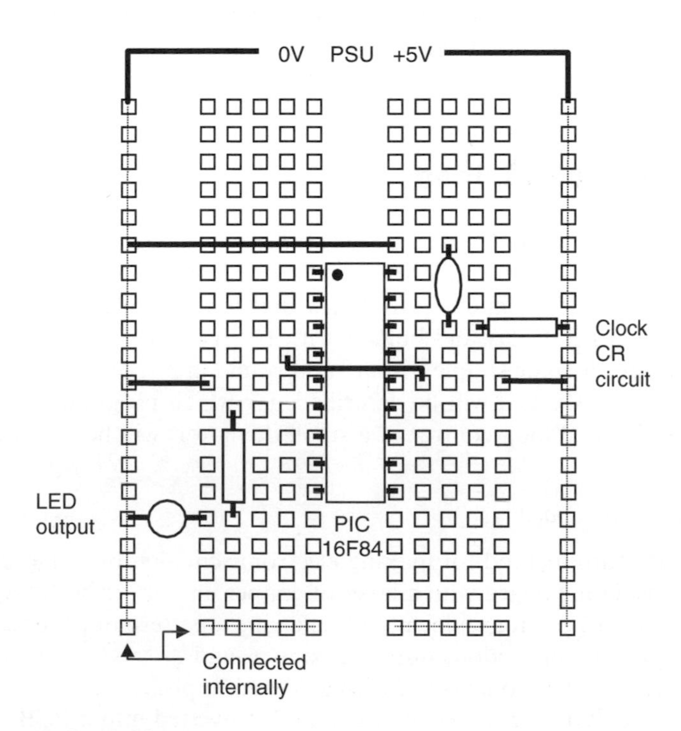

Figure 12.1 Breadboard layout.

	Function	Label	Pin	Pin	Label	Function	
I/O	Port A bit 2	RA2	1	18	RA1	Port A bit 1	I/O
I/O	Port A bit 3	RA3	2	17	RA0	Port A bit 0	I/O
I/O	RA4 or RTCC input	RA4/T0CKI	3	16	OSC1/CLKIN	Crystal or RC oscillator	In
In	Master Clear (Reset)	!MCLR	4	15	OSC2/CLKOUT	Crystal circuit (if used)	Out
In	Supply 0v	V_{SS}	5	14	V_{DD}	Supply +5v	In
I/O	RB0 + Interrupt	RB0/INT	6	13	RB7	Port B bit 7 (+interrupt)	I/O
I/O	Port B bit 1	RB1	7	12	RB6	Port B bit 6 (+interrupt)	I/O
I/O	Port B bit 2	RB2	8	11	RB5	Port B bit 5 (+interrupt)	I/O
I/O	Port B bit 3	RB3	9	10	RB4	Port B bit 4 (+interrupt)	I/O

Figure 12.2 16F84 pin-out.

resistor to form the clock circuit, but we must not forget to connect up the !MCLR (Master Clear) pin to the positive supply, or the chip will not run. The chip could now be programmed to flash the output at a specified rate. As all our circuits are built around the 16F84, it would be useful at this stage to remind ourselves of the chip pin-out (Fig. 12.2).

12.1.3 Powered Prototype Board

Breadboard can be obtained fitted on a base unit containing a +5 V power supply. An illustration of a powered prototype board that is used in combination with ready-made digital plug-in modules to provide switched inputs and LED outputs is shown in Fig. 12.3.

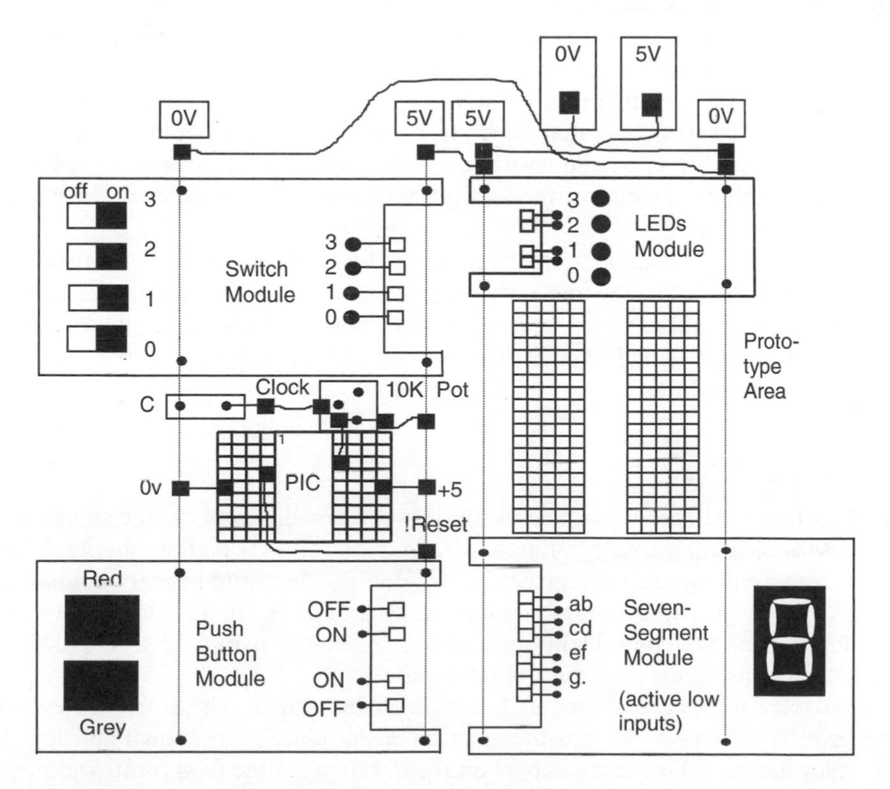

Figure 12.3 Powered prototype board with digital modules.

The modules are small PCBs with power supply pins positioned to plug into the breadboard supply rails, and signal pins that connect to the rows of sockets. A CR clock circuit is shown, with preset pot connected as a variable resistor so that the clock frequency can be adjusted to suit the program. This hardware was used to test the BIN1, BIN2 and BIN3 programs, using two banks of four LEDs. Breadboard is very useful for experimentation with different circuits, but the connections can be unreliable, especially after heavy usage, so a slightly more robust prototyping method may be required, which still avoids designing and building a proper PCB.

12.1.4 Stripboard

Stripboard is a prototyping method which requires no special tools or chemical processing. The components on stripboard are connected via copper tracks laid down in one direction on a 0.1 inch grid of pin holes in an insulating board. The components are soldered in place with their leads through the holes, and the circuit completed using wire links placed on the component side and soldered to the tracks on the copper side. The tracks must be cut where the same strip is used for separate connections in the circuit.

Care is required to avoid 'dry' joints (too little solder) or short circuits between tracks due to solder splashes and whiskers (too much solder!). It can also be quite complicated to work out the connections required in larger circuits, especially if board space is limited. A manual drawing may be used to draft the layout, but a reasonably experienced constructor can build the circuit directly onto the board, with maybe some additional wastage of board area. It is also possible to work out the layout using a standard drawing package, or the drawing tools in a wordprocessor.

An example is shown in Fig. 12.4 of a layout for a motor control board. The stripboard is supplied in sections 300 mm long with 39 longitudinal tracks spaced at the standard 0.1 inch apart to match the pin spacing on ICs and other components. The components are generally placed across the tracks, so that each pin connects with a separate track. The tracks must be cut between the opposing chip pins, and other required positions, using a drill. If the circuit is built at one end of the section, the unused board area can be cut off and used for another circuit.

In this application, a small dc motor is controlled by two power FETs connected to the PIC outputs RA0 and RA1. The motor rotation is monitored by a slotted wheel and optical sensor that feeds back pulses to RA4. Two push buttons allow manual control of the motor operation, and a set of eight DIP switches provide additional inputs to Port B. A similar circuit will be described more fully in Chapter 13.

12.1.5 Drawing Methods

The drawing reprinted here was made using only the drawing tools in the standard wordprocessor. Most general purpose computer drawing packages will allow simple shapes and lines to be drawn and moved around. Shapes for components, and lines for connections on a grid, can be created using suitable scaling. A simplified technique that shows only the essential features of the component connections is shown in Fig. 12.5. This layout was also made using only general purpose drawing tools.

The stripboard tracks are shown as heavy horizontal lines. These are drawn with the 'snap to grid' switched on, so that they can be easily placed with equal spacing. In this example, they are placed 0.2 inches apart on the drawing, so the final result will be printed twice the actual size. This provides more space for labelling.

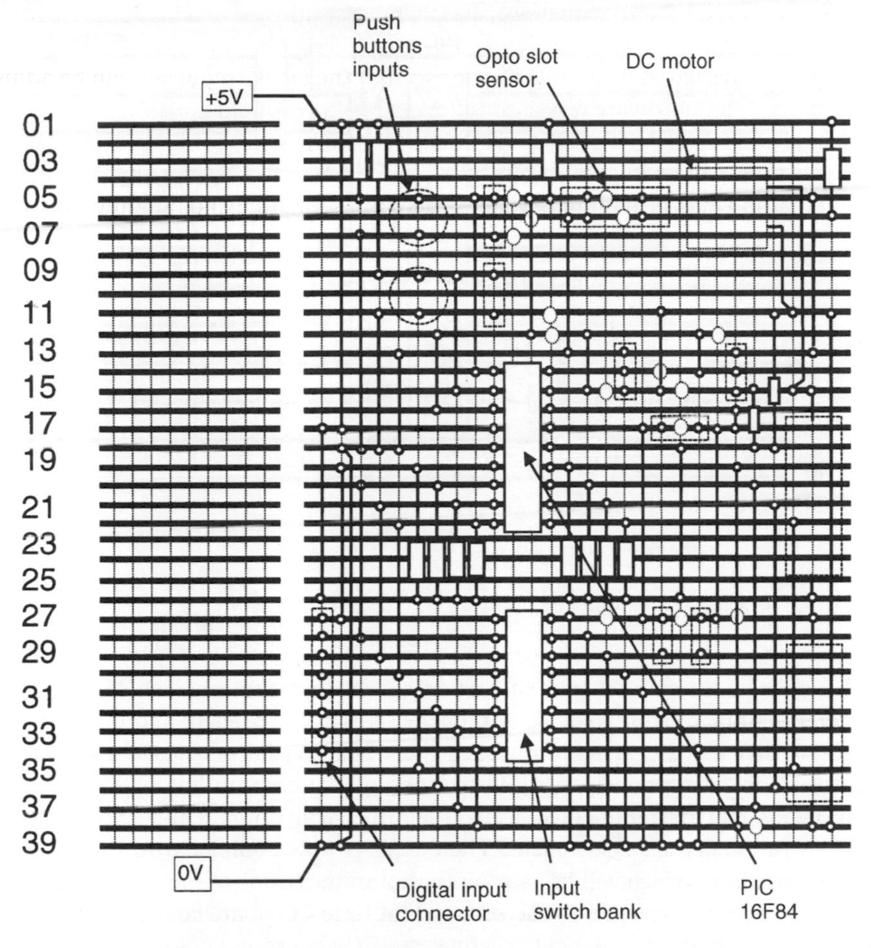

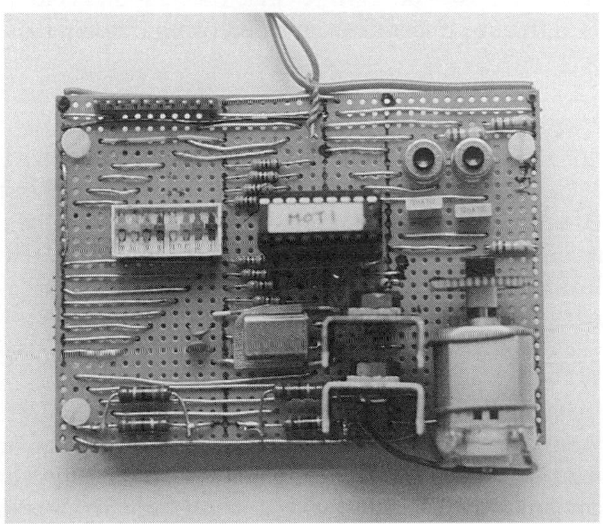

Figure 12.4 Stripboard motor circuit.

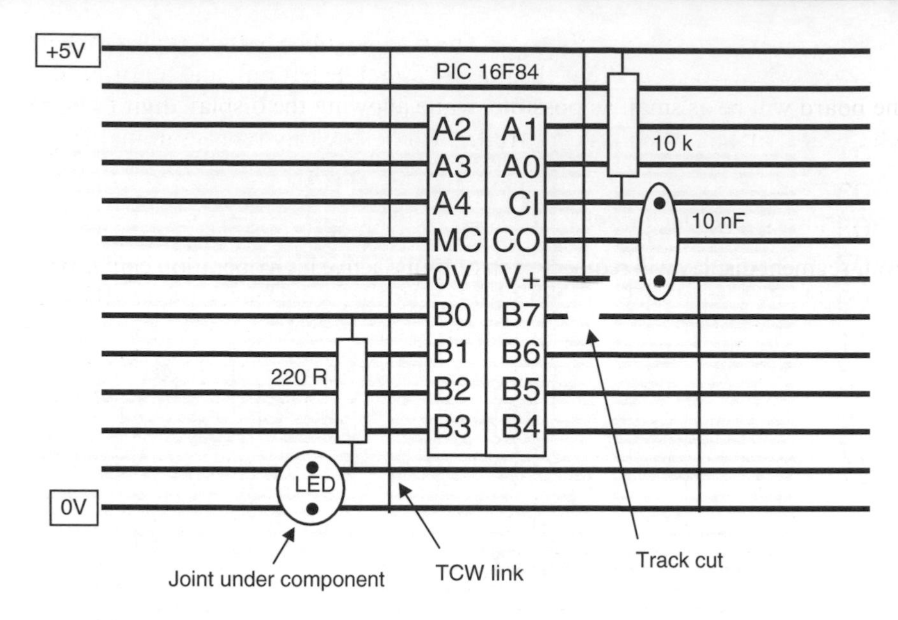

Figure 12.5 Stripboard wiring.

The PIC chip is drawn as a pair of text boxes so that the pins can be labelled. Only two characters can be fitted in here, so abbreviated labels are needed, and the font size must be selected so that the text spacing matches the track spacing as closely as possible. The chip pin positions correspond to the intersections of the chip outline with the track lines. The components can be represented using their physical outlines, and the pin positions shown if they are under the component. The circuit is now completed with vertical lines to represent the wire links which will be used on the component side of the actual board. Ideally, the links should not cross over on the layout, so that bare TCW can be used. To represent the positions where the track must be cut, a white circle (background colour) is drawn over the track line. It is assumed that the tracks are also cut between the chip pins.

12.2 DIZI Demonstration Board

A circuit will be now designed, and a set of programs suggested, to illustrate the programming principles discussed in the previous chapters. The DIZI board will allow the user to experiment with the various features of the PIC hardware and programming techniques with a single hardware module.

12.2.1 Hardware Specification

The microcontroller demonstration board will be suitable for demonstrating a wide range of processes incorporating display, audio, counting, timing and interrupt operations. The board will have a single digit seven-segment display for showing output data in hexadecimal and decimal form, and a low power audio transducer. Manually-operated toggle switches will provide a 4-bit parallel input. Two input push buttons will be available; one to simulate input events to be counted, the other to simulate an external interrupt input. Input and output events must be timed to within 1%. The circuit will be battery powered, with a push button power switch (to ensure that the power cannot be left on), and a power 'on' indicator. The board will be as small as possible, while allowing the display digit to be read, and the audio heard, at 1 m. The microcontroller will be as easily reprogrammable.

12.2.2 Hardware Implementation

The seven-segment display will require seven outputs; active high operation can be provided by a common cathode LED display, and the display decimal point can be used as the power indicator. The audio transducer requires one output; a piezo buzzer was tested for suitability, since its power consumption is low, but the device is generally specified to operate at a fixed frequency. It was found to be satisfactory in its frequency response. A miniature dip switch bank will be used for 4-bit input, and miniature push buttons selected, to conserve space.

Fourteen I/O pins are required: the PIC 16F84 has only 13, so a chip with more I/O could be considered. However, the flash ROM memory is not available in most PIC chips, and the cost would be higher. On the other hand, the audio output and interrupt input could use the same I/O pin at different times. We will assume that this is acceptable, and that the 16F84 will be used. RB0 will be used as the dual function pin, since it is defined as the principal interrupt input, but can also be used as an output. In addition, Port B is preferable for output to the display and audio outputs because its maximum output current rating is 100 mA, compared to 50 mA for Port A.

```
DIZI Project I/O allocation:

     Seven-segment display        Outputs       RB1-RB7
     4-bit switch bank            Inputs        RA0-RA3
     Push button                  Input         RA4
     Push button interrupt        Input         RA0 (dual function)
     Audio transducer             Output        RA0 (dual function)
```

A crystal clock of 4 MHz will be selected for the required timing precision, and to give a 1 μs instruction cycle. The circuit will be powered from two 1.5 V dry cells, giving a 3.0 V supply, with a push button power switch for power conservation, that is, it must be held down to operate the board. A block diagram of the proposed system connections is shown in Fig. 12.6. The inputs and outputs are labelled as for the I/O equates in the program.

12.2.3 Circuit Description

A circuit for the DIZI board is shown in Fig. 12.7. The PIC 16F84 drives an active high (common cathode), low current seven-segment LED display at Port B, RB1–RB7, via a block of 220 R current limiting resistors. RB0 drives an audio sounder when set as an output, but can also be used to detect the 'Interrupt' push button when set as an input and the chip is initialized for this option. To prevent RB0 being shorted to ground if set as an output, the spare 220 R resistor is connected between the push button and RB0. This does not affect the operation of the sounder, which has a relatively high resistance. The flying lead is suggested because this would allow all eight audio frequencies generated by BIN2 program to be heard.

A four-bit DIP switch input is connected to Port A, RA0–RA3, with a push button connected to RA4, which can be used as an external pulse input to the Counter/Timer

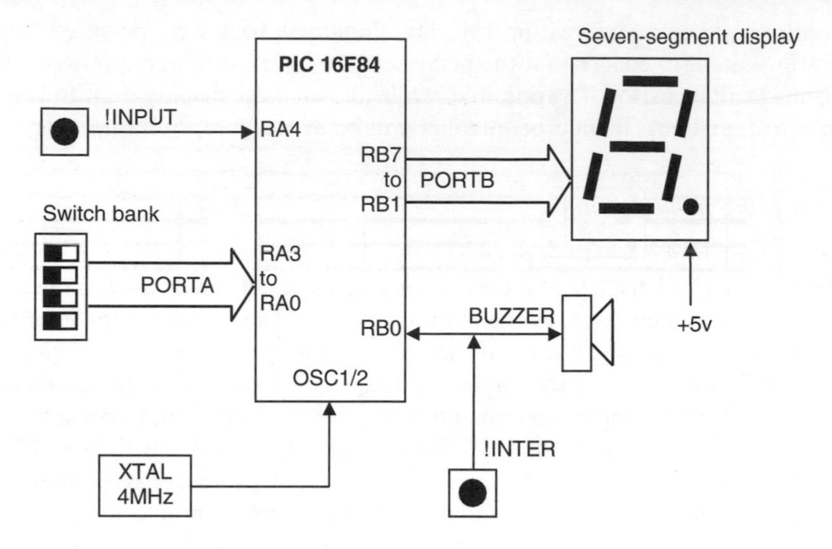

Figure 12.6 DIZI board block diagram.

register RTCC. These operate as active low inputs with 100 K pull-up resistors, as does the interrupt push button. A 4 MHz crystal clock is used so that precise timing operations are possible, with the instruction cycle time equal to 1 μs (±0.01%).

A stripboard layout for the DIZI board is shown in Fig. 12.8. The detail of the component pin connections has been omitted due to the reduced scale of the illustration,

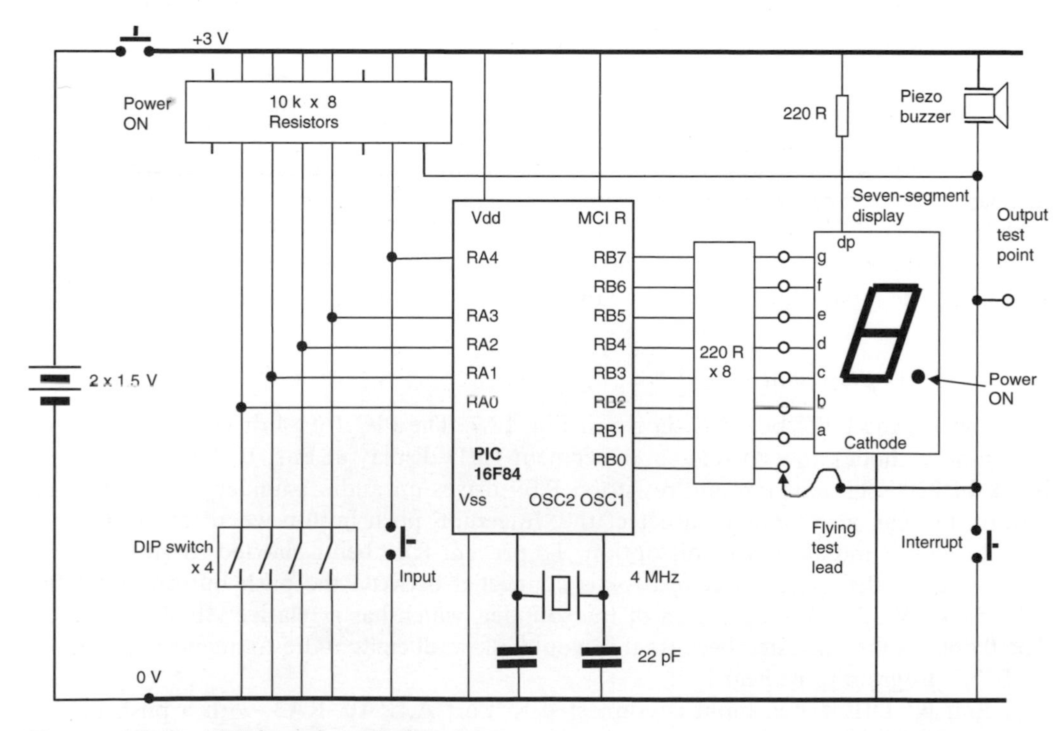

Figure 12.7 DIZI board circuit diagram.

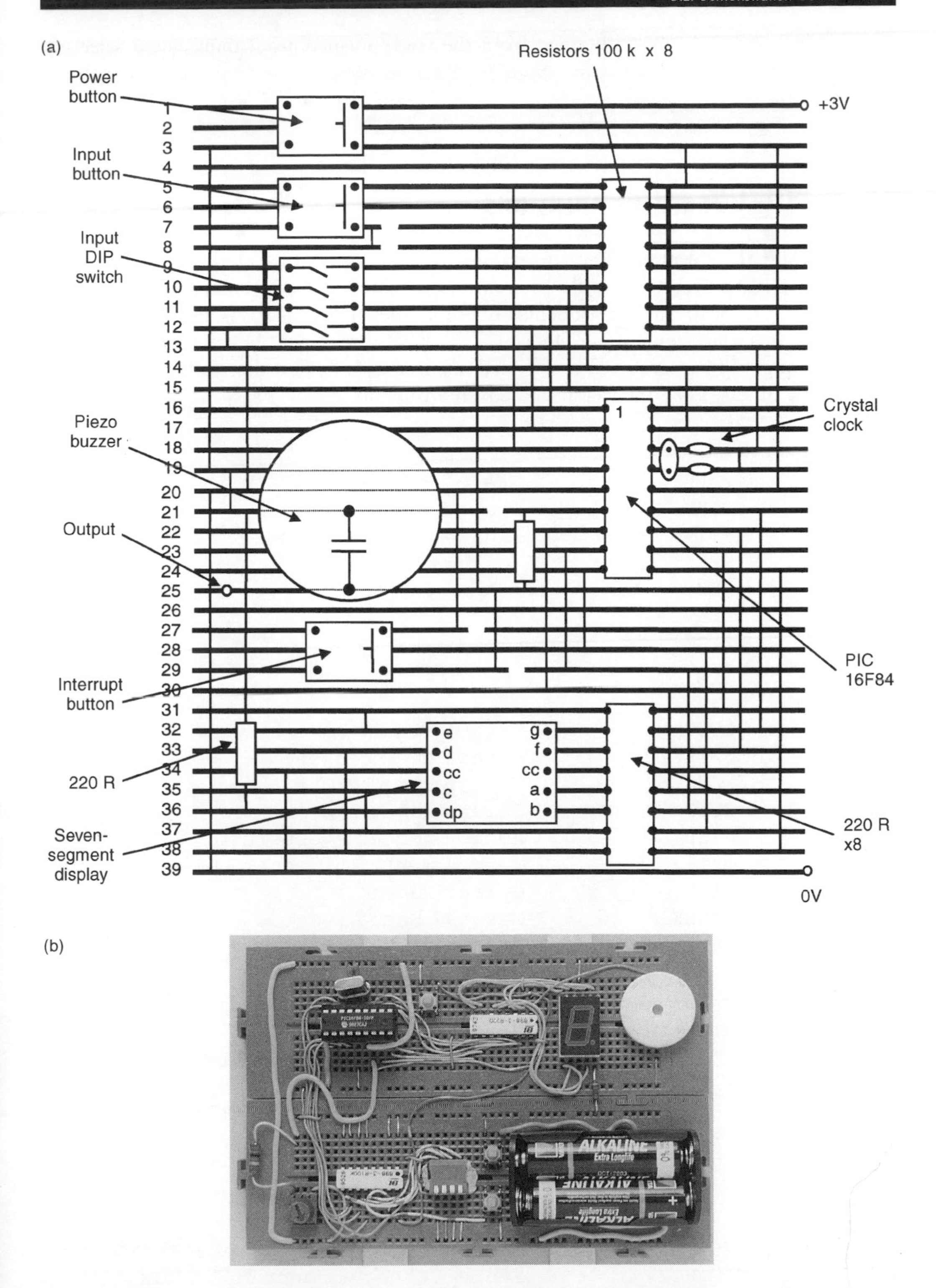

Figure 12.8 (a) Stripboard layout for DIZI circuit; (b) Breadboard prototype of DIZI circuit.

but this information can be obtained from the component pin-out data, when selecting your components. An alternative layout with an on-board power supply is detailed in Appendix B.

12.3 DIZI Board Applications

Most of the earlier demonstration programs can be run on this hardware. The binary output from BINx programs would be seen on the display, but the binary number would not be displayed clearly, as the segments are not in line; also, the LSB is missing. A further set of programs which could be developed for this hardware is listed below. One application in each group will be developed and coded, as an example, and the reader is invited to develop the others, using the techniques covered thus far.

```
DISPLAY
          FLASH1          Flash all Segments
          STEP1           Step through Segments
          HEX1            Binary to Hex Converter
          MESS1           Message Display
          SEC1            1 Second Clock
          REACT1          Reaction Timer
*         DICE1           Electronic Dice

SOUND
          BUZZ1           Output Single Tone
          SWEEP1          Sweep Tone Frequency
*         TONE1           Switch Tone On/Off
          SEL1            Select Tone on Switches
          GEN1            Audio Frequency Generator
          MET1            Metronome
          GIT1            Guitar Tuner
*         SCALE1          Musical Scale
          BELL1           Doorbell Tune

INTERRUPTS
          STEP1           Step through Scale
          STEP2           Step Scale and Display Note
          BUZZ2           Output Tone using TMR0
          REACT2          Reaction Timer using TMR0
          SEC2            One Second Clock using TMR0
          MET2            Metronome using TMR0

EEPROM
          STORE1          Store a display sequence in EEPROM
          STORE2          Store a tone sequence in EEPROM
          LOCK1           Store a code and buzz if matched
```

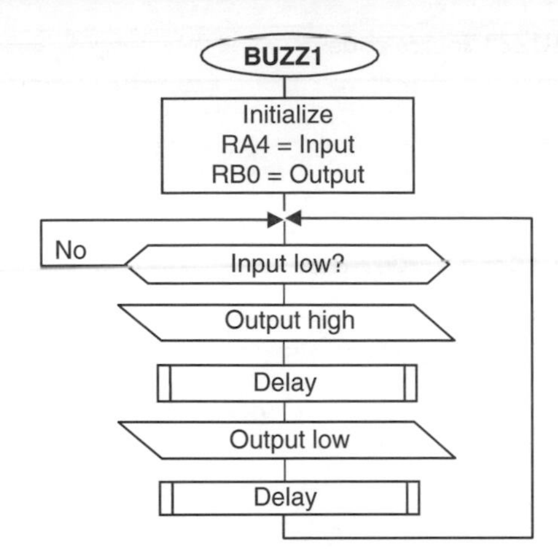

Figure 12.9 BUZZ1 program flowchart.

12.3.1 Program 12.1 BUZZ1

The program (see Fig. 12.9 for flowchart) will generate a single tone at the buzzer when the input button is operated, by toggling the output to the buzzer, with a delay between each change of output state. If a count of 255 is used with a 1 μs instruction cycle time, we have seen that the loop itself will take: $255 \times 3 \times 1 = 765\,\mu s$, which will give a frequency of approximately $1000000/(765 \times 2) = 650\,Hz$.

The frequency is not critical, so we will not add up the additional loop instructions, because they will only make a small difference. The result is well within the audio range, and therefore is suitable. It can be adjusted by simply reducing the count value in the delay loop; 650 Hz is the minimum frequency available. A more precise calculation can be used to obtain a more exact frequency, or the hardware timer can be used.

12.3.2 Program 12.2 DICE1

This program will generate a 'random' number between 1 and 6 at the display when the Input button in pressed. A continuous loop will increment a register from 1 to 6, and back to 1. The loop is stopped when the button is pressed and the number displayed. The display is retained when the button is released. A table is required for the display digit codes. First of all, the allocation of the segments to the pins on the display chip must be checked. Figure 12.10 shows the connections.

The segments must be lit in the combinations necessary to give the required digit display; for instance, segments 'b' and 'c' must be lit for the digit '1' to be displayed. A table is useful here to work out the codes required for output to the display (Table 12.1).

The display is 'active high' in operation. This means a '1' at the output pin will light that segment. This arrangement will also be described as 'common cathode', as all the LED cathodes are connected together at the 'common (0 V)' terminal. A 'common anode' display will therefore operate 'active low'. The binary or hexadecimal code for each digit can be used in the program.

The program represented by the flowchart in Fig. 12.11 uses a GPR as a counter which is continuously decremented from 6 to 0. When the button is pressed, the current number is

Program 12.1 BUZZ1 source code

```
;
;       BUZZ1.ASM                       M. Bates                6/4/99
; ****************************************************************
;
;       Generates an audio tone at Buzzer when the
:       Input button is operated..
;
;       Hardware:          PIC 16F84 DIZI Demo Board
;       Clock:             XTAL 4MHz
;       Inputs:            RA4 : Input (Active Low)
;       Outputs:           RB0 : Buzzer
;       MCLR:              Enabled
;
;       PIC Configuration Settings:
;       WDTimer:           Disable
;       PUTimer:           Enable
;       Interrupts:        Disable
;       Code Protect:      Disable
;
PROCESSOR 16F84 ; Declare PIC device

; Register Label Equates.......................................

PORTA   EQU     05                      ; Port A
PORTB   EQU     06                      ; Port B
Count   EQU     0C                      ; Delay Counter

; Register Bit Label Equates ..................................

Input   EQU     4                       ; Push Button Input RA4
Buzzer  EQU     0                       ; Buzzer Output RB0

; Start Program ***************************************

; Initialize (Default = Input) ...............................

        MOVLW   b'00000000'             ; Define Port B outputs
        TRIS    PORTB                   ; and set bit direction
        GOTO    check

;       Delay Subroutine
...............................................................

delay   MOVLW   0FF                     ; Standard Routine
        MOVWF   Count
down    DECFSZ  Count
        GOTO    down
        RETURN

; Main Loop ...................................................

check   BTFSC   PORTA,Input             ; Check Input Button
        GOTO    check                   ; and wait if not 'on'

        BSF     PORTB,Buzzer            ; Output High
        CALL    delay                   ; run delay subroutine
        BCF     PORTB,Buzzer            ; Output Low
        CALL    delay                   ; run delay subroutine
        GOTO    check                   ; repeat always

        END                             ; Terminate source code
```

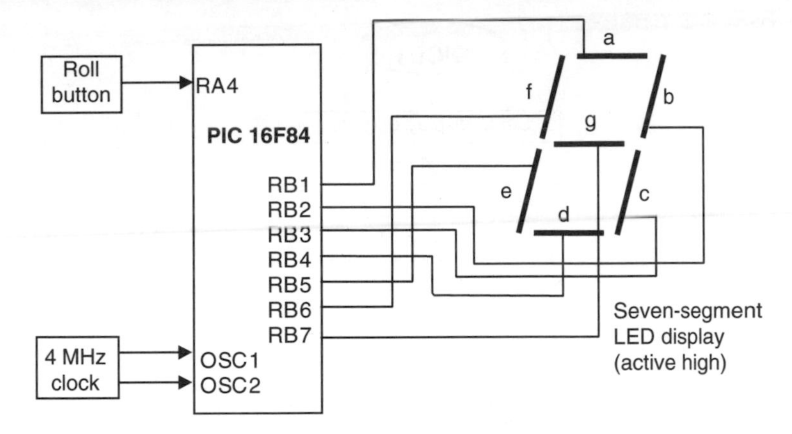

Figure 12.10 Block diagram for DICE1 system.

Table 12.1 DICE1 display encoding table

Displayed digit	Segment code (1 = Segment On)							
	g RB7	f RB6	e RB5	d RB4	c RB3	b RB2	a RB1	HEX (RB0 = 0)
1	0	0	0	0	1	1	0	0C
2	1	0	1	1	0	1	1	B6
3	1	0	0	1	1	1	1	9E
4	1	1	0	0	1	1	0	CC
5	1	1	0	1	1	0	1	DA
6	1	1	1	1	1	0	1	FA

used to select from the table of codes using the method described in Program 9.4. This results in the pseudo-random number code being displayed, and remaining visible until the button is pressed again. Because the number is selected by manually stopping a fast loop, the number cannot be predicted in practice. In the flowchart, the jump destinations have been labelled, and these labels will be used in the program source code. The table subroutine is also named 'table' to match the source code subroutine start label.

12.3.3 Program 12.3 SCALE1

This program will output a musical scale of eight tones. The frequencies (in Hz) for a musical scale from middle C upwards are: 262, 294, 330, 349, 392, 440, 494 and 523. These can be translated into a table of delay counts which give the required tone period, since period, $T = 1/f$, where f = frequency.

The buzzer on the DIZI board is driven from RB0, so this needs to be toggled at a rate determined by the frequency of each tone. We therefore need to use a counter register or the hardware timer to provide a delay corresponding to half the period of each tone. We have seen in Section 9.1.2 how to calculate the delay time for a loop. By producing a formula for the count value, figures were calculated for a half cycle of each tone, which were then placed in the data table in SCALE1.ASM. To keep the program simple, each tone will be output

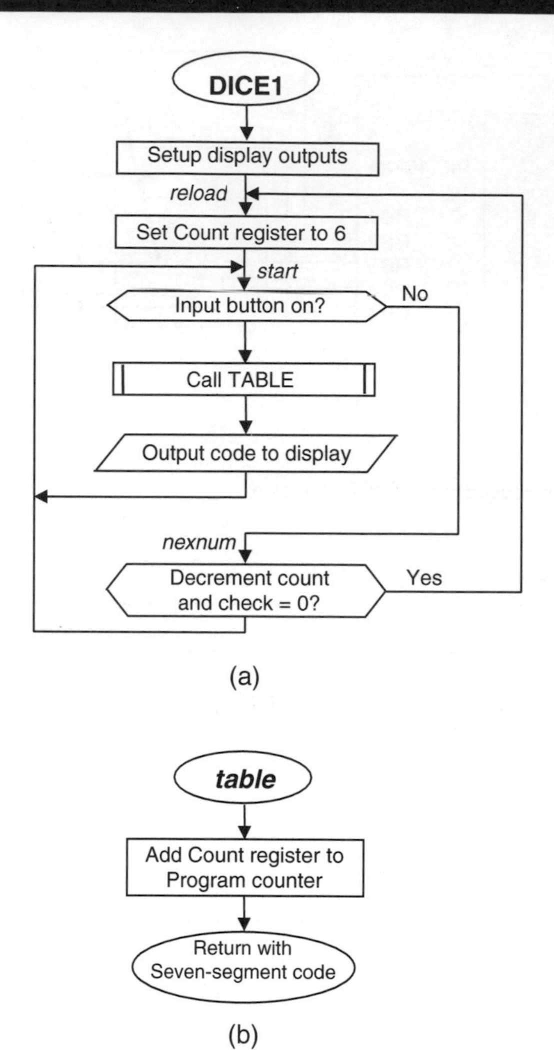

Figure 12.11 DICE1 program flowcharts: (a) main sequence; (b) 'table' routine.

Program 12.2 DICE1 source code

```
; DICE1.ASM              M. Bates                              6/4/99
;
*************************************************************************
;
;       Displays pseudo-random numbers between 1 and 6
;       when a push button is operated.
;
;       Hardware:        PIC 16F84 DIZI Demo Board
;       Clock:           XTAL 4MHz
;       Inputs:          RA4      : Roll (Active Low)
;       Outputs:         RB1-RB7 : 7-Segment LEDs (AH)
;       MCLR:            Enabled
;
                                                          continued...
```

```
;       PIC Configuration Settings:
;       WDTimer:            Disable
;       PUTimer:            Enable
;       Interrupts:         Disable
;       Code Protect:       Disable
;
; Set Processor Options......................................
PROCESSOR 16F84             ; Declare PIC device

; Register Label Equates.....................................
PCL     EQU     02          ; Program Counter
PORTA   EQU     05          ; Port A
PORTB   EQU     06          ; Port B
Count   EQU     0C          ; Counter (1-6)

; Register Bit Label Equates ................................
Roll    EQU     4           ; Push Button Input

; Start Program *********************************************
; Initialize (Default = Input)

        MOVLW   b'00000001'         ; Define RB1-7 outputs
        TRIS    PORTB               ; and set bit direction
        GOTO    reload              ; Jump to main program

; Table subroutine ..........................................

table   MOVF    Count,W             ; Put Count in W
        ADDWF   PCL                 ; Add to Program Counter
        NOP                         ; Skip this location
        RETLW   00C                 ; Display Code for '1'
        RETLW   0B6                 ; Display Code for '2'
        RETLW   09E                 ; Display Code for '3'
        RETLW   0CC                 ; Display Code for '4'
        RETLW   0DA                 ; Display Code for '5'
        RETLW   0FA                 ; Display Code for '6'

; Main Loop .................................................

reload  MOVLW   06                  ; Reset Counter
        MOVWF   Count               ; to 6

start   BTFSC   PORTA,Roll          ; Test Button
        GOTO    nexnum              ; Jump if not pressed
        CALL    table               ; Get Display Code
        MOVWF   PORTB               ; Output Display Code
        GOTO    start               ; start again

nexnum  DECFSZ  Count               ; Decrement and Test Count=0?
        GOTO    start               ; Start again
        GOTO    reload              ; Restart count if zero

        END                         ; Terminate source code
```

for 255 cycles, so we will use another register to count the number of cycles completed during each tone. The scale will then be played over a period of about 5 s. The table of values can be modified later to play a tune in the doorbell program.

Instead of a flowchart, the SCALE1 program source code listing has been annotated with arrows to show the execution sequence. This informal method of analysis can be used on any source code listing to check the program logic prior to simulation. The eight tone frequencies are controlled by the value of 'HalfT', obtained from the program data table at 'getdel'. 'HalfT' is a counter value which will give a delay corresponding to half a cycle of the frequency required when the chip is clocked at 4 MHz. The eight tones are selected in turn by the value of 'TonNum', which is initialized to 8. This is used as the program counter offset in the data table fetch operation. It is decremented in the main loop after each tone has finished to select the next. The 'HalfT' values are thus selected from the bottom of the table upwards.

The tone is generated in the routine 'note', where RB0 is set high, the delay using 'HalfT' runs, RB0 is cleared, and the second half cycle delay executed. No operation instructions (NOP) have been inserted to equalize the tone half cycles duration. RB0 is toggled 255 times using the 'Count' register, which gives a duration of around half a second, depending on which tone is being generated (the lower frequencies take longer).

The main loop thus selects each of the eight values of 'HalfT' in turn, and outputs 255 cycles of each tone. The program is terminated with the SLEEP instruction to stop program execution running into the unused locations following the program.

Program 12.3 SCALE1 source code

```
; SCALE1.ASM                M.Bates                    6/4/99
; ***********************************************************
; Outputs a scale of 8 tones, 255 cycles per tone,
; tone duration of between a half and one second.
; Hardware: PIC 16F84
; XTAL 4MHz, !MCLR to start
; Audio Output: RB0

; Assign Registers ***********************************

PCL      EQU      02                  ; Program Counter
PORTB    EQU      06                  ; Port B for Output
HalfT    EQU      0C                  ; Half Period of Tone
Timer    EQU      0D                  ; Delay Time Counter
Count    EQU      0E                  ; Cycle Count
TonNum   EQU      0F                  ; Tone Number (1-8)

; Initialize Registers

          MOVLW    B'11111110'        ; RB0 set..
          TRIS     PORTB              ; as output
          MOVLW    08                 ; Set intial value of..
          MOVWF    TonNum             ; Tone Number
          GOTO     start              ; Jump to main program



; continued ...
```

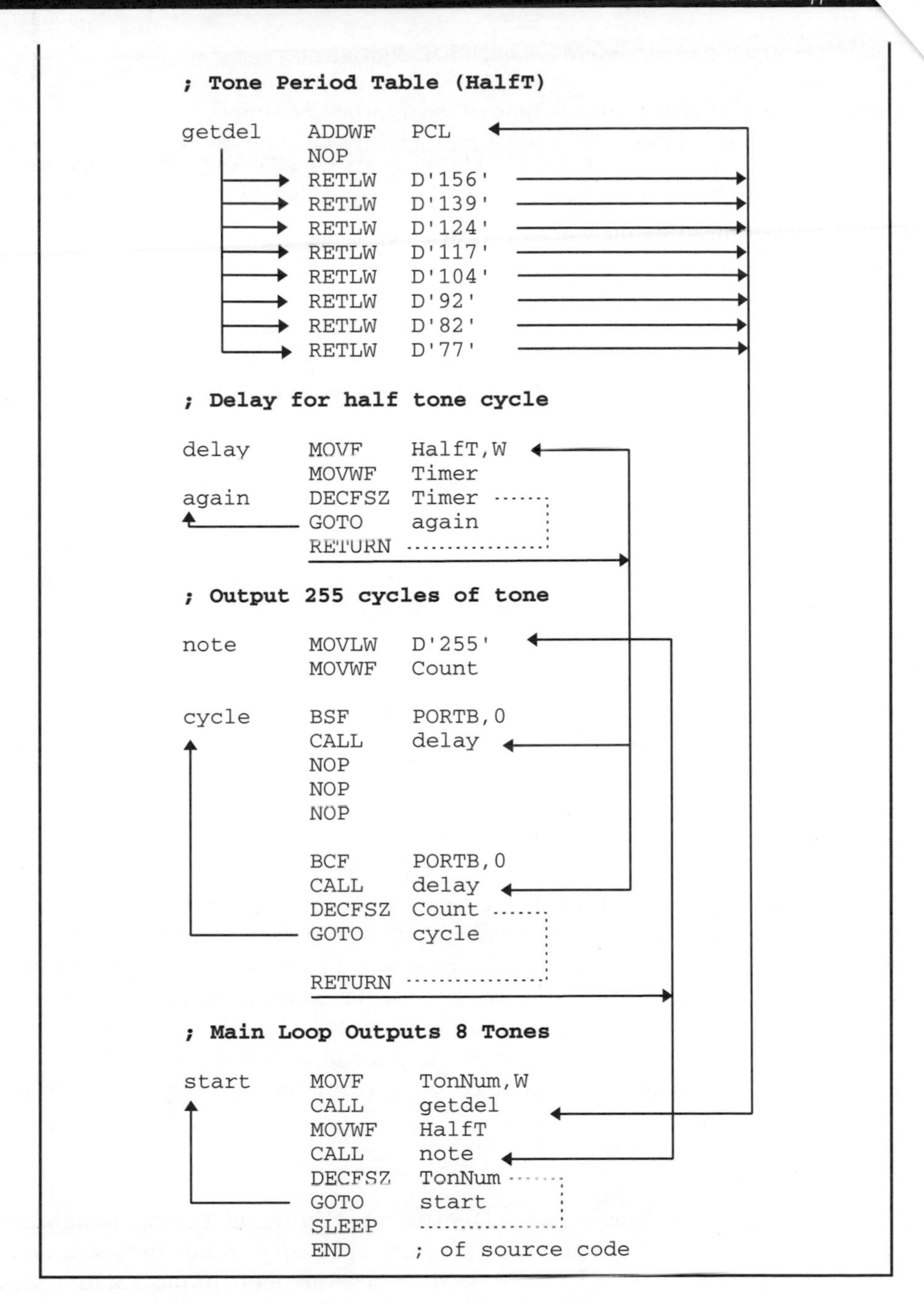

```
; Tone Period Table (HalfT)

getdel     ADDWF    PCL
           NOP
           RETLW    D'156'
           RETLW    D'139'
           RETLW    D'124'
           RETLW    D'117'
           RETLW    D'104'
           RETLW    D'92'
           RETLW    D'82'
           RETLW    D'77'

; Delay for half tone cycle

delay      MOVF     HalfT,W
           MOVWF    Timer
again      DECFSZ   Timer
           GOTO     again
           RETURN

; Output 255 cycles of tone

note       MOVLW    D'255'
           MOVWF    Count

cycle      BSF      PORTB,0
           CALL     delay
           NOP
           NOP
           NOP

           BCF      PORTB,0
           CALL     delay
           DECFSZ   Count
           GOTO     cycle

           RETURN

; Main Loop Outputs 8 Tones

start      MOVF     TonNum,W
           CALL     getdel
           MOVWF    HalfT
           CALL     note
           DECFSZ   TonNum
           GOTO     start
           SLEEP
           END      ; of source code
```

12.3.4 DIZI Application Outlines

Some further information is now provided on the remaining program ideas.

HEX1 Hex Converter

Display the hexadecimal number corresponding to the binary setting of the DIP switch inputs. The input switches select from a table of 16 seven-segment codes which light up the segments in the required pattern for each hex digit display:

```
0,1,2,3,4,5,6,7,8,9,A,b,C,d,E
```

Note that numbers B and D are displayed in lower case on a seven-segment display so that they can be distinguished from 8 and 0 respectively. Use switch input number to select one of 16 codes from a seven-segment table.

MESS1 Message Display

Display a sequence of characters for about 0.5 s each. Most letters of the alphabet can be obtained on the seven-segment display in either upper or lower case, for instance 'HI tHErE'. Output a character code table with delay; the number of characters must be set in a counter, or a termination character used.

SEC1 1 Second Timer

Display an output which increments exactly once per second, from 0 to 9, and then repeats. A table of display codes is required as in the 'Hex Converter'. A 1 s time delay can be achieved using the hardware timer (Chapter 9) and spare register. A 'tick' may be produced at the audio output by pulsing the speaker at each step.

REACT1 Reaction Timer

The user's reaction time is tested by generating a random delay of between 1 and 10 s, outputting a sound, and timing the delay before the Input button is pressed. A number representing the time between the sound and the input is then displayed, say, in multiples of 30 ms.

GEN1 AF Generator

An audio frequency generator outputs frequencies in the range 20 Hz to 20 kHz. The sounder output is toggled up and down with a delay between each operation determined by the frequency required, as in the BUZZ program. For example, for a frequency of 1 kHz, a delay of 1 ms is required, which is 1000 instruction cycles at a cycle time of 1 μs. The information on program timing must be studied in Chapter 10. The delay time, and hence the frequency, can then be incremented using the Input button, and range selection with the input switches might be incorporated, as there are only 255 steps available when using an 8-bit register as the period counter.

MET1 Metronome

Output an audible pulse at a rate set by the DIP switches. The output tick can be adjustable from, say, 1 up to 4 beats per second, using the Interrupt button to step the speed up and down, and the Input button to select up or down. A software loop, or the TMR0 register, can be used to provide the necessary time delays.

BELL1 Doorbell

Plays a tune when the Input button is pressed, using a program look-up table for the tone frequency and duration, since each tone must be played for a suitable time, or number of cycles, as required by the tune. The program can be elaborated by selecting a tune using the DIP switches, and displaying the number of the tune selected.

GIT1 Guitar Tuner

The program will allow the user to step through the frequencies for tuning the strings of a guitar, or other musical instrument using the Input button, or selecting the tone at the DIP switches. The program could be enhanced by displaying the string number to be tuned. The tone frequencies will be generated as for the Doorbell application. The digit display codes would also be required in another table.

12.3.5 LOCK Application

A more complex example is provided in Appendix B. The LOCK program listed demonstrates the use of more advanced features such as analogue to digital conversion and EEPROM memory.

Summary

- Methods of PIC circuit construction available include printed circuit board, breadboard and stripboard.
- PCB design software is available to lay out PIC application boards, but it requires considerable investment of time to master, and PCB etching equipment is then needed.
- Breadboard is reusable, and allows circuits to be prototyped quickly and easily, but is unreliable for complex circuits.
- Stripboard is more reliable, but not reusable. Connections can be laid out on paper, using computer drawing tools or directly onto the board.
- The DIZI demonstration board has a seven-segment display and audio output, with push button input and interrupt and a 4-bit switched input, and can be used for demonstrating a range of simple applications.

Questions

1. State one advantage of (a) breadboard and (b) stripboard compared with a PCB for circuit construction.
2. State the maximum rated current output for (a) Port A pins and (b) Port B pins on the PIC 16F84. How do these ratings simplify the interfacing to this chip compared with standard digital outputs?
3. Explain why a 'common cathode' seven-segment display operates as an active high display. State an input code for (a) all segments off and (b) all segments on.
4. Explain any precautions required in the circuit design if a PIC I/O pin is to be used as both input and output in the same hardware.
5. Outline an algorithm for generating a fixed frequency output, or approximately 1 kHz, from the DIZI board using the hardware timer.
6. Describe a method of outputting a set of codes to drive a seven-segment display of alphanumeric characters at a PIC parallel output.

Activities

1. Confirm by calculation that the values used in the program data table in SCALE1.ASM will give the required delays.
2. Devise a breadboard layout for the BIN circuit in Fig. 6.3. Build the circuit and test the BINx programs.
3. Devise a stripboard layout for the MOT1 circuit in Fig. 10.3, using a VN66 FET and 2 V small dc motor. Build the circuit and test MOT1.
4. Build the DIZI circuit on breadboard, stripboard or PCB and test the following programs: BUZZ1, DICE1, SCALE1.
5. Design and implement any of the programs outlined for the DIZI hardware: HEX1, MESS1, SEC1, REACT1, GEN1, MET1, BELL1, GIT1
6. Investigate how an input from a numeric keypad can be detected. The typical keypad, shown in Fig. 12.12, has 12 keys in four rows of three: 1,2,3; 4,5,6; 7,8,9; #, 0, *. These are connected to seven terminals, and can be scanned in rows and columns. A key pressed is detected as a connection between a row and column. The pull-up resistors ensure that all lines default to logic '1'. If a '0' is applied to one of the column terminals (2,3,4), and a key is pressed, this '0' can be detected at the row terminal (5,6,7,8). If the keypad terminals are connected to Port B of the PIC 16F84, and a '0' output in rotation to the three columns, a key can be detected as a combination of the column selected and the row detected. Terminals 2, 3, and 4 will be set as outputs, and 5, 6, 7 and 8 as inputs.

   A lock function may be implemented by matching an input sequence with a stored sequence of, say, four digits, and switching on an LED output if a match is detected. Design, build and test an electronic lock system using the keypad shown, a PIC 16F84 and an LED to indicate the state of the lock (ON = unlocked). Research the design for the interface to a solenoid operated door lock. Refer to Appendix B for further information.

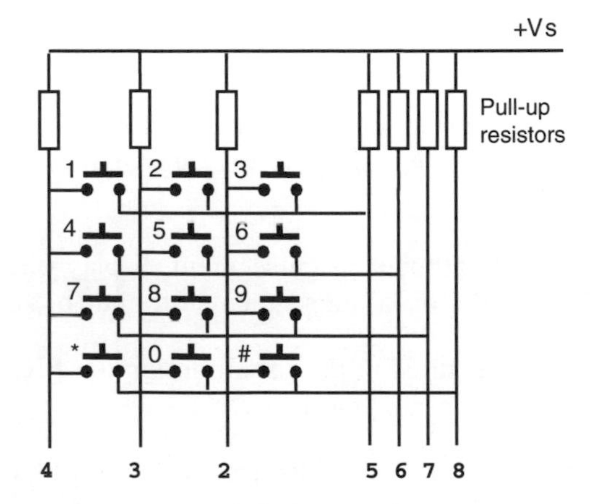

Figure 12.12 Twelve-button keypad connections.

Chapter 13
Motor Applications

There are two main types of control system, open loop and closed loop. An open loop system is essentially manually controlled. For example, a car requires the driver to monitor the direction of travel and correct the steering to follow the road. A closed loop system uses sensors to monitor the system and control the process automatically, once the initial operating selections have been made. An automatic washing machine is an example of a closed loop system that uses fairly simple electromechanical control.

A small, inexpensive, dc motor can be used to demonstrate the use of the PIC microcontroller in a 'real time' control application. Real time systems are those where the time factor is important, and where the dynamic response to control inputs and feedback from sensors is critical.

13.1 Motor Control

Motor output is measured as the shaft speed or position. Open loop control of a motor would consist of simply switching it on and off for a fixed period to position it, or varying the speed, under manual control. There are obvious limitations to open loop control. A dc motor does not start until there is a reasonably large current, due to inertia, 'stiction' and its electromagnetic characteristics. This makes its response 'non-linear', which means that the speed is not directly proportional to the current or voltage supplied. Furthermore, the speed cannot be accurately predicted for any given current, because the load on the shaft will affect it. The final position of the shaft when the motor stops cannot be precisely controlled either. Therefore, if the speed or position of a dc motor is to be controlled accurately, we need sensors to measure these 'output variables', and a control system for the motor drive.

A simple analogue potentiometer can measure position, by converting it to a voltage, or speed can be measured using a tachometer, which produces a voltage that is proportional to the motor speed. These 'transducers' have traditionally been used in analogue motor control systems, where all the signals are continuously variable currents and voltages. It is now more common to use a digital control method, and the microcontroller can be used as the basis of a programmable system in which the control algorithm can be designed to match closely the application requirements. The dynamic (time) response can also be adjusted in the software.

The speed of a dc motor is controlled by the current in the armature, which interacts with the magnetic field produced by the field windings (or permanent magnets in small motors) to produce torque. An analogue control system gives continuous control over the motor current, and a digital to analogue drive converter can be used at the output if the feedback and control is digital. However, the control interface can be simplified if pulse width modulation (PWM) is used, as described in Chapter 10. PWM is a simple and efficient method of converting a digital signal to a proportional drive current. Digital feedback can be obtained from a sensor that detects the shaft rotation. One way of doing this is to use a perforated or sectored disk attached to the shaft and an optical sensor to detect the slots or holes in the disk. The shaft position can be detected by counting pulses, and the speed by measuring their frequency. This signal can be fed directly to a microcontroller running a program that monitors the pulse input, and varies the output to control the speed and/or position of the motor.

13.2 MOTA Board

The block diagram for a motor target board which can demonstrate many of the features of the PIC 16F84 and a range of control methods is shown in Fig. 13.1, and a circuit diagram provided in Fig. 13.2.

A variety of motor control algorithms can be demonstrated: motor on/off, motor forward/reverse, open/closed loop position control, and open/closed loop speed control.

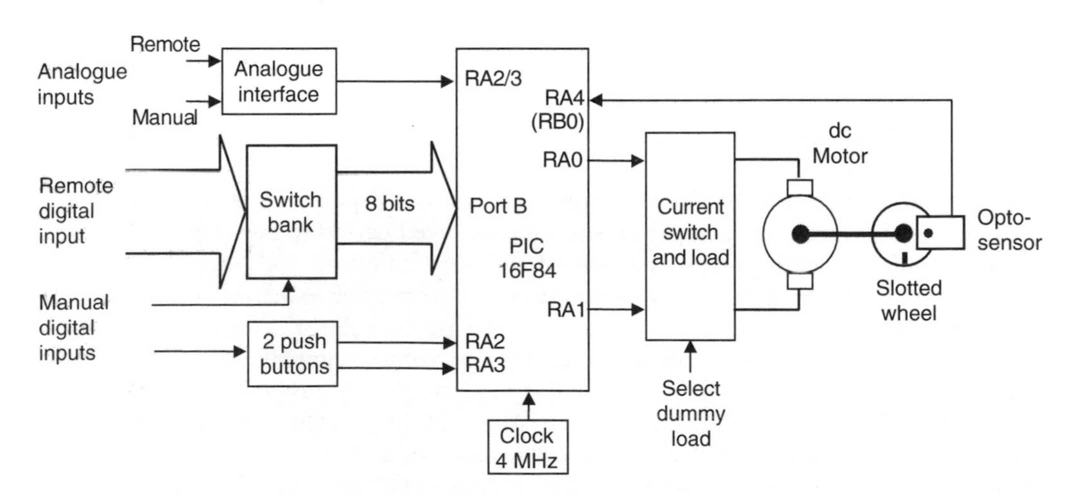

Figure 13.1 Block diagram of MOTA board.

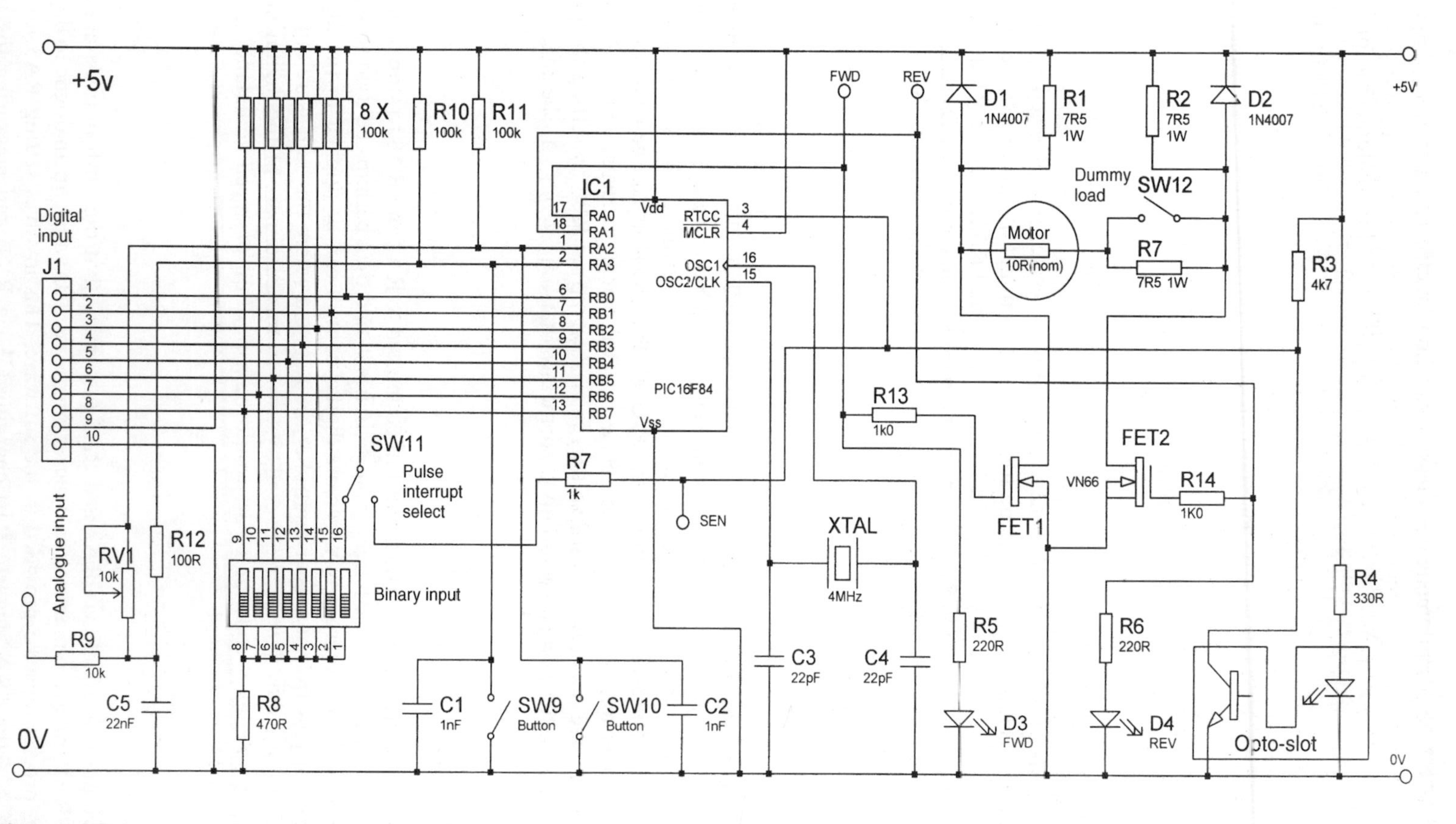

Figure 13.2 Circuit diagram of MOTA board.

13.2.1 MOTA Circuit Description

The MOTA circuit is designed to demonstrate position and speed control with a PIC 16F84 microcontroller. The command inputs can be received from an 8-bit switch bank, a remote 8-bit master controller, two push buttons or an analogue input. The motor can be turned in either direction via bi-directional drive outputs, with LED motor direction indicators. The drive can provide position and speed control, and can be pulse width modulated to control the speed. The shaft speed and position are monitored by a slotted opto-sensor and disk with one slot, feeding back one pulse per revolution to the controller. The PIC is crystal clocked at 4 MHz, for precise feedback measurement.

Motor drive

The small, inexpensive 2 V permanent magnet motor is connected in a passive FET bridge with R1 and R2 as load resistors. The motor current direction, forward or reverse, is controlled by switching on one of two VN66 FETs from RA0 and RA1. A dummy load can be switched in series with the motor to allow testing of closed loop control. LEDs indicate the motor direction. The use of a low quality motor will emphasize the problems that will arise from imperfections in the motor itself.

Opto-sensor

This contains an LED and photodetector mounted either side of the slot in a plastic housing. The perforated disk attached to the motor shaft allows the light to pass through the holes and digital pulses are output from the sensor via a built-in amplifier, which allows the motor speed or position to be monitored by the controller. The sensor input is connected to the T0CKI (RA4) input of the PIC, so that the slots can be counted in the counter/timer register. Alternatively, the pulse interval can be measured at RA4 using the timer. The pulse may also be used to trigger an interrupt at RB0 by setting the pulse interrupt select switch accordingly. This then means that only 7 bits can be read from the binary input.

Switched inputs

The control program can use the push buttons connected to RA2 and RA3 to stop, start or change speed or direction. The binary input switches would then perhaps be used to select the speed or position. Alternatively, remote digital control can be applied to the digital input connector pins from a master controller, in a system where this unit were only one of a set of motor controllers in the overall system. In this case, part of the digital input would be a motor select code, and part would be a position or speed command. Serial commands could also be used.

Analogue input

Analogue input is available via RA2 and RA3 (assuming that the push buttons are not used). Pot RV1 provides a manual analogue setting; to read its value, an analogue to digital conversion process must be provided in the software. This involves setting RA3 low to discharge C5, setting RA2 high to start charging C5 via RV1, and measuring the time until RA3 reaches its input threshold. The time is proportional to the resistance of RV1. An external dc voltage at the analogue input may be read in a similar way, but only values above the threshold of RA3 can be detected, and care must be taken not to exceed the maximum rated voltage on the PIC inputs.

13.3 Control Methods

The PIC 16F84 is crystal clocked at 4 MHz to give an instruction cycle time of 1 μs. No manual reset is required, but the Power On Timer should be enabled during programming to ensure a reliable start.

13.3.1 Open Loop Control

Open loop control of a dc motor has been described in Chapter 10 and a program developed which allows the speed to be controlled manually. This program could be modified to run on the MOTA board. The motor can be driven in either direction by setting RA0 or RA1 high, with both set low to turn the motor off. Either can be pulsed for speed control, but they should not be high together.

Open loop speed control can be implemented in various ways: programmed, push button manual input, manual/remote binary/analogue input. Sequence control can be incorporated in the program; for instance, the speed can be ramped up, held constant and then ramped down over a fixed period of time. The push button inputs could be programmed to run the motor in either direction or increment and decrement the speed in one direction by modifying the delay in a PWM program. The speed could be set at the binary inputs to Port B, either manually at the DIP switches, or with an 8-bit digital input code supplied from a master controller, and analogue control is possible from a manual input (RV1) or from a remote voltage source.

13.3.2 Closed Loop Control

The PIC motor control board has a slotted wheel and opto-sensor to monitor the rotation of the motor. A wheel with only one slot is used for the applications developed here, so that there will then be no possibility of a variation in the inter-slot distance. One pulse per revolution will be then obtained which makes the calculations a little simpler, and provides plenty of time between pulses for the control program to complete its processing tasks. The sensor is connected to RA4 of the PIC. As we know, the 16F84 contains an 8-bit counter/timer register, TMR0, which can be clocked from RA4 or the system clock, which we can use to measure the pulse number, frequency or period.

Closed loop position control involves counting the slots as the shaft turns. This sounds straightforward, but the dynamic characteristics of the motor have to be taken into account. For example, the motor can be switched on from the controller, and the pulses counted, and the motor turned off when a set number of pulses have been counted. However, the motor will probably overshoot the required position due to inertia of the rotor. A simple solution would be to keep counting the slots and turn the motor back by the requisite number of slots. This might have to be repeated several times. More sophisticated control programs will use a mathematical model of the system's physical characteristics, particularly those of the motor, such as inertia, friction and loading. The system performance can then be predicted more accurately, and the control process optimized for accuracy, speed of response and so on.

13.4 Position Control

Position measurement can be obtained simply from counting the number of slots. With only one slot, the position can only be determined to the nearest whole revolution. This may be acceptable if a gearbox is fitted that reduces the angular rotation and speed. For instance, if the gearbox has a reduction ratio of 100:1, the output can be positioned within 1/100 of a revolution. Alternatively, a slotted wheel with more slots can be used. A simple program, which moves the motor by a number of revolutions set on the DIP switches, is outlined in Fig. 13.3.

The hardware timer/counter (TMR0) is used to count the pulses from the sensor. The timer flag (T0IF) is set when the counter rolls over from 255 to 000, so the binary number at the switches has to be complemented (subtracted from 256) to set the count to a value whereby it rolls over after the correct number of pulses. The Option register has to be initialized to select the T0CKI (RA4) pin for input, and to allocate the prescaler to the Watchdog Timer, so that the count is not prescaled. The processor is put to sleep after the move has been completed, so it would have to be reset to repeat the operation.

As mentioned above, the position control achieved with the program may not be accurate. This is because, first, it can only count to the nearest whole revolution, and second, there is no provision for preventing overshoot. One way of achieving better accuracy is for the current position to be continuously compared with the required position, and the motor driven at a speed proportional to the error. The motor will slow down as it approaches the target position. You may find this type of process referred to as PID control, where the response of the system can be tuned to give the best compromise between speed of response, accuracy and overshoot. A simpler process called 'trapezoidal' control

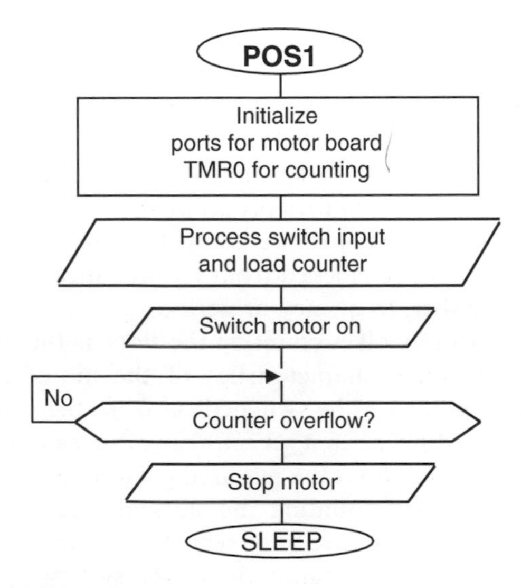

Figure 13.3 Basic position control flowchart.

Program 13.1 POS1 basic position control

```
; ***********************************************************
; POS1.ASM      M. Bates                               7/4/99
; ***********************************************************
;
;       Basic position control program runs motor for
;       number of revs set on binary switch input.
;       Uses TMR0 to count revs, but does not correct
;       for overshoot...
;
;       Hardware:       PIC 16F84 Motor Board

;       Clock:          XTAL 4MHz
;       Inputs:         RB0-RB7      : DIP Switches (High)
;                       T0CKI        : Shaft Sensor (Low)
;       Outputs:        RA0          : Motor (High)

; Set Processor Options.......................................

PROCESSOR 16F84         ; Declare PIC device

; Register Label Equates......................................

PORTA   EQU     05      ; Port A
PORTB   EQU     06      ; Port B
TMR0    EQU     01      ; Counter/Timer
INTCON  EQU     0B      ; Interrupt Control

; Register Bit Label Equates .................................

T0IF    EQU     2       ; Timer Overflow Flag = INTCON,2
motor   EQU     0       ; Motor Output = RA0


; Start Program *******************************************

; Initialize .....................Port B defaults to input

        MOVLW   b'11111100'         ; Port A bit direction code
        TRIS    PORTA               ; Set the bit direction
        MOVLW   b'00100000'         ; Code for Option Register
        OPTION                      ; select T0CKI timer input
        BCF     INTCON,T0IF         ; Clear Timer Overflow Flag

; Main Loop ..................................................

        MOVF    PORTB,W             ; Read switches
        MOVWF   TMR0                ; into counter
        COMF    TMR0                ; and complement

        BSF     PORTA,motor         ; Motor ON
        test    BTFSS INTCON,T0IF   ; Test Overflow Flag
        GOTO    test                ; and wait until set
        BCF     PORTA,motor         ; Motor OFF

        SLEEP                       ; Suspend processor

        END                         ; Terminate source code
```

can also be used. This involves ramping the motor speed up and down at the ends of the move, with a constant speed period in the middle. Because the position control of a dc motor is not very practical without a gearbox, we will look at speed control in more detail. The speed can also be conveniently measured by using an oscilloscope to monitor the pulse period from the sensor; the speed in revs/s will be the reciprocal of the pulse period in seconds.

13.5 Speed Control

A typical application may require the motor board to operate as a slave unit under digital control. A master controller would supply an 8-bit code to set the speed of the motor, with the local controller required to maintain it with a specified degree of precision. The MOTA board allows for such an input, but also for it to be simulated for test purposes using the switch bank. Suppose that the motor is to be controlled to a speed of exactly 50 revs/s (rps), which is about 40% of the nominal full speed. This will produce 50 pulses/s (pps) at RA4 with a single slot. The speed can basically be measured in one of two main ways: (1) counting sensor pulses over a measured time period, or (2) measuring the period between sensor pulses.

13.5.1 Counting Pulses

The accuracy of the speed measurement using this method will depend on the number of slots counted, because the error is always ±1 slot. Thus, if 100 slots are counted, the accuracy will be $1/100 = 1\%$. If the count were made over a period of 1 s at 50 pps, the precision would be $1/50 = 2\%$. Therefore, for 2% accuracy, each count would take 1 s, and speed could only be corrected once per second. This response time is too slow for most practical purposes, so this option will be rejected. It could, however, be used if there were more slots in the disk or the motor were running at high speed.

13.5.2 Measuring Pulse Period

At 50 pps, the test speed, the pulse period will be $1/50\,\text{s} = 20\,\text{ms}$. We therefore need some way of generating a precise 20 ms time interval with which the input period can be compared. This can be done using the TMR0 hardware counter/timer (see Chapter 9). The PIC instruction cycle time is 1 μs with a 4 MHz clock. The counter can therefore be clocked at a maximum 1 MHz (once per instruction cycle). The timer prescaler allows this to be divided by 2, 4, 8, 16, 32, 64, 128, or 256, by setting a 3-bit code in the Option register. A possible prescale factor is calculated as follows:

Now
$$20\,\text{ms} = 20000\,\mu\text{s} = 20000 \text{ instruction cycles}$$

The timer can count 256 cycles, therefore the prescale multiplier required

$$= 20000/256 = 78.125$$

The nearest available prescale multiplier

$$= 128$$

Using this prescale value, the number of counts for 20 ms

$$= 20000/128$$

$$= 156.25$$

$$= 156 \text{ (nearest whole number)}$$

The actual motor period will then be

$$= 156 \times 128 \times 1\,\mu s$$

$$= 19968\,\mu s \text{ (within 1\% of 20 ms)}$$

The longest period then measurable will be

$$= 256 \times 128$$

$$= 32768\,\mu s$$

Motor Period $= 0.032768$ s/rev

Motor Speed $= 1/0.032768$

$$= 30 \text{ rps}$$

Therefore, this method should provide control from the motor maximum speed of about 125 rps down to about 30 rps. Lower speeds may be obtained by increasing the prescale factor in the timer, to give a longer time interval. This could be done 'on the fly' by adding code to allow the push buttons to select the operating speed range.

13.5.3 Motor Control Algorithm 1

Timing diagram

System signals that could provide closed loop control are illustrated in Fig. 13.4. At the start of the control cycle, the motor would be switched on and the timer TMR0 cleared and started. It will then be incremented once per 128 instruction cycles. The sensor and counter can be continuously checked to see whether the counter has reached the set value, or the next pulse has arrived. If the timer times out before the rising edge of the next pulse, the motor is not going fast enough, so the speed must be increased by incrementing the ON time period of the motor. If the pulse arrives first, the motor is going too fast, so the speed must be reduced by decrementing the ON period. If the ON time starts at zero, it will be increased until the motor starts. The speed should then stabilize at the value corresponding to the switch input.

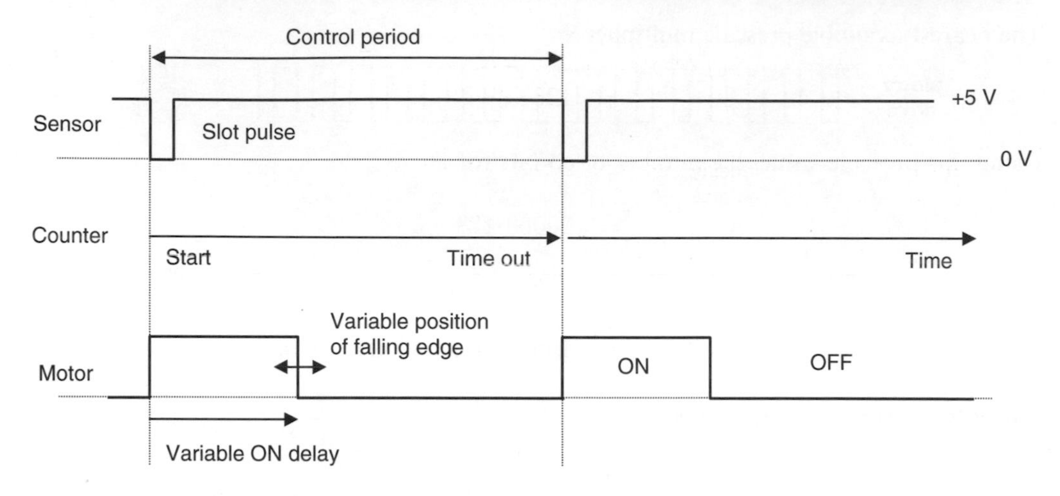

Figure 13.4 Timing diagram for low frequency speed control.

Evaluation

Using this method of control raises a few problems. Note that the drive signal and sensor signal are locked together, with the same period. This means that the drive frequency will be 50 Hz, which is lower than the kind of frequencies normally used. A higher frequency also prevents commutator switching in the motor from interfering with the drive switching. In addition, with this algorithm, the drive will be on for the same part of each revolution, causing uneven wear on the commutator. We will therefore consider an alternative algorithm.

13.5.4 Motor Control Algorithm 2

We would prefer the PWM drive to operate at a higher frequency, so let us consider the delay loop used previously. The proposed algorithm runs the motor using a software delay loop to generate a pulse width modulated drive signal to the motor. The mark/space ratio (MSR) will be adjusted to control the speed. This will be done by waiting for a sensor pulse, starting the hardware timer, and continuously checking whether the timer has finished ('timed out') or the next sensor pulse has arrived first.

Timing diagram

This process is illustrated in Fig. 13.5. The motor drive signal is shown running at a higher frequency than the sensor pulse waveform. It has a variable mark/space ratio, which will be adjusted within the program after the controller has checked if the speed is too high or too low. The hardware timer is used to measure the sensor period by starting an internally clocked count on the falling edge of the sensor pulse, and then continuously checking the time out flag and the sensor for the next pulse.

If the speed is too low, the timer times out first, before the pulse arrives. In this case the speed must be increased for the next timing cycle. On the other hand, if the slot arrives before the timer has timed out, it means that the motor is running too fast, so the speed must be decremented for the next cycle. When the speed is correct, the speed adjustment

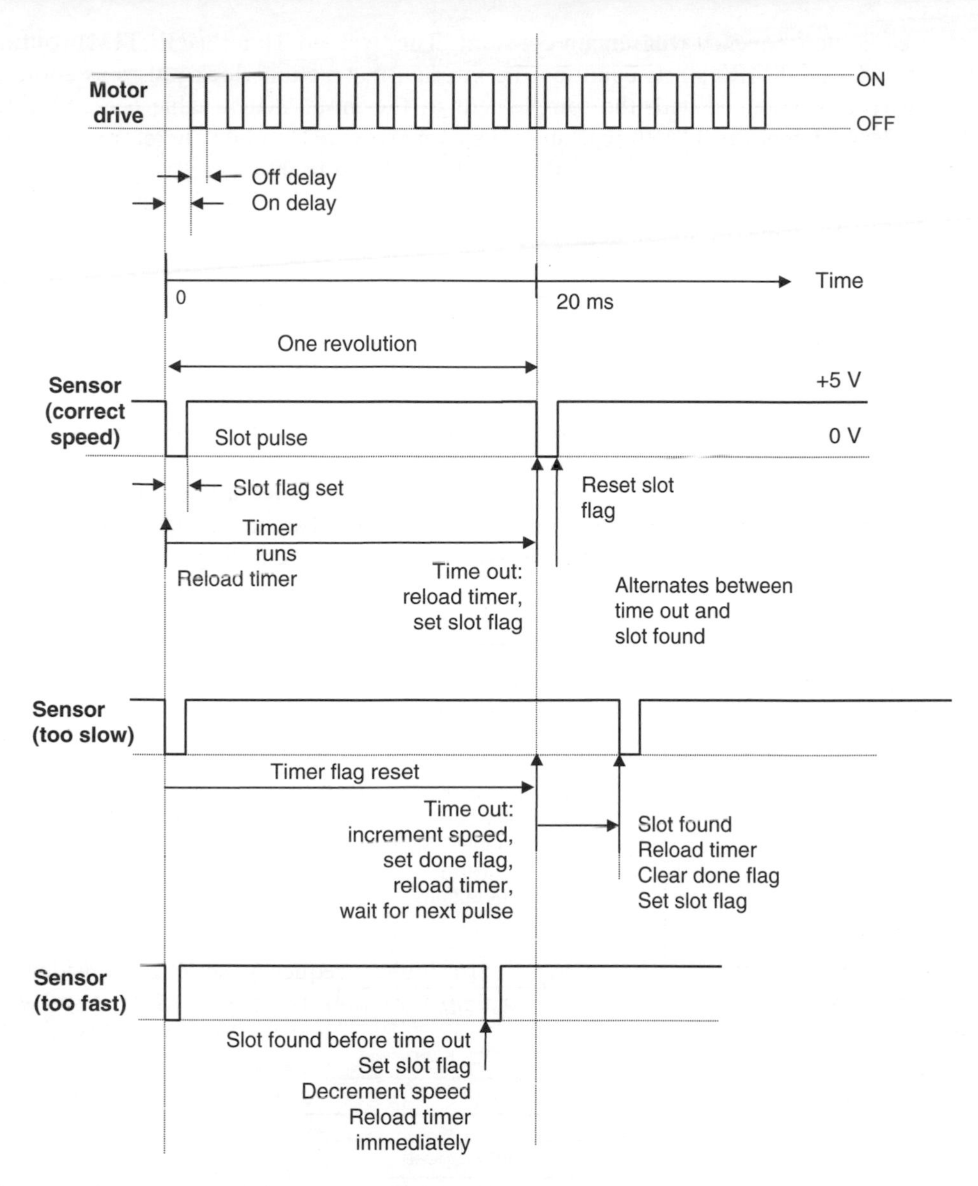

Figure 13.5 Timing diagram for high frequency speed control.

should alternate between incrementing and decrementing. This ideal performance may well be affected by imperfections in the motor. In practice, it has also been found that there was significant 'hunting' in the speed, because of the relatively low sampling rate.

Flowcharts

Figure 13.6(a) shows the top level flow chart for the program. 'Speed' is a general purpose register (GPR) which holds the value for the PWM ON time. The OFF time is derived by complementing this value. The total count for each motor drive cycle is then 255 (FF), which means the frequency will remain constant. The 'Reload Timer' (RELTIM) routine restarts the timer. TMR0 starts counting the internal instruction clock pulses as soon as it is reloaded. It is loaded with the complement of the input switch value because 'time out' is detected when the TMR0 register 'rolls over' from FF to 00. Therefore, the time interval is measured as the count from the loaded value up to 00.

The most complex part of the program is the TESTEM routine, which checks the inputs and modifies the value of 'Speed' accordingly. The requirements are as follows:

1. To start the motor from rest, the MSR must be increased upwards when no slot is detected, so that the motor will eventually start. A fairly high MSR is required to get the rotor moving initially.

2. When a slot is detected before a time out, the timer must be restarted immediately.

3. When the time out occurs first, the speed must be incremented and the timer must be restarted. If a slot then arrives, the timer must be restarted again, so that it can measure the time between the corresponding slot edges. Note that the falling edge of the sensor pulse is used, as shown in Fig. 13.5.

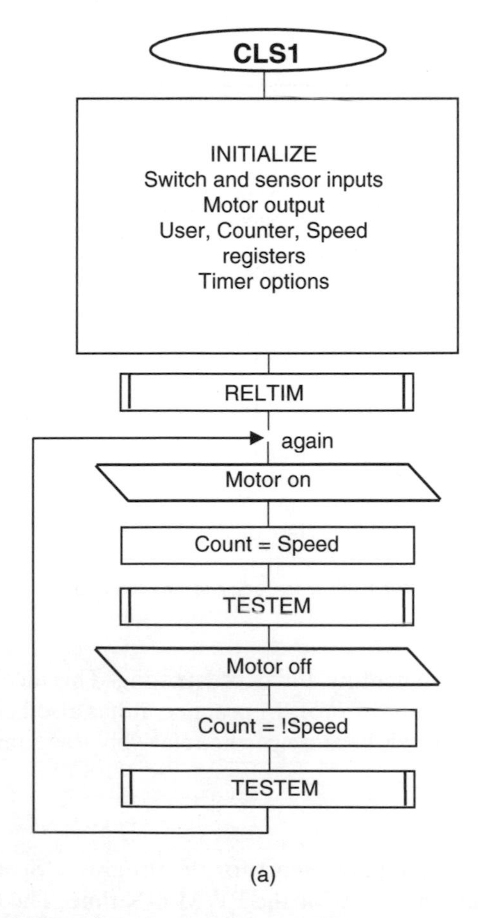

Figure 13.6 (a) General flowchart for program CLS1.

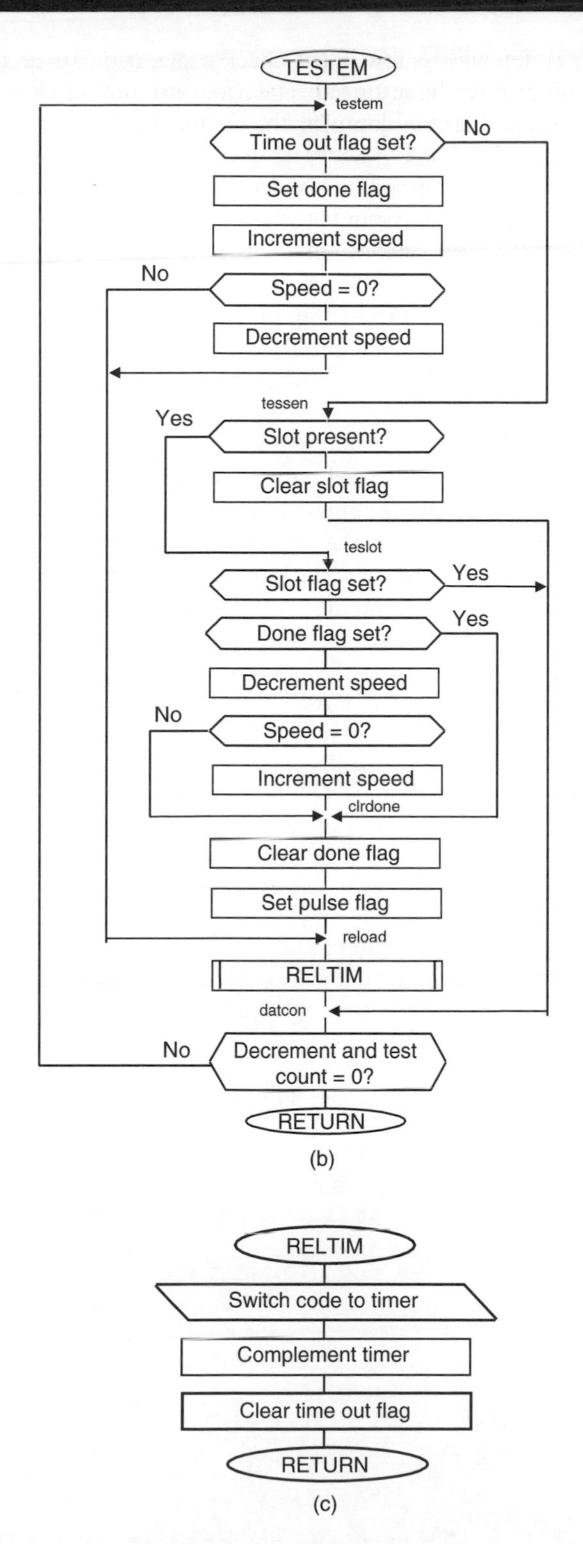

Figure 13.6 (b) Flowchart for TESTEM routine of CLS1. (c) Flowchart for RELTIM routine of CLS1.

4. When the sensor is detected as low in the check cycle, it means that a falling edge has arrived, and the timer must be restarted. The program must then wait for the sensor to go high again before it starts looking for the next slot.

5. The speed must be stopped from rolling over from FF (maximum) to 00 (minimum), so the speed is checked after incrementing and decremented again if it is found to be equal to FE.

To achieve these requirements, user flags have been defined in a GPR to record the fact that the falling edge has been detected and acted upon (flag 'slot'), and another to record the fact that the timer has been reset, to make the program wait for the next slot to restart the timer.

13.5.5 Program Simulation

The source code for the closed loop speed control program is shown as Program 13.2. In order to avoid the need for an input stimulus file, this simulation version of CLS1 loads the timer with the literal value '156' in the subroutine RELTIM, rather than reading Port B switches. For running in the hardware, the comment delimiter on the switch input read to Port B must be removed, and the literal load operation commented out. These options could alternatively be implemented using conditional assembly of the program. The RA4 (sensor) input may be simulated using the asynchronous input window in MPLAB; the equivalent command in MPSIM is 'se RA4'. The result shown in Table 13.1 should be

Program 13.2 Closed loop speed control CLS1

```
; ***********************************************************
; CLS1.ASM                        M. Bates              4/4/99
; ***********************************************************
;
;       Closed Loop DC Motor Speed Control using Pulse
;       Width Modulation (software loop) to control speed
;       and hardware timer to set reference time interval
;
;
;       Hardware:       PIC 16F84 Motor Board
;
;       Clock:          XTAL 4MHz
;       Inputs:         RB0-RB7    : DIP Switches (High)
;                       RA4        : Shaft Sensor (Low)
;       Outputs:        RA0        : Motor (High)
;
;       Configuration Settings:
;
;       WDTimer:        Disable
;       PUTimer:        Enable
;       Interrupts:     Disable
;       Code Protect:   Disable
;

; Set Processor Options......................................
                                                  continued...
```

```
PROCESSOR 16F84            ; Declare PIC device

; Register Label Equates.....................................

PORTA   EQU     05                     ; Port A
PORTB   EQU     06                     ; Port B
TMR0    EQU     01                     ; Counter/Timer
INTCON  EQU     0B                     ; Interrupt Control

Speed   EQU     0C                     ; Counter Pre-load Value
Count   EQU     0D                     ; Delay Counter
Flags   EQU     0E                     ; User Flags

; Register Bit Label Equates ................................

timout  EQU     2                      ; Time-out Flag = TMR0,2
motor   EQU     0                      ; Motor Output = RA0
sensor  EQU     4                      ; Shaft Opto-Sensor = RA4
slot    EQU     0                      ; Slot Found Flag
done    EQU     1                      ; Time Out Done Flag

; Start Program *******************************************

; Initialize ........................Port B defaults to input

        MOVLW   b'11111100'            ; Port A bit direction code
        TRIS    PORTA                  ; Set the bit direction
        MOVLW   b'00000110'            ; Code for Option Register
        OPTION                         ; sets prescale 1:128
        MOVLW   080                    ; Initial value for
        MOVWF   Speed                  ; count pre-load value
        MOVWF   Count                  ; and counter itself
        GOTO    start                  ; Jump to main program

; RELTIM Routine ............................................

; Reloads TMR0 timer/counter register with complement of
; switch input (or dummy value for simulation mode)

reltim  MOVLW   d'156'                 ; Dummy value for timer (sim)
;       MOVF    PORTB,W                ; Input Switches (runtime)

        MOVWF   TMR0                   ; Load Timer with input
        COMF    TMR0                   ; Complement value
        BCF     INTCON,timout          ; Reset 'Time-out' Flag

RETURN


; TESTEM Routine ............................................
```

continued...

```
; Increases speed if timeout detected or
; decreases speed if slot end detected...

testem  BTFSS   INTCON,timout       ; Time Out?
        GOTO    tessen              ; NO: Skip Speed Increment
        BSF     Flags,done          ; Set Time-out Done Flag
        INCFSZ  Speed               ; Test for maximum speed
        GOTO    reload              ; NO: jump to timer reload
        DECF    Speed               ; Decrement again
        GOTO    reload              ; and jump to timer reload

tessen BTFSS    PORTA,sensor        ; Slot Present?
        GOTO    teslot              ; YES: jump to test slot
        BCF     Flags,slot          ; Reset 'Slot' Flag
        GOTO    datcon              ; and continue Count loop

teslot BTFSC    Flags,slot          ; 'Slot' Flag Set?
        GOTO    datcon              ; YES: Skip speed decrement
        BTFSC   Flags,done          ; 'Done' Flag Set?
        GOTO    clrdone             ; YES: Skip speed decrement

        DECFSZ  Speed               ; Test for minimum speed
        GOTO    clrdone             ; NO: continue loop
        INCF    Speed               ; YES: increment again

clrdone BCF     Flags,done          ; Clear 'Done' Flag
setslot BSF     Flags,slot          ; Set 'Slot' Flag

reload  CALL    reltim              ; Reload timer

datcon  DECFSZ  Count               ; Decrement and Test Count
        GOTO    testem              ; Counter not zero yet
        RETURN                      ; End motor cycle if zero

; Main Loop .................................................

start   CALL    reltim              ; Reload timer to start

again   BSF     PORTA,motor         ; Motor ON
        MOVF    Speed,W             ; Put ON delay value
        MOVWF   Count               ; into Counter
        CALL    testem              ; Insert Delay Code

        BCF     PORTA,motor         ; Motor OFF
        MOVF    Speed,W             ; Put ON delay value
        MOVWF   Count               ; into Counter
        COMF    Count               ; and convert to OFF value
        CALL    testem

        GOTO    again               ; Insert Delay Code
        END                         ; Terminate source code
```

Table 13.1 Simulator Test Table for CLS1

After...	*Sensor input*	*Time-out flag*	*Done flag*	*Slot flag*	*Speed register*	*Comment*
start	1	0	0	0	80	Motor stopped
reload timer	1	0	0	0	80	
time-out	1	1	1	0	81	Motor starting up
reload	1	0	1	0	81	
time-out	1	1	1	0	82	
reload	1	0	1	0	82	
slot 1 start and reload	0	0	0	1	82	First slot
slot end	1	0	0	0	82	
time-out	1	1	1	0	83	Increment speed
slot 2 start and reload	0	0	0	1	83	Second slot
slot end	1	0	0	0	83	
time-out	1	1	1	0	84	Increment speed
slot 3 start and reload	0	0	0	1	84	Third slot
slot end	1	0	0	0	84	
time-out	1	1	1	0	85	
etc etc etc						Repeats until up to
slot start and reload	0	0	0	1	XX	Set speed
slot end	1	0	0	0	XX	
time-out	1	1	1	0	XX + 1	Increment speed
slot start and reload	0	0	0	1	XX + 1	slot end
slot end	1	0	0	0	XX + 1	
slot start and reload	0	0	0	1	XX	Decrement speed
slot end	1	0	0	0	XX	
time-out	1	1	1	0	XX + 1	Increment speed
slot start and reload	0	0	0	1	XX + 1	
slot end	1	0	0	0	XX + 1	
slot start and reload	0	0	0	1	XX	Decrement speed
slot end	1	0	0	0	XX	
etc						Speed now stable – repeat increment and decrement

obtained. In MPLAB, the 'modify window' dialogue should be used to modify the timer value, otherwise you will have a long wait for time out!

13.5.6 Hardware Testing

The correct function of the closed loop control program can be tested in the target system by setting the binary input to 156 and checking the actual speed of the motor by measuring the period of the sensor pulse on an oscilloscope; it should be 20 ms. The binary input can then be varied, and the period should vary in proportion, within limits. The transient and start up response can be examined by stalling the motor, and studying the motor response on the oscilloscope as it locks on to the target speed. A dummy load is also provided in series with the motor; when it is switched in, the additional series resistance will reduce the motor in the current, hence its speed. If the closed loop control is working, the drive should compensate and maintain the speed of the motor by increasing the drive MSR.

13.5.7 Evaluation of CLS1

Ideally, the PWM speed control should operate at a frequency above about 15 kHz; the program CLM operates at only about 300 Hz, because of the time required to sample the timer status and sensor input, and to complete the software loop for one drive cycle, which uses a full 8-bit count (256). The frequency could be increased by reducing the total loop count to less than 256, but this will reduce the resolution of the control. Some compromise value could be arrived at by calculation and experimentation, based on the required resolution. A more effective solution is outlined below.

13.6 A 'Real' Application

Attempting closed loop dynamic control using the PIC 16F84, which has only one timer and one interrupt, illustrates its limitations, and why there is a range of more powerful

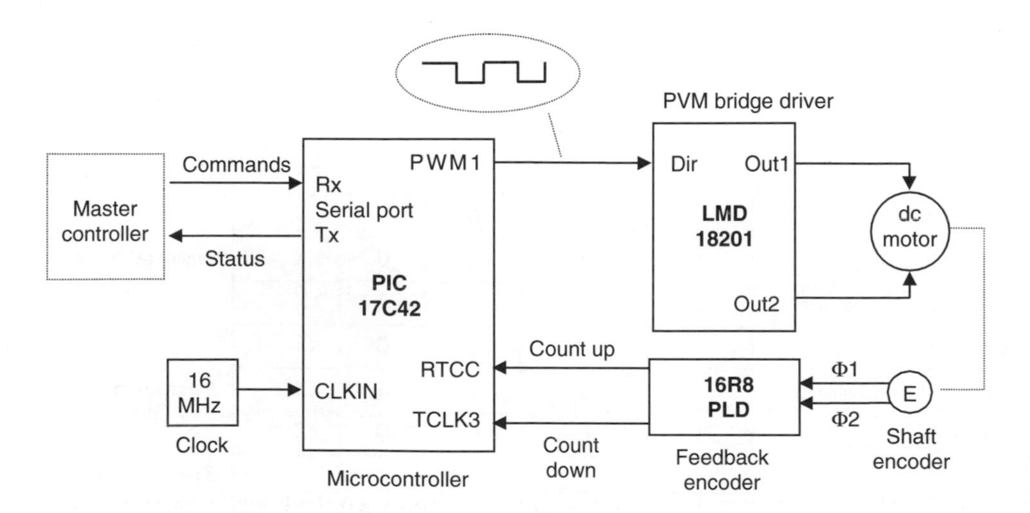

Figure 13.7 Block diagram of servo motor control unit.

Figure 13.8 A servo motor control unit.

processors available in the PIC family. A servo control unit, described in the Microchip Embedded Control Handbook (1993), uses a PIC 17C42 running at 16 MHz, with a dedicated PWM output. It is designed for use in printers and plotters for positioning the print or scan head quickly and precisely. A block diagram is shown in Fig. 13.7.

The motor is driven from a dedicated driver chip (LMD 18201) which requires only a single PWM input. A 50% MSR gives zero output which corresponds to the motor being stationary. The motor can then be driven in either direction by varying the MSR above and below 50%. The driver circuit incorporates a full 'H-bridge' IC, which can supply motor current in either direction with minimal power wastage in the driver chip itself. The motor speed and position are monitored by a shaft encoder which produces two pulse trains. The relative position of the pulses indicates the direction of rotation of the motor. These are fed to a logic circuit which produces separate count up and count down pulses which are counted by 16-bit counters in the PIC to allow the current position of the motor shaft to be calculated. The control software is also far more complex than our demonstration system, with nearly 2000 instructions.

The PIC 17C42 has a serial port which allows commands to the motor to be sent via a single wire, and for the PIC to return information about the actual position, speed and so on, so that, for instance, if the motor is stalled by a mechanical fault, the controller can detect the fault condition. A selection of other PIC chips will be described in the next chapter, so that their features may be compared with the 16F84.

Summary

- A small dc motor under PIC control can be used to demonstrate a range of real-time processes.

- The MOTA demonstration hardware allows motor control via a passive load bidirectional FET bridge driver, with analogue and digital inputs.

- Open loop control can be implemented relatively simply using pulse counting for position control and PWM for speed control.
- Closed loop control requires more complex algorithms that use the pulse feedback to continuously modify the drive output for position and speed control.
- A practical servo typically uses additional hardware, a more powerful microcontroller and complex software for better performance.

Questions

1. Outline an open loop method of controlling the speed of a small dc motor, using a microcontroller. Identify the main hardware components required.
2. Explain how a slotted wheel can be used to provide speed and position feedback from a shaft to a microcontroller.
3. Explain the difference between open and closed loop speed control in a small dc motor.
4. State one problem which might be encountered when attempting to implement accurate dc motor position control using an opto-slot detector.
5. State two alternative methods of measuring the speed of a motor shaft using an opto-slot detector with a crystal clocked microcontroller.

Activities

1. Construct the MOTA board using a method of your choice and devise a test schedule to confirm the correct operation of the hardware prior to fitting the PIC chip.
2. Devise a program to rotate the MOTA board output by exactly 100 revs from its start position. Evaluate the performance of the program in terms of speed of response, accuracy and reliability. What are the characteristics of the motor which affect the performance?
3. Investigate the performance of the program CLS1 in terms of reliability, response time, and range of control (maximum and minimum speeds). Devise a method of loading the motor to test the performance of the controller with varying loads (the speed should be held constant within limits).
4. Modify CLS1 to read the input push buttons on the MOTA board to increase or decrease the set speed.
5. Research a method of analogue to digital conversion at the input of a PIC that does not have an ADC input (see Appendix B). Adapt CLS1 to read an analogue input to the MOTA board and use it to control the speed of the motor.
6. Redesign the MOTA board circuit to use a full bridge motor driver IC instead of the dual FET and passive bridge drive circuit.

Chapter 14
Control Technologies

The PIC 16F84 has been studied in detail here because its architecture and operation are relatively straightforward compared with other processors. A range of other PIC devices, microcontrollers from other manufacturers and conventional 8-, 16- and 32-bit microprocessors operate on the same basic principles, but are generally more complex than the 16F84. To conclude our study then, features of a selection of other PIC devices will be outlined, as well as the standard Intel 8051 microcontroller and a typical 'conventional' microprocessor system based on the Motorola 68000 16-bit CPU. The Intel-based PC processor system has already been introduced in Chapter 1.

14.1 Common Features of PIC Microcontrollers

The main groups of PIC devices are shown in Table 14.1. They are divided into four groups, with three main prefix numbers: the 12XXXX series are 8-pin 'miniature' PICs, the 16XXXX are the 'standard' series and the 17XXXX are the 'high performance' group.

14.1.1 Register and Instruction Set

All PIC microcontrollers use the same basic register and instruction set, modified and extended to take account of the different facilities of each processor. The PIC 17C42, the first high performance PIC developed, had 55 instructions, compared with the 35 of the 16C84. Decrement and Skip if Not Zero (DCFSNZ) is available in this chip, for example, as well as Decrement and Skip if Zero, together with additional data manipulation instructions. The 17C42 has additional ports, timers and special interfaces (serial and

Table 14.1 Main groups of PIC microcontrollers

Series	Features
12CXXX	8-pin devices with minimal features Low cost and small size 33 × 12/14-bit instructions 4 MHz on chip oscillator Analogue to digital converters
16C5X	Original range with minimal features 33 × 12-bit instructions No interrupts
16CXXX	Large range of options, including: 16F84 with flash ROM 35 × 14-bit instructions ADC, serial ports, LCD interface
17CXXX	High performance devices with larger program memory 58 × 16-bit instructions (including multiply) More I/O pins, timers, serial I/O Pulse width modulated outputs

PWM), all requiring additional registers to operate them. A typical application of this chip, a servo-control unit, is outlined in Chapter 13.

14.1.2 Serial In Circuit Programming

The PIC microcontrollers use a common program downloading system, which consists of transferring the program binary code via one of the data pins in serial form when the chip is in programming mode. Programming units usually have a ZIF (zero insertion force) socket which is large enough to accommodate the whole range of PIC chips.

Alternatively, the chip can be programmed in circuit, if the hardware is designed accordingly. This means that the chip can be left in the application circuit at all times, allowing software updating without direct access to the application hardware. The chip can be reprogrammed remotely, after installation in the application hardware. This has obvious advantages for remote sensing systems where the PIC circuit is installed at an inaccessible location.

14.1.3 Power Supplies

Standard digital devices operate with a 5 V supply, and TTL circuits, based on bipolar transistors, must be supplied with a voltage between 4.75 and 5.25 V. This is a problem in battery powered circuits, because 5 V cannot be obtained directly from the available battery voltages. In addition, TTL circuits tend to consume too much power for a battery supply. To overcome this problem, CMOS logic, as used in the PIC, was developed. It is based on FETs, which consume less power than bipolar transistors, and operate at a wider range of supply voltages, while providing the same logic and processing functions.

Its main disadvantage is that it can be damaged more easily by static electricity during assembly and maintenance, but the PIC 16F84 seems, in practice, quite robust!

The PIC 16F84 supply requirements are specified as 2.0–6.0 V, but the supply voltage range for other PIC devices vary somewhat. Nevertheless, most PICs should operate on two or three dry cells (3.0 or 4.5 V) or three or four NiCad (nickel cadmium) batteries (3.6 or 4.8 V).

14.1.4 Microchip Development System

One of the major factors in the success of the PIC microcontroller range must be the availability of the basic development kits at a reasonable cost. The universal availability of the PC compatible computer as a development platform was also important. Professional developers can add the more powerful and expensive items, such as an in-circuit emulator, as required.

14.2 Relatives of the PIC 16F84

Because the PIC range is being expanded and updated all the time, it is impossible to be definitive about the devices available; a representative selection is described in Table 14.2. The original device, which offered the minimum range of features, was the 16C57. It had essentially the same instruction set as the 16F84, but three ports, A, B and C (File Register Addresses 05, 06, 07). It did not have EEPROM and the associated registers present in the 16F84, nor any interrupt facilities.

Other members of the 16XXXX group were developed to offer different combinations of features. The 16C71 provides built-in ADC (analogue to digital converters) inputs for measuring analogue voltage inputs, while the 16C622 has comparator inputs which can detect relative analogue voltages. The 16C65 included serial communication for transfer of commands, data and status information, and multiple interrupt sources. The 16C924 was designed specifically to interface easily with an LCD display. The 17C44 is the successor to the 17C42 which we have already met, with similar high performance features and special interfaces: 58 × 16-bit instructions, multiple I/O ports and interrupt sources, serial port, PWM ports, large memory and high clock rate. More recently, smaller devices have been added. The 12C672 is an 8-pin package with built-in clock and reset, leaving six pins available for I/O, with ADC inputs as an option.

Most of the chips are available with a one time programmable (OTP) memory, and the more expensive EPROM (UV erasable) program memory. The PIC 16F84 is an exception, with its reusable flash ROM.

14.3 Other Microcontrollers

There are several other families of microcontroller available, which can be considered as alternatives to the PIC series.

14.3.1 The Intel 8051 Microcontroller

Originally introduced in 1980, the 8051 can be considered as the original standard microcontroller. As can be seen in the block diagram (Fig. 14.1), it is relatively complex, and has

Table 14.2 Features of selected PIC microcontrollers

	12C672	16C57	16C622	16C65	16C711	16F84	16C924	17C44
Total number of pins	8	28	18	40	18	18	64	40
Number of I/O pins	6	20	13	33	13	13	52	33
Program memory ROM EPROM Flash ROM	2k 	 2k 	 2k 	 4k 	 1k 	 1k	 4k 	 8k
Instruction size (bits)	14	12	14	14	14	14	14	16
Data RAM	128	72	128	192	68	68	176	454
Data EEPROM	–	–	–	–	–	64	–	–
Max clock speed (MHz)	4	20	20	20	20	10	8	25
Hardware timers	1	1	1	3	1	1	3	4
ADC inputs	4	–	–	–	4	–	5	–
Serial I/O	–	–	–	yes	–	–	yes	yes
Interrupt sources	–	–	4	11	4	4	9	11
Brown out protection	–	–	yes	yes	yes	–	–	–
Comparator inputs/PWM Outputs	–	–	–	2	–	–	1	4
Number of instructions	33	33	35	35	35	35	35	58
Special feature	Small and low cost	Minimum specification	Analog comparators	Serial communication	Analog/ digital converters	Flash ROM	LCD display output	High performance

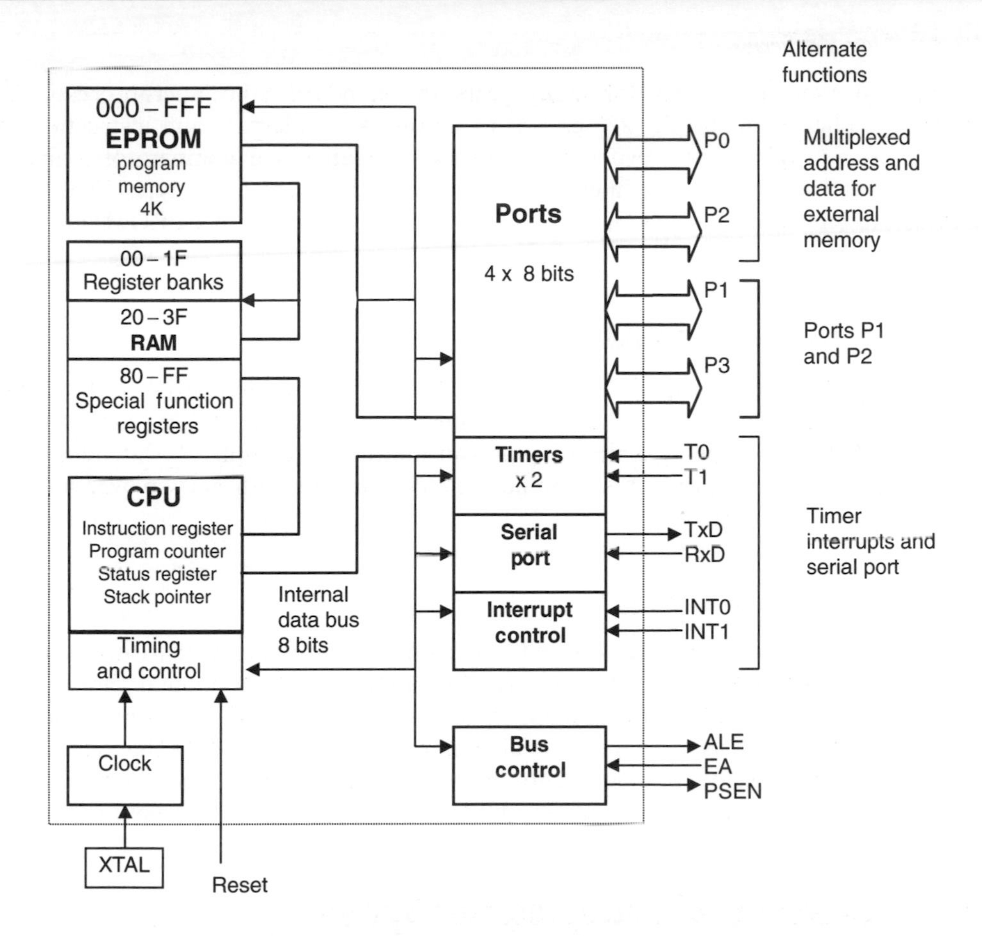

Figure 14.1 Block diagram of the 8051 microcontroller.

the features to be used for a wide range of applications: multiple parallel ports, timers and interrupts, and a serial port. The 8051 can be used as conventional processor, as well as a microcontroller; it can access external memory using Port 0 and Port 2, which can be set up as multiplexed data and address lines. Some of Port 1 and Port 3 pins also have a dual purpose, providing connection to the timers, serial port and interrupts.

14.3.2 Philips 8XCXXX Series

Philips Components now supply a range of microcontrollers related to the 80C51. Most devices in this series are available with either external ROM, internal ROM that is pre-programmed to order by the manufacturer, or internal EPROM for user programming. For example, the 80C31 is a version of the 80C51 without internal ROM; the application program is stored in a separate EPROM. As usual, a range of related microcontrollers is available with different features, so that the device can be selected to match the application as closely as possible to obtain the required functions at minimum cost. At the upper end of the Philips series is the 8XC552 which has 8 k ROM/EPROM, 256 bytes of RAM, PWM outputs, ADC inputs, two types of serial port, and various different types of I/O lines – an all-singing, all-dancing microcontroller, with a price to match.

14.3.3 Zilog Z8 Series

The Z80 microprocessor was for many years the standard 8-bit microprocessor for industrial applications, and the Z8 series is based on its architecture and instruction set. These microcontrollers have two counter/timers, two analogue comparators, internal reset and watchdog timers, six interrupt levels and a choice of clock types. The range offers controllers with 14, 24 or 32 I/O pins, 1, 2, or 4 k of program EPROM and clock speeds up to 12 MHz.

14.3.4 SGS-Thomson ST62 Series

This range is designed principally for appliance, automotive and industrial control applications which require multiple analogue sensing inputs. All the devices have at least seven 8-bit A/D (analogue to digital) inputs and multiple interrupts. A watchdog timer allows recovery from program crashes in hostile environments with noisy power supplies, heat, vibration, and so on.

14.3.5 Motorola MC68HC11 Series

This series is based on the architecture and instruction set of the standard Motorola 68000 microprocessor, which is discussed below. The MC68HC11A1P is a popular choice, offering 8 k program ROM, 512 bytes of data, 256 bytes of static RAM, 16-bit multifunction timer, synchronous and asynchronous serial communications, eight A/D channels and 38 I/O pins. It can operate in microcontroller or microprocessor mode with external memory accesses via multiplexed address and data busses at the multipurpose port pins.

14.4 Conventional Microprocessor System

Now that we are familiar with the features of a typical microcontroller, our overview can be extended by looking briefly at a conventional microprocessor system. Here, the CPU, memory and I/O chips are separate, and linked together by system address, data and control busses. We have already met the Intel CPU-based PC architecture, so we will look briefly at the other most commonly used processor type, the Motorola 68000 (M68K), which has been in widespread use in both home computers and industrial applications for many years. It is also the CISC processor most often used in education and training because of its relatively regular architecture.

A typical development and training system is shown in Fig. 14.2. The 68000 Target Board is the conventional equivalent of the microcontroller, with EPROM, RAM and Port chips on board which would typically measure about 150 × 200 mm. When used for training, the 68000 board can be connected to an applications board that has a range of peripheral transducers, such as switches, LEDs and PWM controlled motor and shaft opto-sensor. The applications board is controlled by the 68000 CPU via a standard 68230 parallel interface/timer (PI/T) that has three 8-bit ports, of which Port A provides data transfer and Port B the individual control and data lines.

The program for the 68000 is prepared on a host PC, in a similar way to the PIC programs. Assembly language source code is written using a text editor and converted to machine code by an assembler program. Alternatively, the source code can be written in

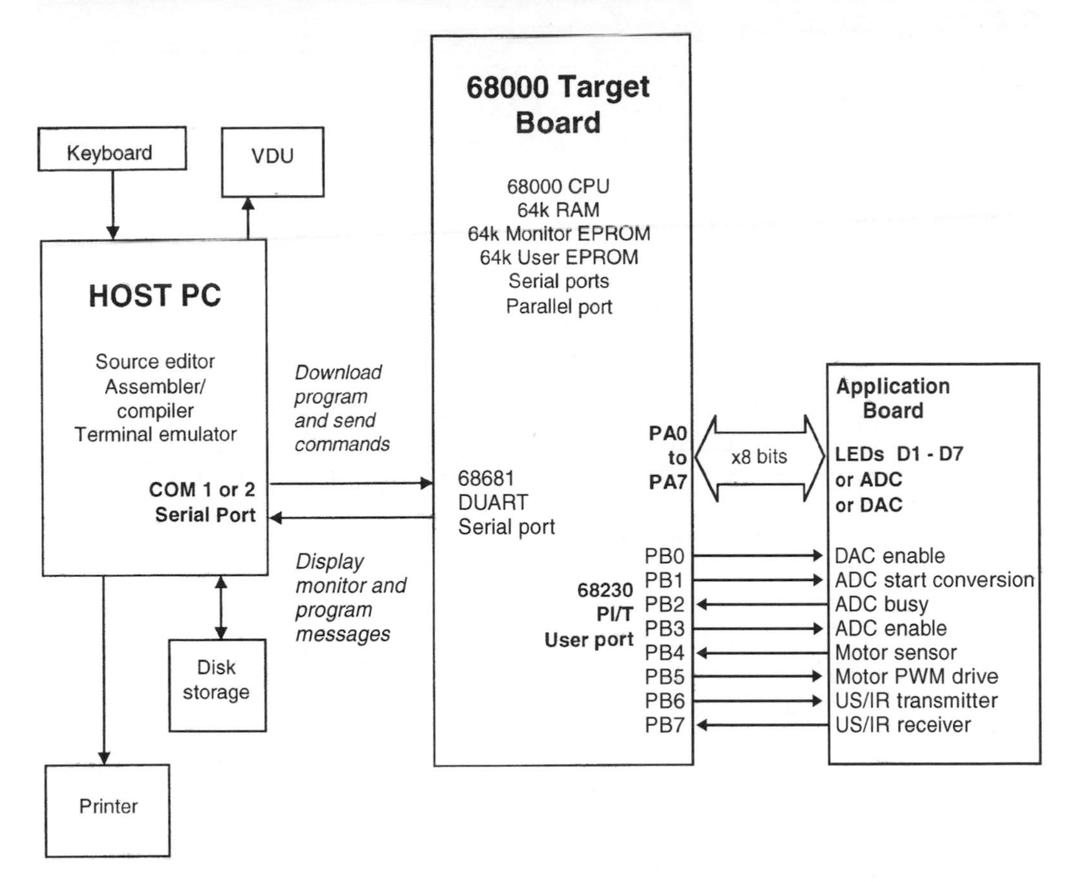

Figure 14.2 M68000 microprocessor demonstration system.

the high-level language, typically standard 'C'. A compiler then converts the source code initially into assembler code, which is then assembled. The machine code program created is downloaded via the PC serial port to the serial port of the target board and hence into its RAM block. A 'terminal emulator' utility is used for downloading, which also allows the target board to use the PC screen and keyboard to control its monitor program (minimal operating system). The monitor commands are used to run and debug the program. In single step or trace mode, the 68000 can display its register contents on the PC screen. The PC provides the keyboard, screen, disk storage and printer during program development. Once an application is up and running on the 68000 board, a user interface may no longer be required, or if it is, a simple keypad and display may be sufficient. At this stage, the program can be blown into EPROM to run independently and the PC disconnected.

The block diagram of the M68000 target board is shown in Fig. 14.3, so that it can be compared with the PIC 16F84 internal architecture. Note that in the PIC, the internal architecture of the processor is visible in the manufacturer's block diagram (data sheet, Fig. 3.1), whereas in the 68000 system, it is concealed within the CPU. Therefore, to see all the details of the 68000 system, both the internal CPU architecture and the circuit board diagram must be studied.

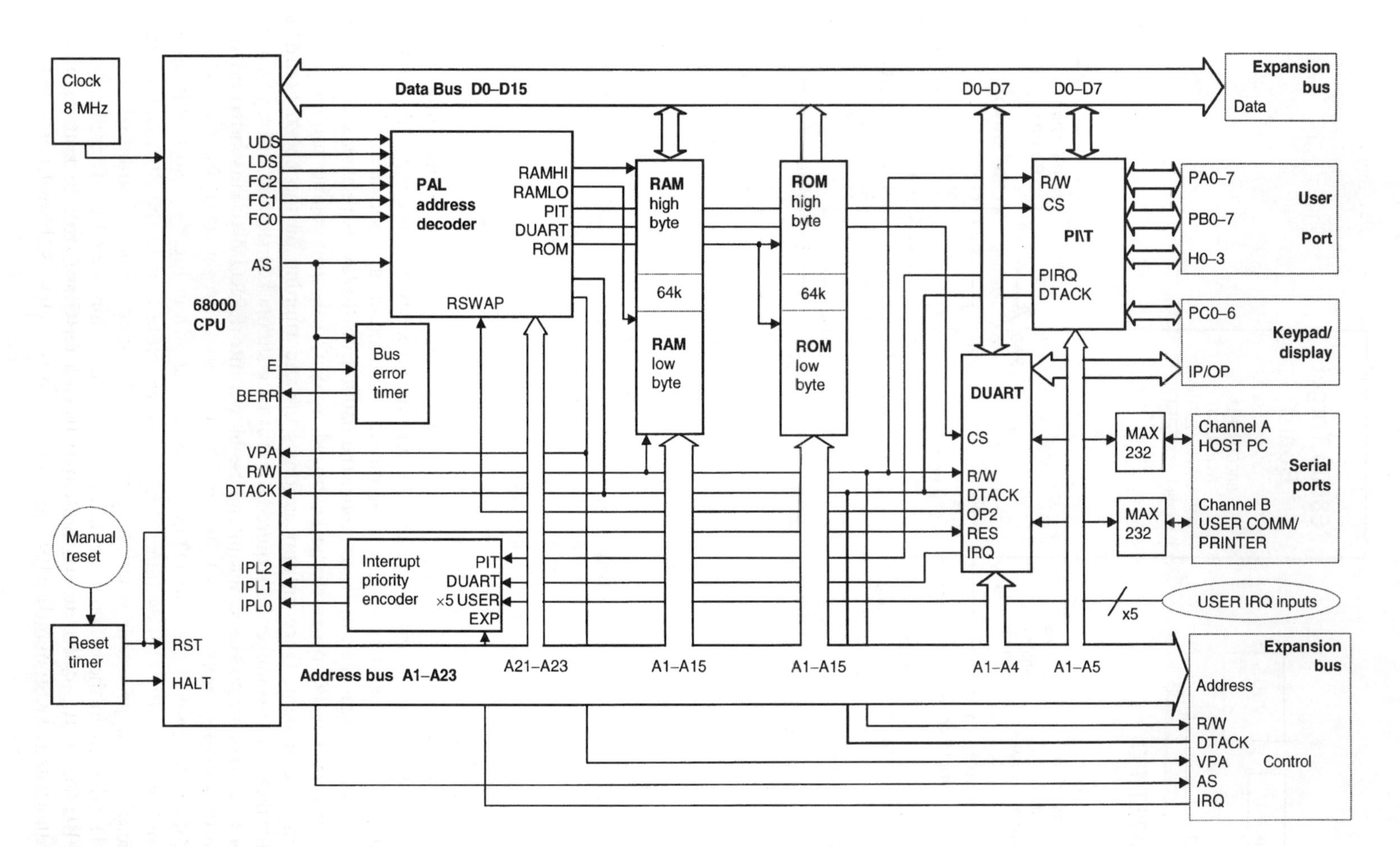

Figure 14.3 Block diagram of M68000 target board.

The conventional microprocessor system is controlled by the CPU driven at 8 MHz. The RAM holds the user data and program, and the EPROM stores the monitor and communications program that allows the program in RAM to be downloaded, run and debugged.

14.4.1 Program Execution

The lowest memory locations, 0000–0FFF, are used by the system control software in the 68000, so the user's machine code program is typically stored in RAM at a set of locations from address $1000 ($ = hex). Assuming it has already been downloaded to RAM using the monitor commands in EPROM, the program is started by issuing a command from the terminal (PC), specifying the start address, if necessary; for example, 'G $1000' to start at the default user program origin.

The program is executed by the CPU fetching each 16-bit code in turn from the memory into the CPU via the data bus. The program codes are found in the memory by the CPU sending out the addresses in sequence, starting with $1000, from its program counter register. The address is 'decoded' by the address decoding logic that is contained partly in the PAL (programmable array logic) chip and partly in the memory chips themselves. In this way, each individual memory location (1 byte) can be accessed for reading and writing.

The 68000 system has a 16-bit data bus, and 24-bit addressing giving a maximum address space of 16 Mb (megabytes). In this minimal system, only a small fraction of this memory space is used; the EPROM and RAM blocks are all 64 k bytes, installed as pairs of identical chips that store the upper and lower byte of the 16-bit data word. The 68000 instructions are of variable length, 2-, 4- or 6-bytes long, stored in adjacent locations.

The PAL address decoder chip generates the system chip select (CS) signals from the address lines A21, A22, and A23, which 'enable' the memory or I/O device that is to be accessed. The lower order address lines are fed directly to the memory chips to select an individual location within the memory array. The Read/Write (R/W) controls the data direction for the data transfer between the CPU and the other chips.

14.4.2 Ports

The parallel interface/timer (PI/T) chip connects the 68000 board to the application board. In a 'real' application, this board would be replaced by the hardware required by that particular application, and a keypad and display connected, if required. The DUART serial port allows the program to be downloaded to RAM from the host PC. The MAX 232 chips are line drivers that boost the signal power on the serial connection between the host and the target board.

The ports can request service from the processor by using the interrupt signal (PIRQ, IRQ) so that more important data transfers can be completed quickly. The Interrupt Priority Encoder converts the active interrupt line number into a 3-bit code which identifies the device requesting service to the CPU, which then runs a corresponding interrupt service routine.

14.4.3 Bus Control

The 68000 has what is called an 'asynchronous bus'. This means that the memory or I/O data transfer is not completed unless the CPU receives a Data Acknowledge (DTACK) signal from the peripheral chip. The 68000 I/O chips (PI/T and DUART) are designed to provide DTACK, but memory chips are not processor-specific in their design, so the DTACK for the memory access cycles must be generated by the PAL decoder. If not received within a preset time, the Bus Error Timer generates an error signal to the CPU and the bus cycle is aborted, and an error reported to the terminal.

The busses and most of the system signals are available at an expansion connector so that additional devices can be attached or data passed to another processor. The address decoder generates a VPA signal which allows external devices to be added to the decoding system.

14.5 M68k Application Example

Figure 14.4 shows a block diagram of the M68000 board used as a supervisory controller, making use of its multiple I/O facilities and the keyboard and display option. The motor control board is designed to operate six motors under closed loop speed control. Each motor has its own microcontroller, and could form part of a robot control system, for example.

In a robot controller, the command to move to a particular position, given as a set of three-dimensional co-ordinates, could be received via the serial port. The 68000 would work out the move required for each motor, and move them simultaneously to produce

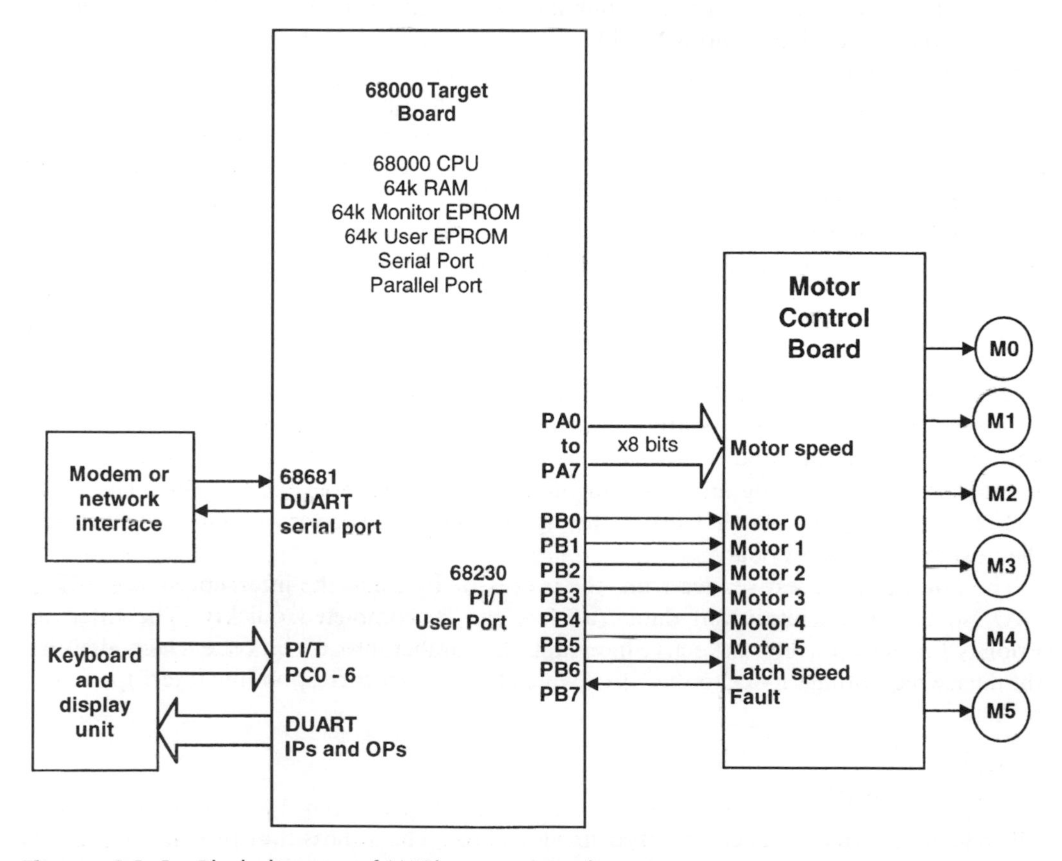

Figure 14.4 Block diagram of M68k motor control system.

a smooth path for the robot end effector. The processor would then select each motor in turn and download its speed setting, using the 'Latch Speed' to signal each individual motor controller to store the current value. The motors could then be run at different controlled speeds, while the 68000 board monitors the 'fault' input, which allows each motor controller to signal a mismatch between the programmed speed and that actually attained. Motor operating sequences could be stored in the RAM block, while the communication and control firmware would be resident in ROM.

14.6 Overview of Control Technologies

Microprocessor and microcontrollers fit into a range of technologies that may be used to implement control systems and include:

- electromechanical relays
- programmable logic controllers (PLCs)
- microcontrollers
- standard microprocessor boards
- dedicated microprocessor systems
- Wintel/PC based controllers
- mini/mainframe computers

We will conclude our study of microcontroller principles with an overview of this range of related technologies, of which the microcontroller is part. In fact, microcontrollers are literally a part of some of these devices; the PLC, for example, contains a microprocessor or microcontroller at its heart.

14.6.1 Relay Control

The relay is the simplest form of controller – a small input current through a coil controls (electromagnetically) a set of contacts which can switch a load (motor, heater, pump, etc.) on and off. Thus a low power input circuit controls a high power load circuit. This reduces power consumption on the control side, and is inherently safer because the control circuits, with which the operator has contact, can be electrically isolated (separated) from the power circuits. In addition, relays can be combined together to make simple sequence control circuits, as illustrated below.

Machine control

A simple system for controlling a machine tool is illustrated in Fig. 14.5. The control system is designed to prevent the motor starting immediately when the power is turned on, and to prevent the motor running unless the cutting fluid pump is on and the guard is closed. There

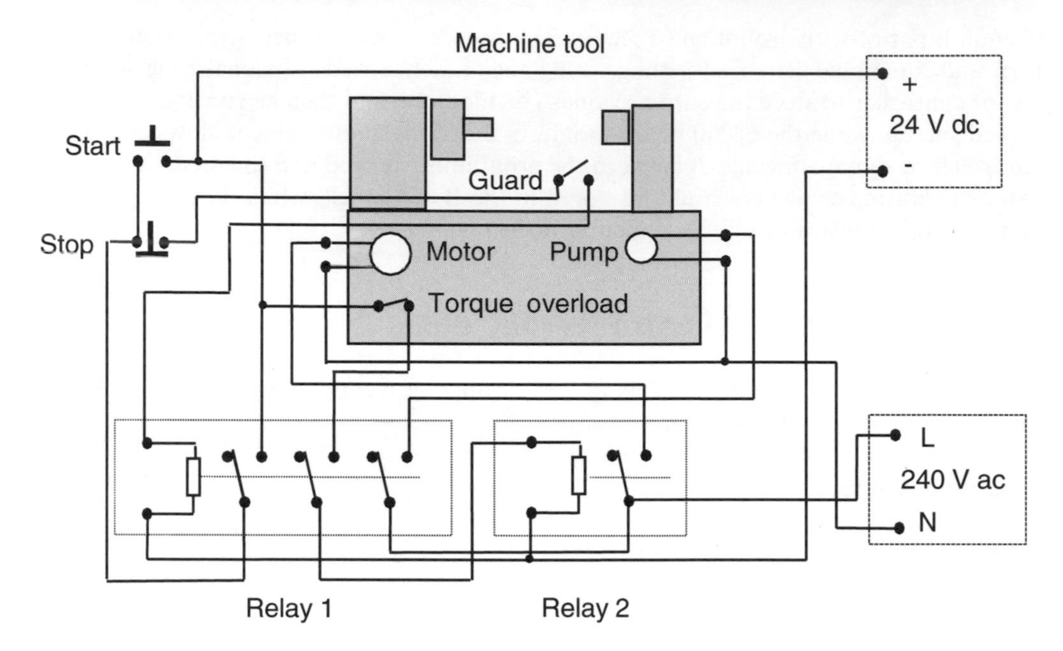

Figure 14.5 Relay machine control system.

is also a torque overload sensor which disables the machine if the tool jams or the motor is stalled for some other reason. Relay 2 (motor) is controlled from relay 1 (control). The relay coils and control switching are operated at 24 V. The motor and pump use the 240 V supply, which in most machines would, in fact, be a three-phase supply. If the wiring is followed through, it can be seen that the machine will be safely operated via the push buttons as required.

Sequence control

Another familiar example of a sequence control system is a domestic washing machine. The main controller is a motorized rotary switch, which operates multiple contacts in the required sequence to open valves (filling), switch on motors (washing, spinning and pumping) and heaters. It operates in conjunction with switched sensors (level, temperature) and safety interlocks (door switch), to give the wash sequence. Thus, electromechanical devices provide a relatively straightforward solution for simple control applications, or ones where the environment is hostile to electronics, such as inside a washing machine. However, switches and relays are unreliable because of the moving contacts. Solid state logic offers a method of implementing more complex control functions, but programmable logic controllers (PLCs) are now more often used for sequential control.

14.6.2 PLC Control

The PLC is a self-contained sequence controller, based on a microprocessor, but with all the electronics and switched interfaces built in. The PLC can be programmed to act like a set of relays to give a particular output sequence in response to switched inputs, which can be manual inputs or derived from sensors. It is suitable for controlling systems where motors,

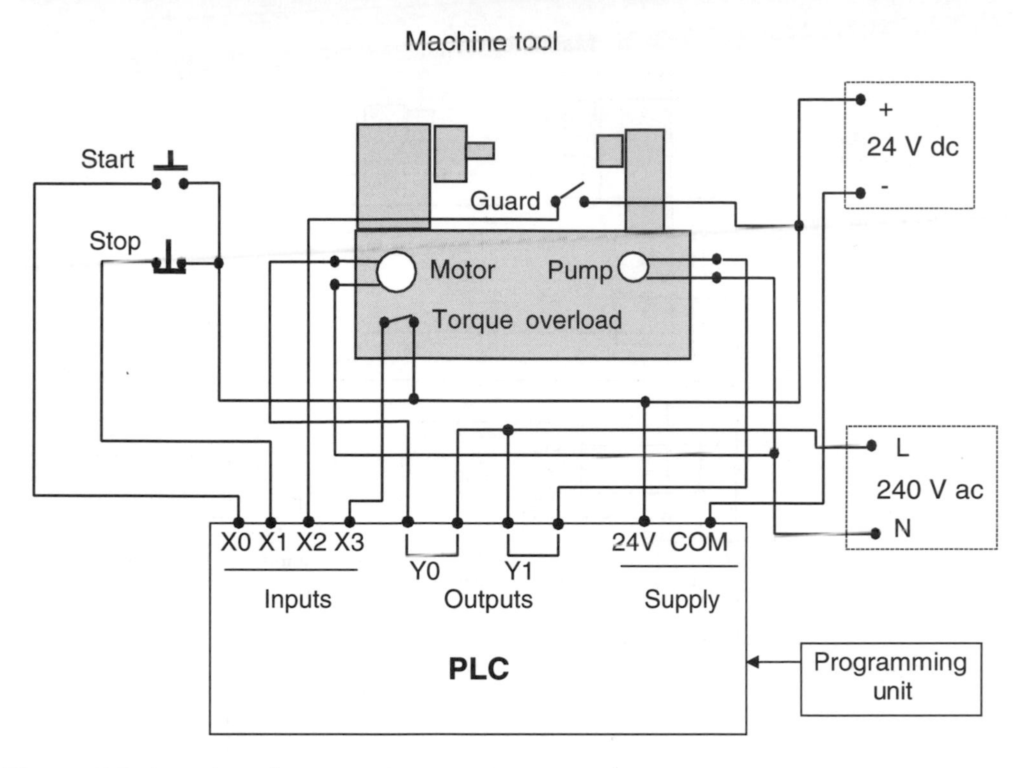

Figure 14.6 PLC machine control system.

heaters, valves and other loads must be switched directly from a power supply. The same machine tool seen in the previous example is now shown under PLC control in Fig. 14.6.

The PLC has inputs labelled X0, X1, X2 and X3. These inputs are detected as 'on' when connected to 24 V via system switches. The PLC is programmed to operate the outputs, labelled Y0 and Y1, according to the input sequence. The program can be written in 'ladder logic' form, which allows the control program to be specified as if the PLC contained the relay system shown in Fig. 14.5. The outputs are simple switched contacts, and require two connections to complete a load circuit. They are designed to handle high power loads operating with mains voltage, or, more usually, 415 V three-phase supplies. If necessary, the PLC outputs can control external contactors if the load current exceeds the PLC output contact rating. The control and load circuits are electrically isolated from each other, for safety, reliability and ease of use.

14.6.3 Microcontroller

This brings us back to microcontrollers, which are now available in increasing variety and at reducing cost. They can be used in small, self-contained systems, or in sub-circuits or interfaces within a larger control system. They are now widely used in domestic appliances, computer peripherals, smart cards and other consumer products, as well as more specialist areas, such as production systems. Microprocessors and microcontrollers are thus often used together as part of a more complex system. In fact, the PIC was originally developed as a general purpose programmable interface controller, hence the name. The keyboard

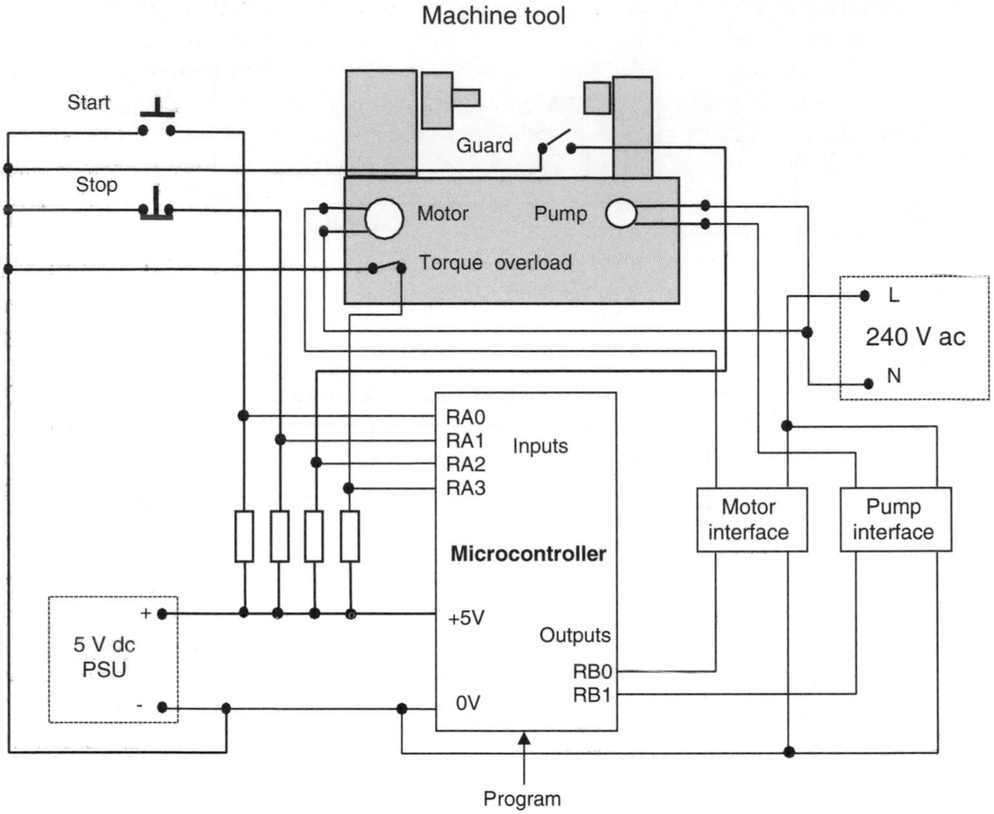

Figure 14.7 Microcontroller machine control system.

controller in a PC is a microcontroller, which scans the keys and sends the code to the PC via a serial link. Printers also use microcontrollers; even a basic dot matrix printer may have two of them! For comparison with other control technologies, Fig. 14.7 shows the machine tool operated by a microcontroller.

As we know, the microcontroller uses signal levels around 5 V, so the input switches have to be connected with pull-up resistors. The microcontroller can then be programmed to operate the output loads via suitable interfaces that allow its outputs to switch the high power motors. Solid-state relays are useful here, as they are designed to operate with 5 V inputs and have no moving parts. They operate as a high current electronic switch, using FET type outputs. The microcontroller would have to be programmed in its native language, which takes time to learn, as you now know! This is why the PLC was developed, with a microcontroller and interfaces built into a self-contained unit, and using a simplified programming method.

14.7 Control Application Design

One of the disadvantages of the microcontroller and the standard microprocessor board such as the M68k target board, is that the hardware is designed to be multi-purpose, and has not been designed for any specific application. This means that there are likely to be features of the hardware which are not utilized by a particular application, and the user will be paying for unused facilities. Thus, the advantage of a conventional microprocessor is that the hardware can be designed at component level to meet the requirements of the application exactly. For example, the amount of memory, the type and number of I/O ports can be tailored precisely to the needs of the application. The decision on which type of hardware to use must therefore be made by balancing the ongoing cost of unused features against the extra development costs of a tailor-made design. Existing expertise will be another factor; it takes time for the design engineer to become familiar with any new technology.

The PC itself can also be used as a controller, with a standard operating system, graphics, disk storage, communications and printing. One of the great advantages of the modular design of the PC is that special interface cards can be fitted for control applications. The PC can thus be turned into an oscilloscope, logic analyser, controller or data logger (and you can still play your favourite game if the boss doesn't catch you!). It is also a universal platform for running design software, both for mechanical and electronic CAD. Thus, you can design hardware, run a control application, write the support documentation and sell your product on the Internet, all using the same machine! Its role as a development platform for PIC microcontroller designs is only one of many uses.

As part of this rapidly evolving field, microcontroller based designs will increasingly replace circuits which use discrete components to implement control applications. The PIC family is at the forefront of that technology, and is therefore of vital interest to all students of engineering. Learning about the PIC is important – and fun too!

Summary

- The PIC family of microcontrollers share a common basic architecture and instruction set.
- Each PIC microcontroller type has different hardware features: memory size and type, interfaces, number of ports, etc.
- Other microcontroller families are often based on conventional processor architectures.
- The Motorola 68000 (M68k) CISC microprocessor has been widely used in education and industry, and has a relatively straightforward architecture, but a complex instruction set.
- The M68k system has a 16-bit data bus and 24-bit address bus, and can access up to 16 Mb of memory directly, using asynchronous bus control.
- Simple sequence control can be implemented using electromechanical relays.
- PLCs have built-in interfacing and user friendly programming methods to simplify sequence control system design.
- The PC can be used as a universal design, control, monitoring, instrumentation, IT applications and networking hardware platform.

Questions

1. Summarize the common features of the PIC range of microcontrollers.
2. Describe the differences between the PIC 12C672 and the PIC 16F84.
3. Select a PIC chip which is most suited for use with an LCD display.
4. State the main differences between the PIC 12XXXX, 16XXXX and 17XXXX devices.
5. Outline the differences between the PIC 16F84 and the Intel 8051 microcontrollers.
6. Explain briefly how a machine can be switched on via a relay. Why is this safer than using a manually operated mains switch?
7. Explain briefly the advantages of using a PLC compared with a microprocessor system in control applications.
8. State one advantage and one disadvantage of designing microprocessor hardware at component level.
9. What are the advantages of using the PC as a controller, compared with a microcontroller?

Activities

1. Study the information on the range of PIC microcontrollers available in a current catalogue, or the 'Microchip' website. Compare the cost and features of one chip from the 12XXXX and 17XXXX series with the PIC 16F84.
2. Take the casing off an old printer and obtain a circuit diagram or block diagram, if possible; see if you can identify a microcontroller in the circuit.
3. Study the catalogue description of the MC68HC11 microcontroller and compare its features with the PIC 16F84.
4. Draw a flowchart for the control program to be loaded into the microcontroller in Fig. 14.7, if the machine is to operate as follows:

   When the 'Start' button is operated, the machine motor and pump should start only if the guard switch is closed and the torque overload not tripped. If the guard switch opens, both should stop. If the torque overload goes active, the motor should be disabled but the coolant flow maintained. The 'Stop' should switch off both motor and pump, and the machine must not restart until the 'Start' button is used, and the 'Stop' button is closed. The 'Stop' button is mechanically retained until manually reset.

   Write the program for the PIC 16F84 and test in the simulator.

5. Devise a block diagram of a domestic washing machine, controlled by a microcontroller. Show the interface blocks between the switched actuators and sensors and the microcontroller. Write a description of the operating sequence of the machine, and devise a flowchart for the control sequence, constructed so that it could be implemented in PIC assembly language.

Appendix A
PIC 16F84 Data Sheet

This appendix contains an edited version of the PIC 16F84 data sheet. The full data sheet can be downloaded from the Microchip Technology website at www.microchip.com, along with up-to-date information on the PIC range, support software and other user services.

A.1 PIC 16F8X Data Summary

Devices included in this data sheet:

- PIC 16F83
- PIC 16CR83
- PIC 16F84
- PIC 16CR84
- Extended voltage range devices available (PIC 16LF8X, PIC 16LCR8X)

High performance RISC CPU features:

- Only 35 single word instructions to learn
- All instructions single cycle (400 ns @ 10 MHz) except for program branches which are two-cycle
- Operating speed:
  dc – 10 MHz clock input
  dc – 400 ns instruction cycle

Device	Memory			Freq Max.
		Data		
	Flash	RAM	EEPROM	
PIC 16F83	512 words	36	64	10 MHz
PIC 16CR83	512 words	36	64	10 MHz
PIC 16F84	1 K-words	68	64	10 MHz
PIC 16CR84	1 K-words	68	64	10 MHz

F = flash; CR = ROM

- 14-bit wide instructions
- 8-bit wide data path
- 15 special function hardware registers
- Eight-level deep hardware stack
- Direct, indirect and relative addressing modes
- Four interrupt sources:
  – External RB0/INT pin
  – TMT0 timer overflow
  – PORTB ⟨7 : 4⟩ interrupt on change
  – Data EEPROM write complete
- 1 000 000 data memory EEPROM ERASE/WRITE cycles
- EEPROM Data Retention >40 years

Peripheral features:

- 13 I/O pins with individual direction control
- High current sink/source for direct LED drive
  – 25 mA sink max. per pin
  – 20 mA source max. per pin
- TMR0: 8-bit timer/counter with 8-bit programmable prescaler

Pin diagram:

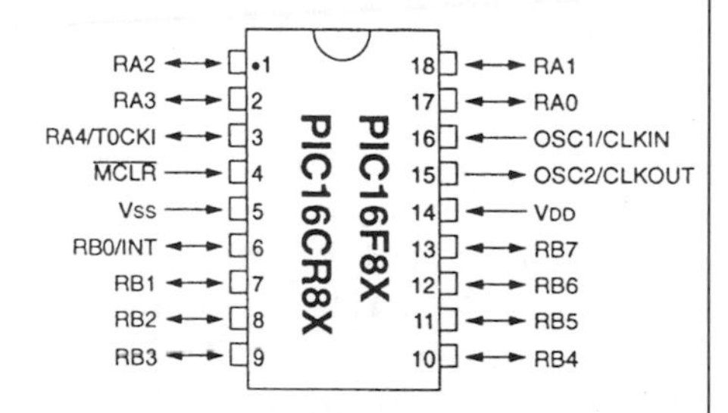

Special microcontroller features:

- Power-on Reset (POR)
- Power-up Timer (PWRT)
- Oscillator Start-up Timer (OST)
- Watchdog Timer (WDT) with its own on-chip RC oscillator for reliable operation
- Code-protection
- Power saving SLEEP mode
- Selectable oscillator options
- Serial in-system programming – via two pins (ROM devices support only Data EEPROM programming)

CMOS technology:

- Low-power, high-speed CMOS flash/EEPROM technology
- Fully static design
- Wide operating voltage range:
  – Commercial: 2.0–6.0 V
  – Industrial: 2.0–6.0 V
- Low power consumption:
  – <2 mA typical @ 5 V, 4 MHz
  – 15 μA typical @ 2 V, 32 kHz
  – <1 μA typical standby current @ 2 V

A.2 General Description

The PIC 16F8X is a group in the PIC 16CXX family of low-cost, high-performance, CMOS, fully-static, 8-bit microcontrollers. This group contains the following devices:

- PIC 16F83
- PIC 16CR83
- PIC 16F84
- PIC 16CR84.

All PIC 16/17 microcontrollers employ an advanced RISC architecture. PIC 16CXX devices have enhanced core features, eight-level deep stack, and multiple internal and external interrupt sources. The separate instruction and data buses of the Harvard architecture allow a 14-bit wide instruction word with a separate 8-bit wide data bus. The two stage instruction pipeline allows all instructions to execute in a single cycle, except for program branches (which require two cycles). A total of 35 instructions (reduced instruction set) are available. In addition, a large register set is used to achieve a very high performance level.

PIC 16F8X microcontrollers typically achieve a 2:1 code compression and up to a 2:1 speed improvement (at 10 MHz) over other 8-bit microcontrollers in their class.

The PIC 16F8X has up to 68 bytes of RAM, 64 bytes of data EEPROM memory, and 13 I/O pins. A timer/counter is also available.

The PIC 16CXX family has special features to reduce external components, thus reducing costs, enhancing system reliability and reducing power consumption. There are four oscillator options, of which the single pin RC oscillator provides a low-cost solution, the LP oscillator minimizes power consumption, XT is a standard crystal, and the HS is for high speed crystals. The SLEEP (power-down) mode offers power saving. The users can wake the chip from sleep through several external and internal interrupts and resets.

A highly reliable Watchdog Timer with its own on-chip RC oscillator provides protection against software lock-up.

The devices with flash program memory allow the same device package to be used for prototyping and production. In-circuit reprogrammability allows the code to be updated without the device being removed from the end application. This is useful in the development of many applications where the device may not be easily accessible, but the prototypes may require code updates. This is also useful for remote applications where the code may need to be updated (such as rate information).

A simplified block diagram of the PIC 16F8X is shown in Fig. A.1.

The PIC 16F8X fits perfectly in applications ranging from high speed automotive and appliance motor control to low-power remote sensors, electronic locks, security devices and smart cards. The flash/EEPROM technology makes customization of application programs (transmitter codes, motor speeds, receiver frequencies, security codes, etc.) extremely fast and convenient. The small footprint packages make this microcontroller series perfect for all applications with space limitations. Low-cost, low-power, high performance, ease of use and I/O flexibility make the PIC 16F8X very versatile even in areas where no microcontroller use has been considered before (e.g., timer functions, serial communication, capture and compare, PWM functions and co-processor applications).

The serial in-system programming feature (via two pins) offers flexibility of customizing the product after complete assembly and testing. This feature can be used to serialize a product, store calibration data, or program the device with the current firmware before shipping.

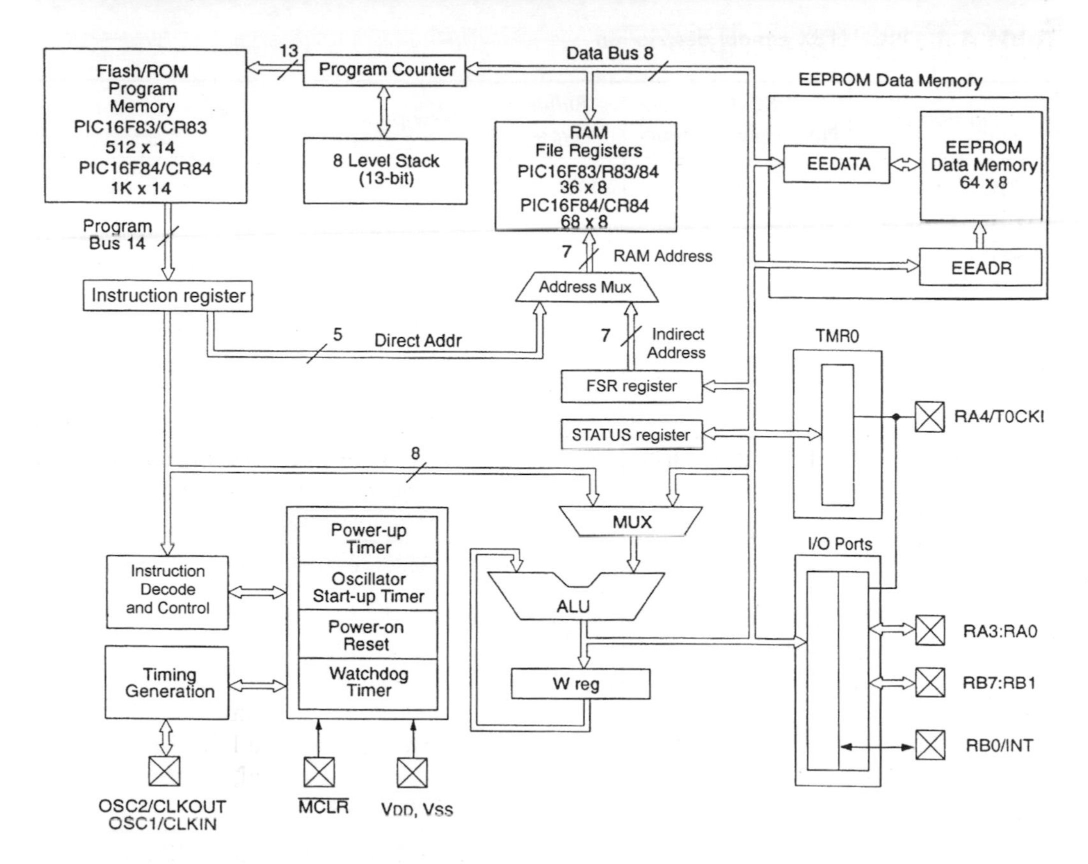

Figure A.1 PIC 16F8X block diagram.

A.3 Architectural Overview

The high performance of the PIC 16CXX family can be attributed to a number of architectural features commonly found in RISC microprocessors. To begin with, the PIC 16CXX uses a Harvard architecture. This architecture has the program and data accessed from separate memories. So the device has a program memory bus and a data memory bus. This improves bandwidth over traditional von Neumann architecture where program and data are fetched from the same memory (accesses over the same bus). Separating program and data memory further allows instructions to be sized differently than the 8-bit wide data word. PIC 16CXX opcodes are 14-bits wide, enabling single word instructions. The full 14-bit wide program memory bus fetches a 14-bit instruction in a single cycle. A two-stage pipeline overlaps fetch and execution of instructions. Consequently, all instructions execute in a single cycle (400 ns @ 10 MHz) except for program branches.

The PIC 16F83 and PIC 16CR83 address 512 × 14 of program memory, and the PIC 16F84 and PIC 16CR84 address 1 k × 14 program memory. All program memory is internal.

The PIC 16CXX can directly or indirectly address its register files or data memory. All special function registers, including the program counter, are mapped in the data memory. An orthogonal (symmetrical) instruction set makes it possible to carry out any operation on

Table A.1 PIC 16F8X pin-out description

Pin name	DIP No.	SOIC No.	I/O/P type	Buffer type	Description
OSC1/CLKIN	16	16	I	ST/CMOS‡	Oscillator crystal input/external clock source input.
OSC2/CLKOUT	15	15	O	—	Oscillator crystal output. Connects to crystal or resonator in crystal oscillator mode. In RC mode, OSC2 pin outputs CLKOUT which has 1/4 the frequency of OSC1, and denotes the instruction cycle rate.
$\overline{\text{MCLR}}$	4	4	I/P	ST	Master clear (reset) input/programming voltage input. This pin is an active low reset to the device.
					PORTA is a bi-directional I/O port.
RA0	17	17	I/O	TL	
RA1	18	18	I/O	TTL	
RA2	1	1	I/O	TTL	
RA3	2	2	I/O	TTL	
RA4/TOCKI	3	3	I/O	ST	Can also be selected to be the clock input to the TMR0 timer/counter. Output is open drain type.
					PORTB is a bi-directional I/O port. PORTB can be software programmed for internal weak pull-up on all inputs.
RB0/INT	6	6	I/O	TTL/ST*	RB0/INT can also be selected as an external interrupt pin.
RB1	7	7	I/O	TTL	
RB2	8	8	I/O	TTL	
RB3	9	9	I/O	TTL	
RB4	10	10	I/O	TTL	Interrupt on charge pin.
RB5	11	11	I/O	TTL	Interrupt on change pin.
RB6	12	12	I/O	TTL/ST†	Interrupt on change pin. Serial programming clock.
RB7	13	13	I/O	TTL/ST†	Interrupt on change pin. Serial programming data.
V_{SS}	5	5	Pi	—	Ground reference for logic and I/O pins.
V_{DD}	14	14	P	—	Positive supply for logic and I/O pins.

I = input; O = output; I/O = input/output; P = power; — = not used; TTL = TTL input; ST = Schmitt Trigger input.

* This buffer is a Schmitt Trigger input when configured as the external interrupt.

† This buffer is a Schmitt Trigger input when used in serial programming mode.

‡ This buffer is a Schmitt Trigger input when configured in RC oscillator mode and a CMOS input otherwise.

any register using any addressing mode. This symmetrical nature and lack of 'special optimal situations' make programming with the PIC 16CXX simple yet efficient. In addition, the learning curve is reduced significantly.

PIC 16CXX devices contain an 8-bit ALU and Working register. The ALU is a general purpose arithmetic unit. It performs arithmetic and Boolean functions between data in the Working register and any register file.

The ALU is 8-bits wide and capable of addition, subtraction, shift and logical operations. Unless otherwise mentioned, arithmetic operations are two's complement in nature. In two-operand instructions, typically one operand is the Working register (W register), and the other operand is a file register or an immediate constant. In single operand instructions, the operand is either the W register or a file register.

The W register is an 8-bit working register used for ALU operations. It is not an addressable register.

Depending on the instruction executed, the ALU may affect the values of the Carry (C), Digit Carry (DC), and Zero (Z) bits in the STATUS register. The C and DC bits operate as a borrow and digit borrow out bit, respectively, in subtraction.

A simplified block diagram for the PIC 16F8X is shown in Fig. A.1, its corresponding pin description is shown in Table A.1.

A.3.1 Clocking Scheme/Instruction Cycle

The clock input (from OSC1) is internally divided by four to generate four non-overlapping quadrature clocks namely Q1, Q2, Q3 and Q4. Internally, the Program Counter (PC) is incremented every Q1, the instruction is fetched from the program memory and latched into the instruction register in Q4. The instruction is decoded and executed during the following Q1 through Q4. The clocks and instruction execution flow is shown in Fig. A.2.

A.3.2 Instruction Flow/Pipelining

An 'Instruction Cycle' consists of four Q cycles (Q1, Q2, Q3 and Q4). The instruction fetch and execute are pipelined such that fetch takes one instruction cycle while decode and execute takes another instruction cycle. However, due to the pipelining, each instruction effectively executes in one cycle. If an instruction causes the Program Counter to change (e.g., GOTO) then two cycles are required to complete the instruction (Example A.1).

A fetch cycle begins with the Program Counter (PC) incrementing in Q1.

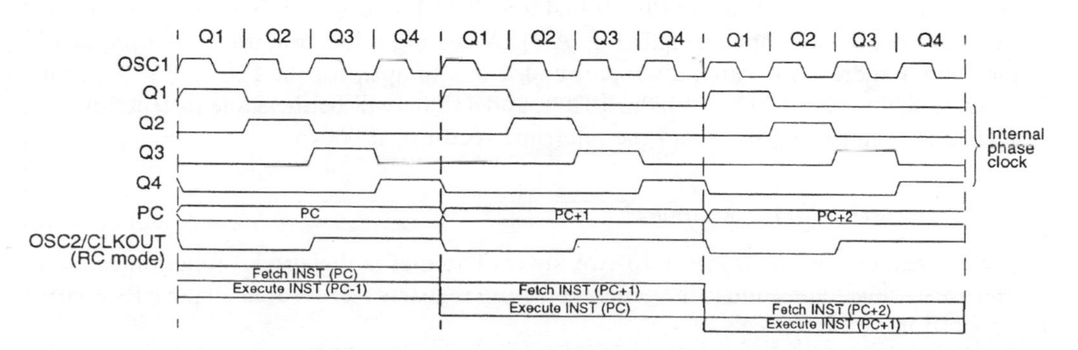

Figure A.2 Clock/instruction cycle.

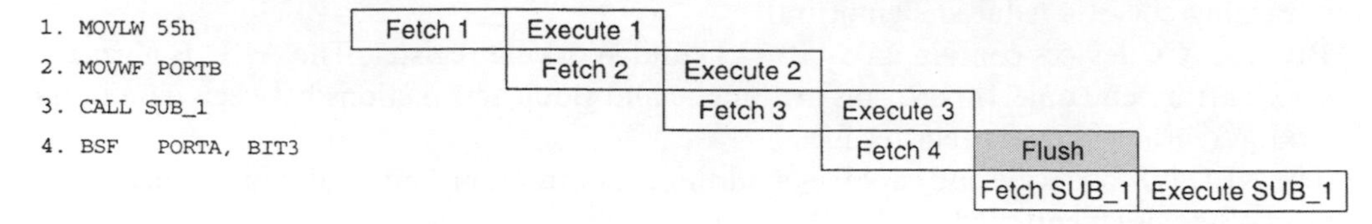

Example A.1 Instruction pipeline flow.

In the execution cycle, the fetched instruction is latched into the 'Instruction Register' in cycle Q1. This instruction is then decoded and executed during the Q2, Q3, and Q4 cycles. Data memory is read during Q2 (operand read) and written during Q4 (destination write).

A.4 Memory Organization

There are two memory blocks in the PIC 16F8X. These are the program memory and the data memory. Each block has its own bus, so that the access to each block can occur during the same oscillator cycle.

The data memory can be broken down further into the general purpose RAM and the special function registers (SFRs). The operation of the SFRs that control the 'core' are described here. The SFRs used to control the peripheral modules are described in the section discussing each individual peripheral module.

The data memory area also contains the data EEPROM memory. This memory is not directly mapped into the data memory, but is indirectly mapped. That is, an indirect address pointer specifies the address of the data EEPROM memory to ready/write. The 64 bytes of data EEPROM memory have the address range 0h–3Fh. More details on the EEPROM memory can be found in Section A.7.

A.4.1 Program Memory Organization

The PIC 16FXX has a 13-bit program counter capable of addressing an 8 k × 14 program memory space. For the PIC 16F83 and PIC 16CR83, the first 512 × 14 (0000h–01FFh) are physically implemented. For the PIC 16F84 and PIC 16CR84, the first 1 k × 14 (0000h–03FFh) are physically implemented (Fig. A.3). Accessing a location above the physically implemented address will cause a wraparound. For example, for the PIC 16F84 locations 20h, 420h, 820h, C20h, 1020h, 1420h, 1820h, and 1C20h will be the same instruction.

The reset vector is at 0000h and the interrupt vector is at 0004h.

A.4.2 Data Memory Organization

The data memory is partitioned into two areas. The first is the special function registers (SFR) area, while the second is the general purpose registers (GPR) area. The SFRs control the operation of the device.

Portions of data memory are banked. This is for both the SFR area and the GPR area. The GPR area is banked to allow greater than the 116 bytes of general purpose RAM. The

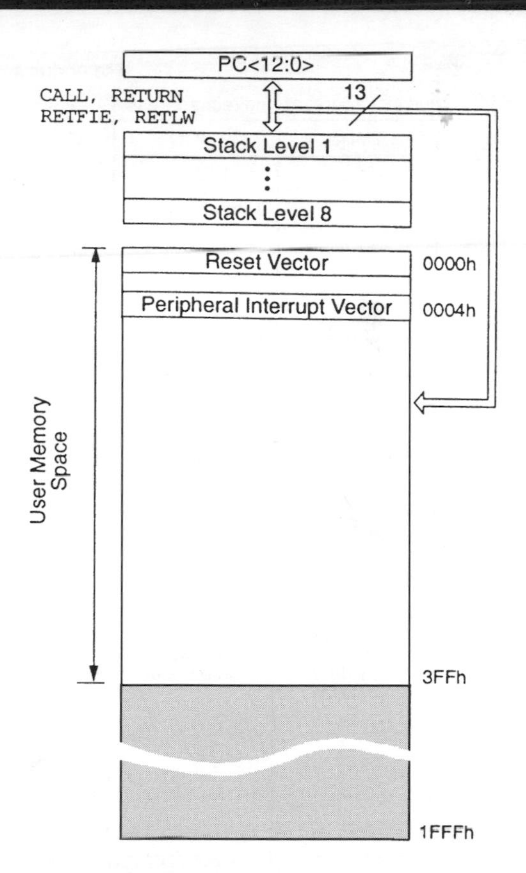

Figure A.3 Program memory map and stack.

banked areas of the SFR are for the registers that control the peripheral functions. Banking requires the use of control bits for bank selection. These control bits are located in the STATUS Register. Figure A.4 shows the data memory map organization.

Instructions MOVWF and MOVF can move values from the W register to any location in the register file ('F'), and vice-versa.

The entire data memory can be accessed either directly using the absolute address of each register file or indirectly through the File Select Register (FSR) (Section A.4.5). Indirect addressing uses the present value of the RP1:RP0 bits for access into the banked areas of data memory.

Data memory is partitioned into two banks which contain the general purpose registers and the special function registers. Bank 0 is selected by clearing the RP0 bit (STATUS⟨5⟩). Setting the RP0 bit selects bank 1. Each bank extends up to 7Fh (128 bytes). The first 12 locations of each bank are reserved for the special function registers. The remainder are general purpose registers implemented as static RAM.

General purpose register file

All devices have some amount of GPR area. Each GPR is 8 bits wide and is accessed either directly or indirectly through the FSR (Section A.4.5).

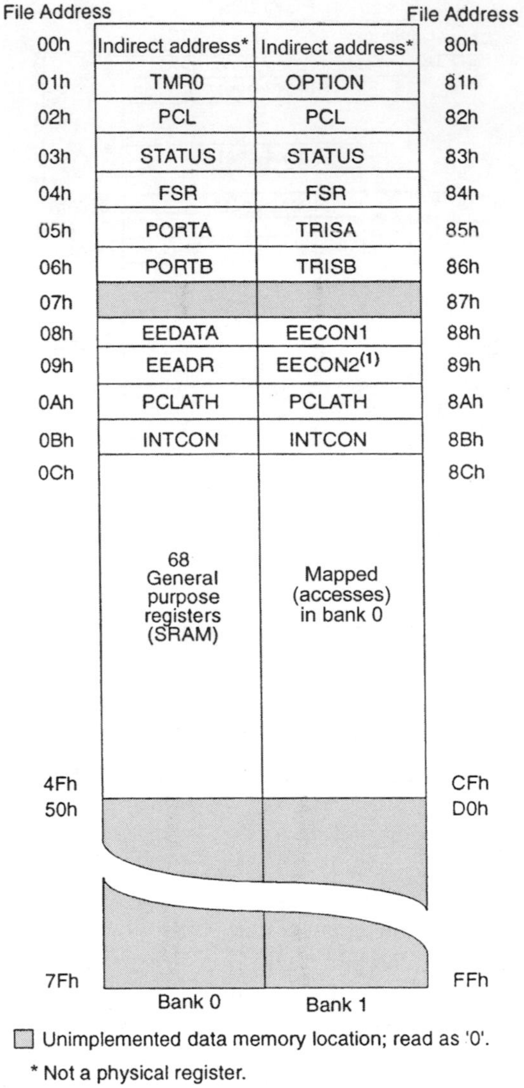

File Address	Bank 0	Bank 1	File Address
00h	Indirect address*	Indirect address*	80h
01h	TMR0	OPTION	81h
02h	PCL	PCL	82h
03h	STATUS	STATUS	83h
04h	FSR	FSR	84h
05h	PORTA	TRISA	85h
06h	PORTB	TRISB	86h
07h			87h
08h	EEDATA	EECON1	88h
09h	EEADR	EECON2(1)	89h
0Ah	PCLATH	PCLATH	8Ah
0Bh	INTCON	INTCON	8Bh
0Ch–4Fh	68 General purpose registers (SRAM)	Mapped (accesses) in bank 0	8Ch–CFh
50h–7Fh			D0h–FFh

▢ Unimplemented data memory location; read as '0'.

* Not a physical register.

Figure A.4 Register file map.

Special function registers

The SFR (Fig. A.4 and Table A.2) are used by the CPU and peripheral functions to control the device operation. These registers are static RAM.

The SFR can be classified into two sets, core and peripheral. Those associated with the core functions are described in this section. Those related to the operation of the peripheral features are described in the section for that specific feature.

Status register

The STATUS register contains the arithmetic status of the ALU, the RESET status and the bank select bit for data memory.

Table A.2 Register file summary

Address	Name	Bit 7	Bit 6	Bit 5	Bit 4	Bit 3	Bit 2	Bit 1	Bit 0	Value on power-on reset	Value on all other resets‡
Bank 0											
00h	INDF	Uses contents of FSR to address data memory (not a physical register)								---- ----	---- ----
01h	TMR0	8-bit real-time clock/counter								xxxx xxxx	uuuu uuuu
02h	PCL	Low order 8 bits of the Program Counter (PC)								0000 0000	0000 0000
03h	STATUS†	IRP	RP1	RP0	$\overline{\text{TO}}$	$\overline{\text{PD}}$	Z	DC	C	0001 1xxx	000q quuu
04h	FSR	Indirect data memory address pointer 0								xxxx xxxx	uuuu uuuu
05h	PORTA	—	—	—	RA4/T0CKI	RA3	RA2	RA1	RA0	---x xxxx	---u uuuu
06h	PORTB	RB7	RB6	RB5	RB4	RB3	RB2	RB1	RB0/INT	xxxx xxxx	uuuu uuuu
07h		Unimplemented location, read as '0'								---- ----	---- ----
08h	EEDATA	EEPROM data register								xxxx xxxx	uuuu uuuu
09h	EEADR	EEPROM address register								xxxx xxxx	uuuu uuuu
0Ah	PCLATH	—	—	—	Write buffer for upper 5 bits of the PC*					---0 0000	---0 0000
0Bh	INTCON	GIE	EEIE	T0IE	INTE	RBIE	T0IF	INTF	RBIF	0000 000x	0000 000u
Bank 1											
80h	INDF	Uses contents of FSR to address data memory (not a physical register)								---- ----	---- ----
81h	OPTION	$\overline{\text{RBPU}}$	INTEDG	T0CS	T0SE	PSA	PS2	PS1	PS0	1111 1111	1111 1111
82h	PCL	Low order 8 bits of Program Counter (PC)								0000 0000	0000 0000
83h	STATUS†	IRP	RP1	RP0	$\overline{\text{TO}}$	$\overline{\text{PD}}$	Z	DC	C	0001 1xxx	000q quuu
84h	FSR	Indirect data memory address pointer 0								xxxx xxxx	uuuu uuuu
85h	TRISA	—	—	—	PORTA data direction register					---1 1111	---1 1111
86h	TRISB	PORTB data direction register								1111 1111	1111 1111
87h		Unimplemented location, read as '0'								---- ----	---- ----
88h	EECON1	—	—	—	EEIF	WRERR	WREN	WR	RD	---0 x000	---0 q000
89h	EECON2	EEPROM control register 2 (not a physical register)								---- ----	---- ----
0Ah	PCLATH	—	—	—	Write buffer for upper 5 bits of the PC*					---0 0000	---0 0000
0Bh	INTCON	GIE	EEIE	T0IE	INTE	RBIE	T0IF	INTF	RBIF	0000 000x	0000 000u

x = unknown; u = unchanged; - = unimplemented read as '0'; q = value depends on condition.
* The upper byte of the program counter is not directly accessible. PCLATH is a slave register for PC⟨12 : 8⟩. The contents of PCLATH can be transferred to the upper byte of the program counter, but the contents of PC⟨12 : 8⟩ is never transferred to PCLATH.
† The $\overline{\text{TO}}$ and $\overline{\text{PD}}$ status bits in the STATUS register are not affected by a $\overline{\text{MCLR}}$ reset.
‡ Other (non power-up) resets include: external reset through $\overline{\text{MCLR}}$ and the Watchdog Timer Reset.

As with any register, the STATUS register can be the destination for any instruction. If the STATUS register is the destination for an instruction that affects the Z, DC or C bits, then the write to these three bits is disabled. These bits are set or cleared according to device logic. Furthermore, the $\overline{\text{TO}}$ and $\overline{\text{PD}}$ bits are not writable. Therefore, the result of an instruction with the STATUS register as destination may be different than intended.

For example, `CLRF STATUS` will clear the upper-three bits and set the Z bit. This leaves the STATUS register as `000u u1uu` (where u = unchanged).

Only the `BCF`, `BSF`, `SWAPF` and `MOVWF` instructions should be used to alter the STATUS register because these instructions do not affect any status bit.

R/W-0	R/W-0	R/W-0	R-1	R-1	R/W-x	R/W-x	R/W-x
IRP	RP1	RP0	$\overline{TO}$	$\overline{PD}$	Z	DC	C
bit7							bit0

R = Readable bit
W = Writable bit
U = Unimplemented bit, read as '0'
- n = Value at POR reset

bit 7: **IRP**: Register Bank Select bit (used for indirect addressing)
0 = Bank 0, 1 (00h – FFh)
1 = Bank 2, 3 (100h – 1FFh)
The IRP bit is not used by the PIC16F8X. IRP should be maintained clear.

bit 6-5: **RP1:RP0**: Register Bank Select bits (used for direct addressing)
00 = Bank 0 (00h – 7Fh)
01 = Bank 1 (80h – FFh)
10 = Bank 2 (100h – 17Fh)
11 = Bank 3 (180h – 1FFh)
Each bank is 128 bytes. Only bit RP0 is used by the PIC16F8X. RP1 should be maintained clear.

bit 4: **$\overline{TO}$**: Time-out bit
1 = After power-up, CLRWDT instruction, or SLEEP instruction
0 = A WDT time-out occurred

bit 3: **$\overline{PD}$**: Power-down bit
1 = After power-up or by the CLRWDT instruction
0 = By execution of the SLEEP instruction

bit 2: **Z**: Zero bit
1 = The result of an arithmetic or logic operation is zero
0 = The result of an arithmetic or logic operation is not zero

bit 1: **DC**: Digit carry/$\overline{\text{borrow}}$ bit (for ADDWF and ADDLW instructions) (For $\overline{\text{borrow}}$ the polarity is reversed)
1 = A carry-out from the 4th low order bit of the result occurred
0 = No carry-out from the 4th low order bit of the result

bit 0: **C**: Carry/$\overline{\text{borrow}}$ bit (for ADDWF and ADDLW instructions)
1 = A carry-out from the most significant bit of the result occurred
0 = No carry-out from the most significant bit of the result occurred
Note: For $\overline{\text{borrow}}$ the polarity is reversed. A subtraction is executed by adding the two's complement of the second operand. For rotate (RRF, RLF) instructions, this bit is loaded with either the high or low order bit of the source register.

Figure A.5 Status register (address 03h, 83h).

Note the following.

1. The IRP and RP1 bits (STATUS⟨7 : 6⟩) are not used by the PIC 16F8X and should be programmed as cleared. Use of these bits as general purpose R/W bits is *not* recommended, since this may affect upward compatibility with future products.
2. The C and DC bits operate as a $\overline{\text{borrow}}$ and digit borrow out bit, respectively, in subtraction.
3. When the STATUS register is the destination for an instruction that affects the Z, DC or C bits, then the write to these three bits is disabled. The specified bit(s) will be updated according to device logic.

OPTION register

The OPTION register is a readable and writable register which contains various control bits to configure the TMR0/WDT prescaler, the external INT interrupt, TMR0, and the weak pull-ups for PORTB.

Note: When the prescaler is assigned to the WDT (PSA = '1'), TMRO has a 1 : 1 prescaler assignment.

R/W-1	R/W-1	R/W-1	R/W-1	R/W-1	R/W-1	R/W-1	R/W-1
$\overline{RBPU}$	INTEDG	T0CS	T0SE	PSA	PS2	PS1	PS0
bit7							bit0

R = Readable bit
W = Writable bit
U = Unimplemented bit, read as '0'
- n = Value at POR reset

bit 7: **$\overline{\text{RBPU}}$**: PORTB Pull-up Enable bit
1 = PORTB pull-ups are disabled
0 = PORTB pull-ups are enabled (by individual port latch values)

bit 6: **INTEDG**: Interrupt Edge Select bit
1 = Interrupt on rising edge of RB0/INT pin
0 = Interrupt on falling edge of RB0/INT pin

bit 5: **T0CS**: TMR0 Clock Source Select bit
1 = Transition on RA4/T0CKI pin
0 = Internal instruction cycle clock (CLKOUT)

bit 4: **T0SE**: TMR0 Source Edge Select bit
1 = Increment on high-to-low transition on RA4/T0CKI pin
0 = Increment on low-to-high transition on RA4/T0CKI pin

bit 3: **PSA**: Prescaler Assignment bit
1 = Prescaler assigned to the WDT
0 = Prescaler assigned to TMR0

bit 2-0: **PS2:PS0**: Prescaler Rate Select bits

Bit Value	TMR0 Rate	WDT Rate
000	1 : 2	1 : 1
001	1 : 4	1 : 2
010	1 : 8	1 : 4
011	1 : 16	1 : 8
100	1 : 32	1 : 16
101	1 : 64	1 : 32
110	1 : 128	1 : 64
111	1 : 256	1 : 128

Figure A.6 Option register (address 81h).

INTCON register

The INTCON register is a readable and writable register which contains the various enable bits for all interrupt sources.

Note: Interrupt flag bits get set when an interrupt condition occurs regardless of the state of its corresponding enable bit or the global enable bit, GIE (INTCON⟨7⟩).

A.4.3 Program Counter: PCL and PCLATH

The Program Counter (PC) is 13-bits wide. The low byte is the PCL register, which is a readable and writable register. The high byte of the PC (PC⟨12 : 8⟩) is not directly readable nor writable and comes from the PCLATH register. The PCLATH (PC latch high) register is a holding register for PC⟨12 : 8⟩. The contents of PCLATH are transferred to the upper byte of the program counter when the PC is loaded with a new value. This occurs during a `CALL`, `GOTO` or a write to PCL. The high bits of PC are loaded from PCLATH as shown in Fig. A.8.

Computed GOTO

A computed GOTO is accomplished by adding an offset to the program counter (`ADDWF PCL`). When doing a table read using a computed GOTO method, care should be exercised if the table location crosses a PCL memory boundary (each 256 word block).

R/W-0	R/W-0	R/W-0	R/W-0	R/W-0	R/W-0	R/W-0	R/W-x
GIE	EEIE	T0IE	INTE	RBIE	T0IF	INTF	RBIF
bit7							bit0

R = Readable bit
W = Writable bit
U = Unimplemented bit, read as '0'
- n = Value at POR reset

bit 7: **GIE:** Global Interrupt Enable bit
1 = Enables all un-masked interrupts
0 = Disables all interrupts

Note: For the operation of the interrupt structure, please refer to Section 8.5.

bit 6: **EEIE**: EE Write Complete Interrupt Enable bit
1 = Enables the EE write complete interrupt
0 = Disables the EE write complete interrupt

bit 5: **T0IE**: TMR0 Overflow Interrupt Enable bit
1 = Enables the TMR0 interrupt
0 = Disables the TMR0 interrupt

bit 4: **INTE**: RB0/INT Interrupt Enable bit
1 = Enables the RB0/INT interrupt
0 = Disables the RB0/INT interrupt

bit 3: **RBIE**: RB Port Change Interrupt Enable bit
1 = Enables the RB port change interrupt
0 = Disables the RB port change interrupt

bit 2: **T0IF**: TMR0 overflow interrupt flag bit
1 = TMR0 has overflowed (must be cleared in software)
0 = TMR0 did not overflow

bit 1: **INTF**: RB0/INT Interrupt Flag bit
1 = The RB0/INT interrupt occurred
0 = The RB0/INT interrupt did not occur

bit 0: **RBIF**: RB Port Change Interrupt Flag bit
1 = When at least one of the RB7:RB4 pins changed state (must be cleared in software)
0 = None of the RB7:RB4 pins have changed state

Figure A.7 INTCON register (address 0Bh, 8Bh).

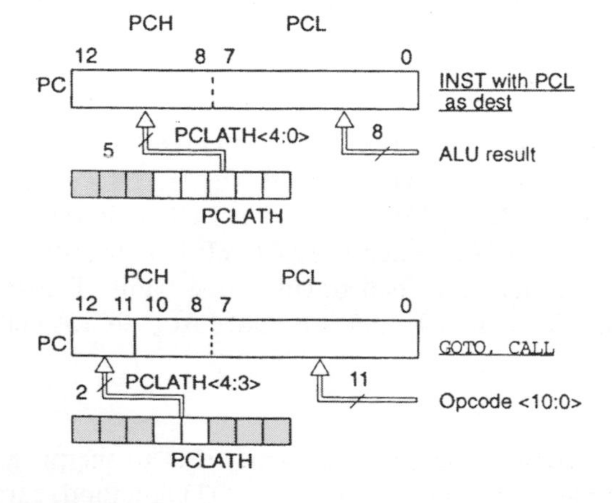

Figure A.8 Loading of PC.

Program memory paging

The PIC 16F84 and PIC 16CR84 have 1 k of program memory. The CALL and GOTO instructions have an 11-bit address range. This 11-bit address range allows a branch within a 2 k program memory page size. For future PIC 16F8X program memory expansion, there must be another two bits to specify the program memory page. These paging bits come from the PCLATH⟨4 : 3⟩ bits (Fig. A.8). When doing a CALL or a GOTO instruction, the user must ensure that these page bits (PCLATH⟨4 : 3⟩) are programmed to the desired program memory page. If a CALL instruction (or interrupt) is executed, the entire 13-bit PC is 'pushed' onto the stack (see next section). Therefore, manipulation of the PCLATH⟨4 : 3⟩ is not required for the return instructions (which 'pops' the PC from the stack).

Note. The PIC 16F8X ignores the PCLATH⟨4 : 3⟩ bits, which are used for program memory pages 1, 2 and 3 (0800h–1FFFh). The use of PCLATH⟨4 : 3⟩ as general purpose R/W bits is not recommended since this may affect upward compatibility with future products.

A.4.4 Stack

The PIC 16FXX has an 8 deep × 13-bit wide hardware stack (Fig. A.3). The stack space is not part of either program or data space and the stack pointer is not readable or writable.

The entire 13-bit PC is 'pushed' onto the stack when a CALL instruction is executed or an interrupt is acknowledged. The stack is 'popped' in the event of a RETURN, RETLW or a RETFIE instruction execution. PCLATH is not affected by a push or a pop operation.

Note. There are no instruction mnemonics called push or pop. These are actions that occur from the execution of the CALL, RETURN, RETLW, and RETFIE instructions, or the vectoring of an interrupt address.

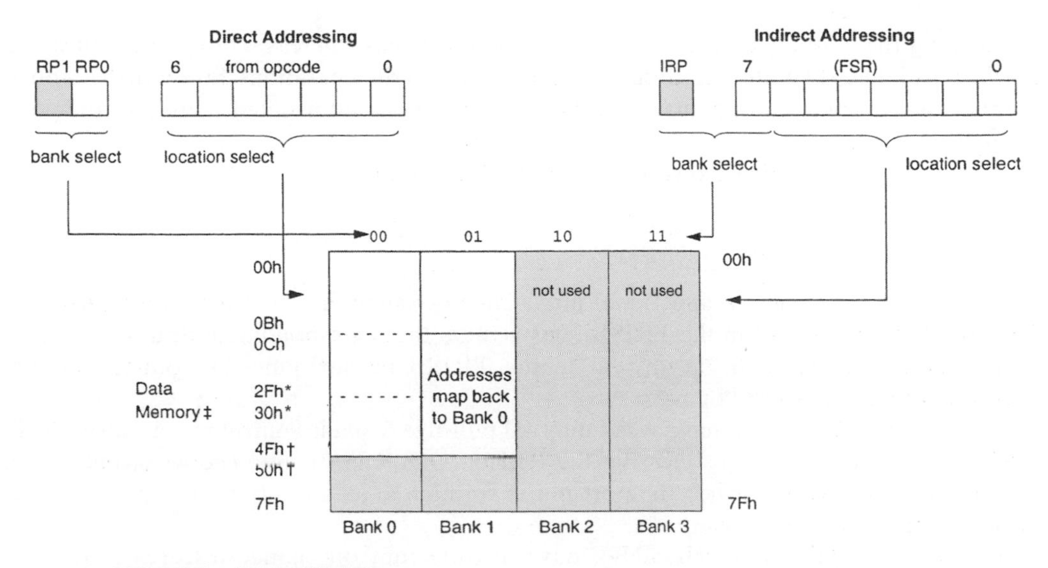

Figure A.9 Direct/indirect addressing.

The stack operates as a circular buffer. That is, after the stack has been pushed eight times, the ninth push overwrites the value that was stored from the first push. The 10th push overwrites the second push (and so on).

If the stack is effectively popped nine times, the PC value is the same as the value from the first pop.

Note. There are no status bits to indicate stack overflow or stack underflow conditions.

A.4.5 Indirect addressing: INDF and FSR registers

The INDF register is not a physical register. Addressing INDF actually addresses the register whose address is contained in the FSR register (FSR is a *pointer*). This is indirect addressing.

A.5 I/O Ports

The PIC 16F8X has two ports, PORTA and PORTB. Some port pins are multiplexed with an alternate function for other features on the device.

A.5.1 PORTA and TRISA registers

PORTA is a 5-bit wide latch. RA4 is a Schmitt Trigger input and an open drain output. All other RA port pins have TTL input levels and full CMOS output drivers. All pins have data direction bits (TRIS registers) which can configure these pins as output or input.

Setting a TRISA bit (=1) will make the corresponding PORTA pin an input, that is, put the corresponding output driver in a hi-impedance mode. Clearing a TRISA bit (=0) will make the corresponding PORTA pin an output, that is, put the contents of the output latch on the selected pin.

Reading the PORTA register reads the status of the pins whereas writing to it will write to the port latch. All write operations are read-modify-write operations. So a write to a port implies that the port pins are first read, then this value is modified and written to the port data latch.

The RA4 pin is multiplexed with the TMR0 clock input.

A.5.2 PORTB and TRISB registers

PORTB is an 8-bit wide bi-directional port. The corresponding data direction register is TRISB. A '1' on any bit in the TRISB register puts the corresponding output driver in a hi-impedance mode. A '0' on any bit in the TRISB register puts the contents of the output latch on the selected pin(s).

Each of the PORTB pins has a weak internal pull-up. A single control bit can turn on all the pull-ups. This is done by clearing the $\overline{\text{RBPU}}$ (OPTION⟨7⟩) bit. The weak pull-up is automatically turned off when the port pin is configured as an output. The pull-ups are disabled on a power-on reset.

Four of PORTB's pins, RB7:RB4, have an interrupt on change feature. Only pins configured as inputs can cause this interrupt to occur (i.e., any RB7:RB4 pin configured as an output is excluded from the interrupt on change comparison). The pins value in input mode are compared with the old value latched on the last read of PORTB. The 'mismatch' outputs of the pins are OR'ed together to generate the RB port change interrupt.

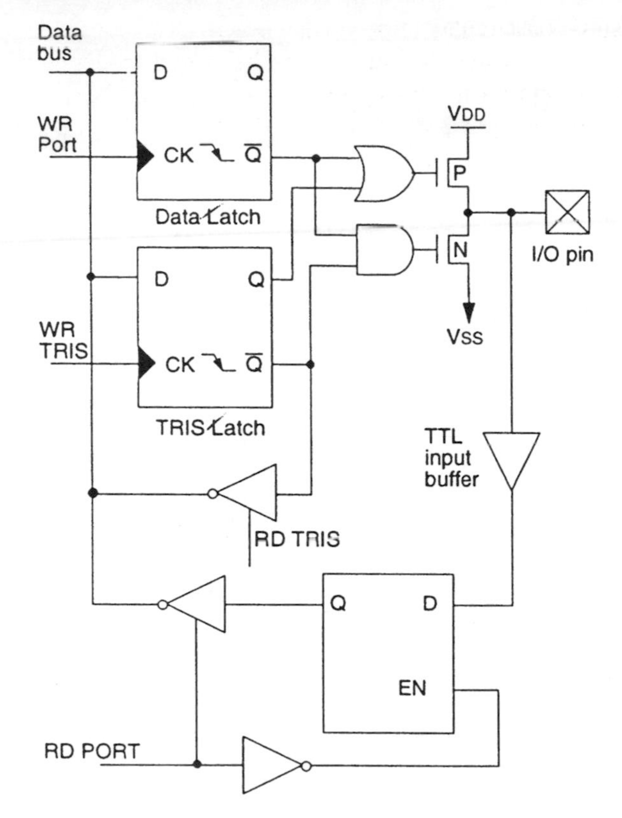

Figure A.10 Block diagram of pins RA3:RA0.

This interrupt can wake the device from SLEEP. The user, in the interrupt service routine, can clear the interrupt in the following manner:

(a) read (or write) PORTB. This will end the mismatch condition;
(b) clear flag bit RBIF.

A mismatch condition will continue to set the RBIF bit. Reading PORTB will end the mismatch condition, and allow the RBIF bit to be cleared.

This interrupt on mismatch feature, together with software configurable pull-ups on these four pins allow easy interface to a key pad and make it possible for wake-up on key-depression.

The interrupt on change feature is recommended for wake-up on key depression operation and operations where PORTB is only used for the interrupt on change feature. Polling of PORTB is not recommended while using the interrupt on change feature.

A.5.3 I/O Programming Considerations

Bi-directional I/O ports

Any instruction which writes, operates internally as a read followed by a write operation. The `BCF` and `BSF` instructions, for example, read the register into the CPU, execute the bit

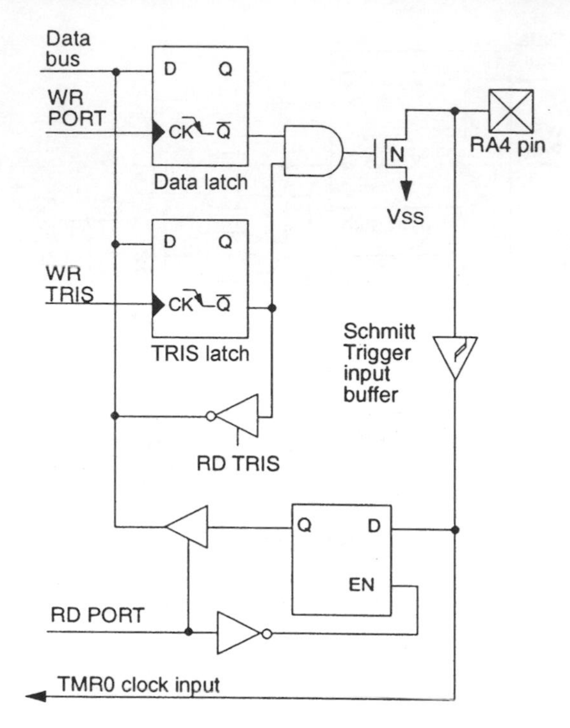

Figure A.11 Block diagram of pin RA4.

operation and write the result back to the register. Caution must be used when these instructions are applied to a port with both inputs and outputs defined. For example, a `BSF` operation on bit 5 of PORTB will cause all eight bits of PORTB to be read into the CPU. Then the `BSF` operation takes place on bit 5 and PORTB is written to the output latches. If another bit of PORTB is used as a bi-directonal I/O pin (i.e., bit 0) and it is defined as an input at this time, the input signal present on the pin itself would be read into the CPU and rewritten to the data latch of this particular pin, overwriting the previous content. As long as the pin stays in the input mode, no problem occurs. However, if bit 0 is switched into output mode later on, the content of the data latch is unknown.

Reading the port register, reads the values of the port pins. Writing to the port register writes the value to the port latch. When using read-modify-write instructions (i.e., `BCD`, `BSF`, etc.) on a port, the value of the port pins is read, the desired operation is done to this value, and this value is then written to the port latch.

A pin actively outputting a Low or High should not be driven from external devices at the same time in order to change the level on this pin ('wired-or', 'wired-and'). The resulting high output current may damage the chip.

Successive operations on I/O ports

The actual write to an I/O port happens at the end of an instruction cycle, whereas for reading, the data must be valid at the beginning of the instruction cycle. Therefore, care must be exercised if a write followed by a read operation is carried out on the same I/O port. The

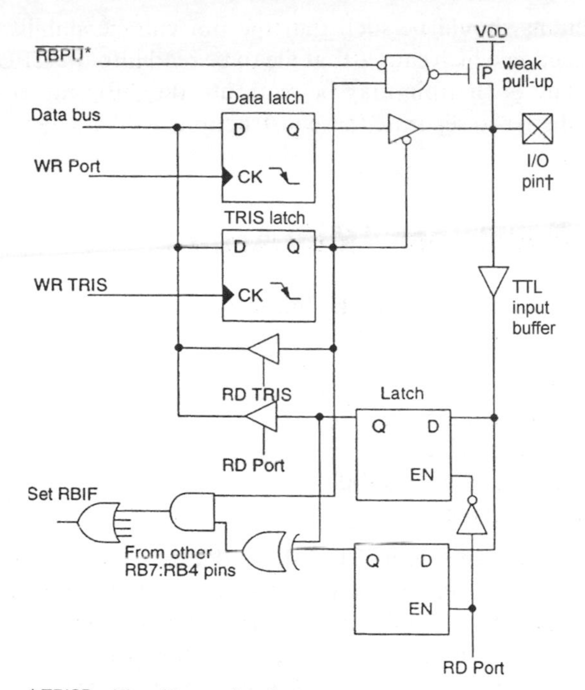

Figure A.12 Block diagram of pins RB7:RB4.

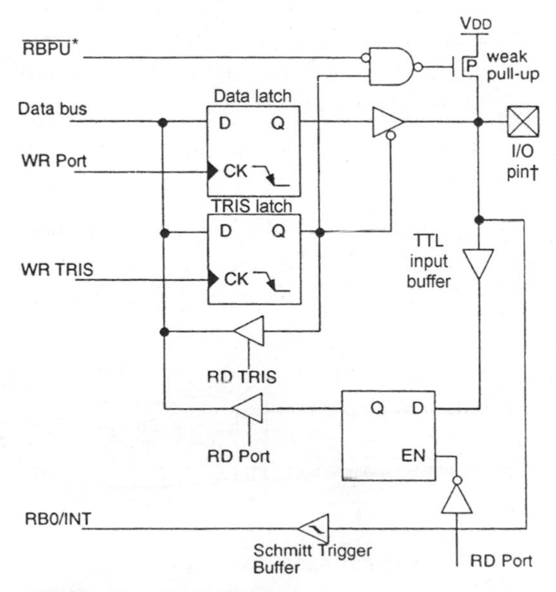

Figure A.13 Block diagram of pins RB3:RB0.

sequence of instructions should be such that the pin voltage stabilizes (load dependent) before the next instruction which causes that file to be read into the CPU is executed. Otherwise, the previous state of that pin may be read into the CPU rather than the new state. When in doubt, it is better to separate these instructions with a `NOP` or another instruction not accessing this I/O port.

A.6 Timer0 Module and TMR0 Register

The Timer0 module timer/counter has the following features:

- 8-bit timer/counter;
- readable and writable;
- 8-bit software programmable prescaler;
- internal or external clock select;
- interrupt on overflow from FFh to 00h;
- edge select for external clock.

Timer mode is selected by clearing the T0CS bit (OPTION⟨5⟩). In timer mode, the Timer0 module (Fig. A.14) will increment every instruction cycle (without prescaler). If the TMR0 register is written, the increment is inhibited for the following two cycles. The user can work around this by writing an adjusted value to the TMR0 register.

Counter mode is selected by setting the T0CS bit (OPTION⟨5⟩). In this mode TMR0 will increment either on every rising or falling edge of pin RA4/T0CKI. The incrementing edge is determined by the T0 source edge select bit, T0SE (OPTION⟨4⟩). Clearing bit T0SE selects the rising edge.

The prescaler is shared between the Timer0 Module and the Watchdog Timer. The prescaler assignment is controlled, in software, by control bit PSA (OPTION⟨3⟩). Clearing bit PSA will assign the prescaler to the Timer0 Module. The prescaler is not readable or writable. When the prescaler (Section A.6.3) is assigned to the Timer0 Module, the prescale value $(1:2, 1:4, \ldots, 1:256)$ is software selectable.

A.6.1 TMR0 interrupt

The TMR0 interrupt is generated when the TMR0 register overflows from FFh to 00h. This overflow sets the T0IF bit (INTCON⟨2⟩). The interrupt can be masked by clearing enable bit T0IE (INTCON⟨5⟩). The T0IF bit must be cleared in software by the Timer0 Module

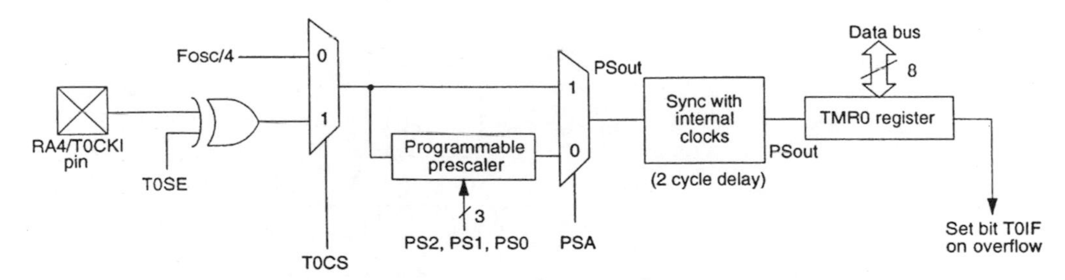

Notes: (1) Bits T0CS, T0SE, PS2, PS1, PS0 and PSA are located in the OPTION register.
(2) The prescaler is shared with the Watchdog Timer (Fig. A.17).

Figure A.14 TMR0 block diagram.

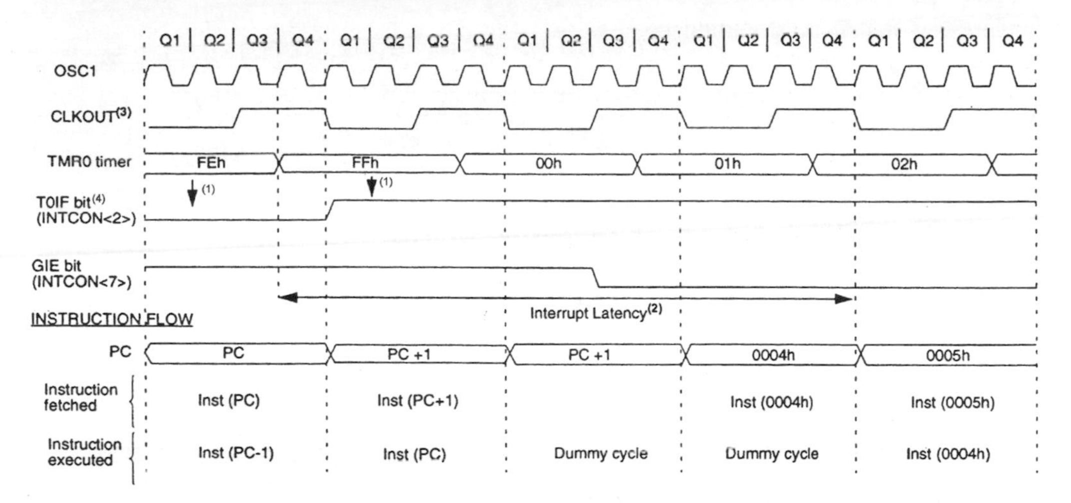

Figure A.15 TMR0 interrupt timing.

interrupt service routine before re-enabling this interrupt. The TMR0 interrupt cannot wake the processor from SLEEP since the timer is shut off during SLEEP.

A.6.2 Using TMR0 with external clock

When an external clock input is used for TMR0, it must meet certain requirements. The external clock requirement is due to internal phase clock (Tosc) synchronization. Also, there is a delay in the actual incrementing of the TMR0 register after synchronization.

External clock synchronization

When no scaler is used, the external clock output is the same as the presenter output. The synchronization of pin RA4/T0CKI with the internal phase clocks is accomplished by sampling the prescaler output on the Q2 and Q4 cycles of the internal phase clocks (Fig. A.16). Therefore, it is necessary for T0CKI to be high for at least $2T_{osc}$ (plus a small RC delay) and low for at least $2T_{osc}$ (plus a small RC delay). Refer to the electrical specification of the desired device.

When a prescaler is used, the external clock input is divided by an asynchronous ripple counter type prescaler so that the prescaler output is symmetrical. For the external clock to meet the sampling requirement, the ripple counter must be taken into account. Therefore, it is necessary for T0CKI to have a period of at least $4T_{osc}$ (plus a small RC delay) divided by the prescaler value. The only requirement on T0CKI high and low time is that they do not violate the minimum pulse width requirement of 10 ns.

TMR0 increment delay

Since the prescaler output is synchronized with the internal clocks, there is a small delay from the time the external clock edge occurs to the time the Timer0 Module is actually

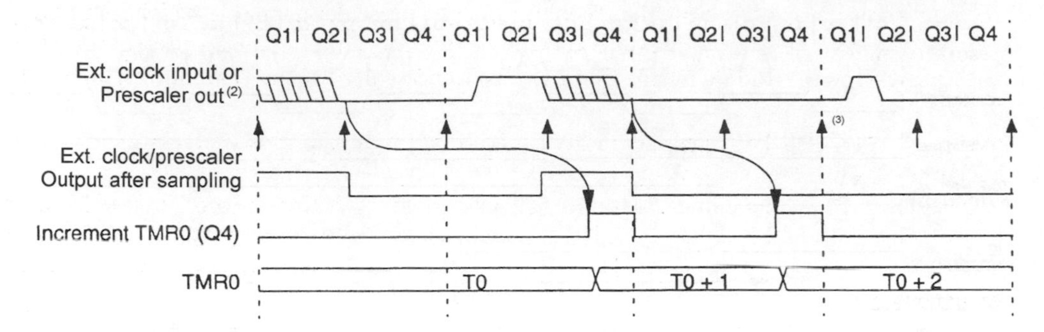

Notes: (1) Delay from clock input change to TMR0 increment is 3Tosc to 7Tosc (duration of Q = Tosc). Therefore, the error in measuring the interval between two edges on TMR0 input = ± 4Tosc max.
(2) External clock if no prescaler selected, Prescaler output otherwise.
(3) The arrows ↑ indicate where sampling occurs. A small clock pulse may be missed by sampling.

Figure A.16 Timer0 timing with external clock.

incremented. Figure A.16 shows the delay from the external clock edge to the time incrementing.

A.6.3 Prescaler

An 8-bit counter is available as a prescaler for the Timer0 Module, or as a postscaler for the Watchdog Timer (Fig. A.17). For simplicity, this counter is being referred to as 'prescaler'

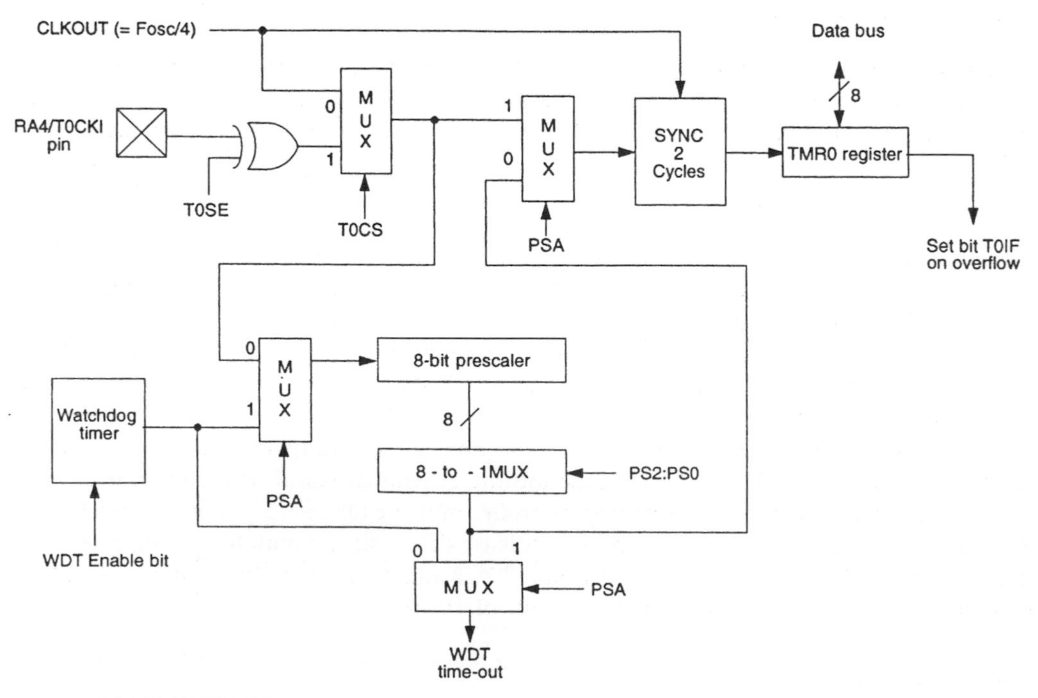

Note: T0CS, T0SE, PSA, PS2:PS0 are bits in the OPTION register.

Figure A.17 Block diagram of the TMR0/WDT prescaler.

throughout this data sheet. Note that there is only one prescaler available which is mutually exclusive between the Timer0 Module and the Watchdog Timer. Thus, a prescaler assignment for the Timer0 Module means that there is no prescaler for the Watchdog Timer, and vice-versa.

The PSA and PS2:PS0 bits (OPTION⟨3 : 0⟩) determine the prescaler assignment and prescale ratio.

When assigned to the Timer0 Module, all instructions writing to the Timer0 Module (e.g., `CLRF 1`, `MOVWF 1`, `BSF 1, x` ... etc.) will clear the prescaler. When assigned to WDT, a `CLRWDT` instruction will clear the prescaler along with the Watchdog Timer. The prescaler is not readable or writable.

A.7 Data EEPROM Memory

The EEPROM data memory is readable and writable during normal operation (full VDD range). This memory is not directly mapped in the register file space. Instead it is indirectly addressed through the special function registers. There are four SFRs used to read and write this memory. These registers are:

- EECON1
- EECON2
- EEDATA
- EEADR.

EEDATA holds the 8-bit data for read/write, and EEADR holds the address of the EEPROM location being accessed. PIC 16F8X devices have 64 bytes of data EEPROM with an address range from 0h to 3Fh.

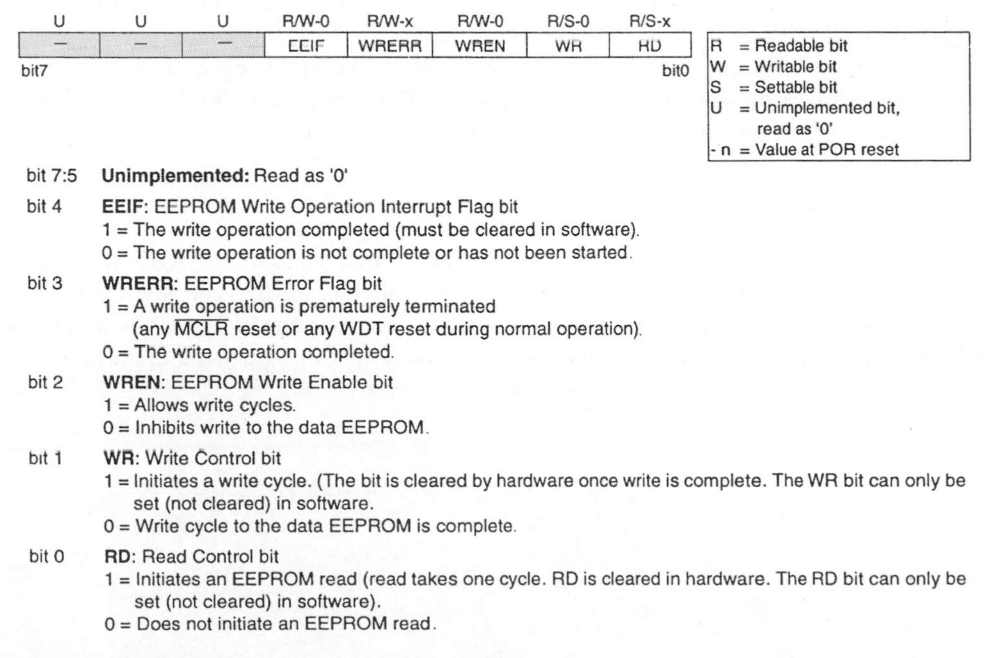

U	U	U	R/W-0	R/W-x	R/W-0	R/S-0	R/S-x
—	—	—	EEIF	WRERR	WREN	WR	RD
bit7							bit0

R = Readable bit
W = Writable bit
S = Settable bit
U = Unimplemented bit, read as '0'
- n = Value at POR reset

bit 7:5 **Unimplemented:** Read as '0'

bit 4 **EEIF**: EEPROM Write Operation Interrupt Flag bit
1 = The write operation completed (must be cleared in software).
0 = The write operation is not complete or has not been started.

bit 3 **WRERR**: EEPROM Error Flag bit
1 = A write operation is prematurely terminated
(any $\overline{\text{MCLR}}$ reset or any WDT reset during normal operation).
0 = The write operation completed.

bit 2 **WREN**: EEPROM Write Enable bit
1 = Allows write cycles.
0 = Inhibits write to the data EEPROM.

bit 1 **WR**: Write Control bit
1 = Initiates a write cycle. (The bit is cleared by hardware once write is complete. The WR bit can only be set (not cleared) in software.
0 = Write cycle to the data EEPROM is complete.

bit 0 **RD**: Read Control bit
1 = Initiates an EEPROM read (read takes one cycle. RD is cleared in hardware. The RD bit can only be set (not cleared) in software).
0 = Does not initiate an EEPROM read.

Figure A.18 EECON1 register (address 88h).

The EEPROM data memory allows byte read and write. A byte write automatically erases the location and writes the new data (erase before write). The EEPROM data memory is rated for high erase/write cycles. The write time is controlled by an on-chip timer. The write-time will vary with voltage and temperature as well as from chip to chip. Please refer to AC specifications for exact limits.

When the device is code protected, the CPU may continue to read and write the data EEPROM memory. The device programmer can no longer access this memory.

A.7.1 EEADR register

The EEADR register can address up to a maximum of 256 bytes of data EEPROM. Only the first 64 bytes of data EEPROM are implemented.

The upper two bits are address decoded. This means that these two bits must always be '0' to ensure that the address is in the 64 byte memory space.

A.7.2 EECON1 and EECON2 registers

EECON1 is the control register with five low order bits physically implemented. The upper-three bits are non-existent and read as '0's.

Control bits RD and WR initiate read and write, respectively. These bits cannot be cleared, only set, in software. They are cleared in hardware at completion of the read or write operation. The inability to clear the WR bit in software prevents the accidental, premature termination of a write operation.

The WREN bit, when set, will allow a write operation. On power-up, the WREN bit is clear. The WRERR bit is set when a write operation is interrupted by a $\overline{\text{MCLR}}$ reset or a WDT time-out reset during normal operation. In these situations, following reset, the user can check the WRERR bit and rewrite the location. The data and address will be unchanged in the EEDATA and EEADR registers.

Interrupt flag bit EEIF is set when write is complete. It must be cleared in software.

EECON2 is not a physical register. Reading EECON2 will read all '0's. The EECON2 register is used exclusively in the Data EEPROM write sequence.

A.7.3 Reading the EEPROM data memory

To read a data memory location, the user must write the address to the EEADR register and then set control bit RD (EECON1⟨0⟩). The data is available, in the very next cycle, in the EEDATA register; therefore it can be read in the next instruction. EEDATA will hold this value until another read or until it is written to by the user (during a write operation).

```
BCF      STATUS, RP0     ;   Bank 0
MOVLW    CONFIG__ADDR    ;
MOVWF    EEADR           ;   Address to read
BSF      STATUS, RP0     ;   Bank 1
BSF      EECON1, RD      ;   EE Read
BCF      STATUS, RP0     ;   Bank 0
MOVF     EEDATA, W       ;   W = EEDATA
```

Example A.2 Data EEPROM read

```
            BSF      STATUS, RP0     ;   Bank 1
            BCF      INTCON, GIE     ;   Disable INTs.
            BSF      EECON1, WREN    ;   Enable Write
Required    MOVLW    55h             ;
Sequence    MOVWF    EECON2          ;   Write 55h
            MOVLW    AAh             ;
            MOVWF    EECON2          ;   Write AAh
            BSF      EECON1, WR      ;   Set WR bit
                                     ;     begin write
            BSF      INTCON,GIE      ;   Enable INTs.
```

Example A.3 Data EEPROM write.

A.7.4 Writing to the EEPROM data memory

To write an EEPROM data location, the user must first write the address to the EEADR register and the data to the EEDATA register. Then the user must follow a specific sequence to initiate the write for each byte.

The write will not initiate if the above sequence is not exactly followed (write 55h to EECON2, write AAh to EECON2, then set WR bit) for each byte. We strongly recommend that interrupts be disabled during this code segment.

In addition, the WREN bit in EECON1 must be set to enable write. This mechanism prevents accidental writes to data EEPROM due to errant (unexpected) code execution (i.e., lost programs). The user should keep the WREN bit clear at all times, except when updating EEPROM. The WREN bit is not cleared by hardware.

After a write sequence has been initiated, clearing the WREN bit will not affect this write cycle. The WR bit will be inhibited from being set unless the WREN bit is set.

At the completion of the write cycle, the WR bit is cleared in hardware and the EE Write Complete Interrupt Flag bit (EEIF) is set. The user can either enable this interrupt or poll this bit. EEIF must be cleared by software.

A.8 Special Features of the CPU

What sets a microcontroller apart from other processors are special circuits to deal with the needs of real time applications. The PIC 16F8X has a host of such features intended to maximize system reliability, minimize cost through the elimination of external components, provide power saving operating modes and offer code protection. These features are:

- OSC Selection
- Reset
  - Power-on Reset (POR)
  - Power-up Timer (PWRT)
  - Oscillator Start-up Timer (OST)
- Interrupts
- Watchdog Timer (WDT)
- SLEEP
- Code protection
- ID locations
- In-circuit serial programming

The PIC 16F8X has a Watchdog Timer which can be shut off only through configuration bits. It runs off its own RC oscillator for added reliability. There are two timers that offer necessary delays on power-up. One is the Oscillator Start-up Timer (OST), intended to keep the chip in reset until the crystal oscillator is stable. The other is the Power-up Timer (PWRT), which provides a fixed delay of 72 ms (nominal) on power-up only. This design keeps the device in reset while the power supply stabilizes. With these two timers on-chip, most applications need no external reset circuitry.

SLEEP mode offers a very low current power-down mode. The user can wake-up from SLEEP through external reset, Watchdog Timer time-out or through an interrupt. Several oscillator options are provided to allow the part to fit the application. The RC oscillator option saves system cost while the LP crystal option saves power. A set of configuration bits are used to select the various options.

A.8.1 Configuration bits

The configuration bits can be programmed (read as '0') or left unprogrammed (read as '1') to select various device configurations. These bits are mapped in program memory location 2007h.

Address 2007h is beyond the user program memory space and it belongs to the special test/configuration memory space (2000h–3FFFh). This space can only be accessed during programming.

A.8.2 Oscillator configurations

Oscillator types

The PIC 16F8X can be operated in four different oscillator modes. The user can program two configuration bits (FOSC1 and FOSC0) to select one of these four modes:

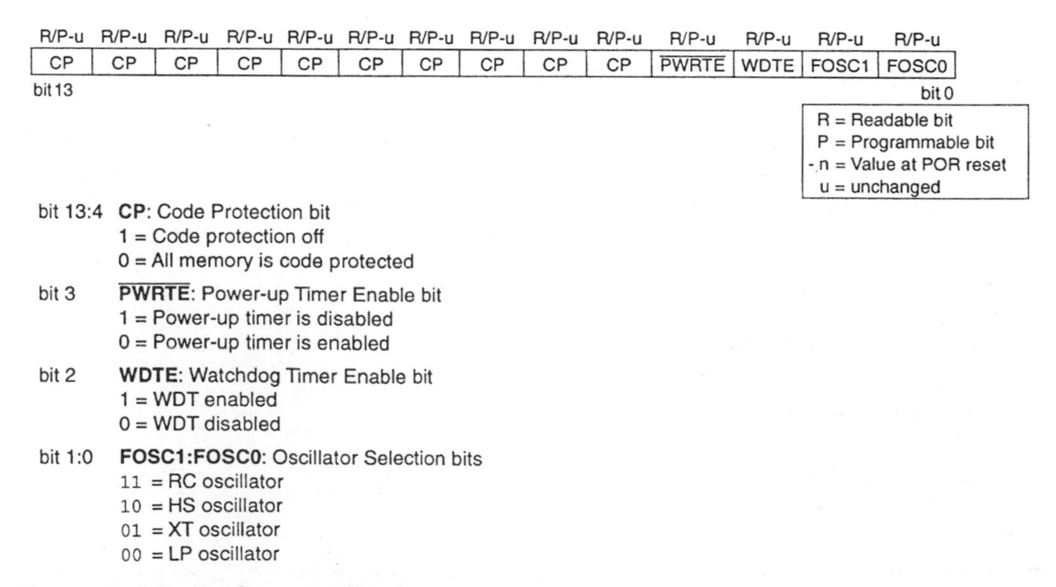

Figure A.19 Configuration word.

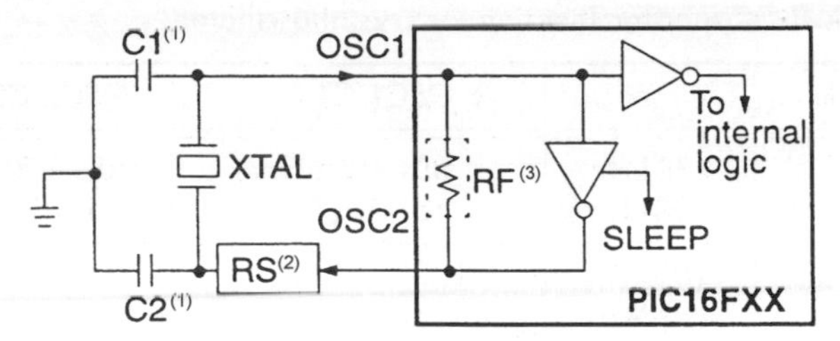

Figure A.20 Crystal/ceramic resonator operation (HS, XT or LP OSC configuration.

1. LP Low Power Crystal
2. XT Crystal/Resonator
3. HS High Speed Crystal/Resonator
4. RC Resistor/Capacitor

Crystal oscillator/ceramic resonators

In XT, LP or HS modes a crystal or ceramic resonator is connected to the OSC1/CLKIN and OSC2/CLKOUT pins to establish oscillation (Fig. A.20).

The PIC 16F8X oscillator design requires the use of a parallel cut crystal. Use of a series cut crystal may give a frequency out of thc crystal manufacturer's specifications. When in XT, LP or HS modes, the device can have an external clock source to drive the OSC1/CLKIN pin (Fig. A.21).

External crystal oscillator circuit

Either a prepackaged oscillator can be used or a simple oscillator circuit with TTL gates can be built. Prepackaged oscillators provide a wide operating range and better stability. A well-designed crystal oscillator will provide good performance with TTL gates. Two types of crystal oscillator circuits are available; one with series resonance, and one with parallel resonance.

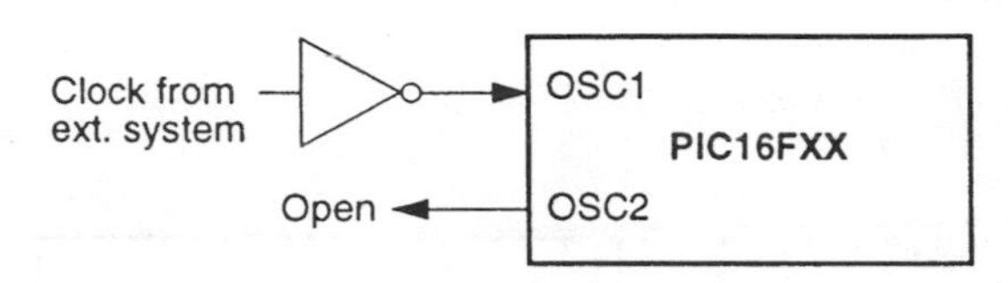

Figure A.21 External clock input operation (HS, XT or LP OSC configuration).

RC oscillator

For timing insensitive applications the RC device option offers additional cost savings. The RC oscillator frequency is a function of the supply voltage, the resistor (Rext) values,

Table A.3 Capacitor selection for crystal oscillator

Mode	Freq	OSC1/C1	OSC2/C2
LP	32 kHz 200 kHz	68–100 pF 15–33 pF	68–100 pF 15–33 pF
XT	100 kHz 2 MHz 4 MHz	100–150 pF 15–33 pF 15–33 pF	100–150 pF 15–33 pF 15–33 pF
HS	4 MHz 10 MHz	15–33 pF 15–33 pF	15–33 pF 15–33 pF

Note. Higher capacitance increases the stability of oscillator but also increases the start-up time. These values are for design guidance only. Rs may be required in HS mode as well as XT mode to avoid overdriving crystals with low drive level specification. Since each crystal has its own characteristics, the user should consult the crystal manufacturer for appropriate values of external components. For $V_{DD} > 4.5V$, $C1 = C2 = 30\,pF$ is recommended.

capacitor (Cext) values, and the operating temperature. In addition to this, the oscillator frequency will vary from unit to unit due to normal process parameter variation. Furthermore, the difference in lead frame capacitance between package types also affects the oscillation frequency, especially for low Cext values. The user needs to take into account variation due to tolerance of the external R and C components. Figure A.22 shows how an R/C combination is connected to the PIC 16F8X. For Rext values below 2.2 kΩ, the oscillator operation may become unstable, or stop completely. For very high Rext values (e.g., 1 MΩ), the oscillator becomes sensitive to noise, humidity and leakage. Thus, we recommend keeping Rext between 5 kΩ and 100 kΩ.

The oscillator frequency, divided by 4, is available on the OSC2/CLKOUT pin, and can be used for test purposes or to synchronize other logic (see Fig. A.2 for waveform).

Note. When the device oscillator is in RC mode, do not drive the OSC1 pin with an external clock or you may damage the device.

A.8.3 Reset

The PIC 16F8X differentiates between kinds of reset:

- Power-on Reset (POR)
- $\overline{\text{MCLR}}$ reset during normal operation

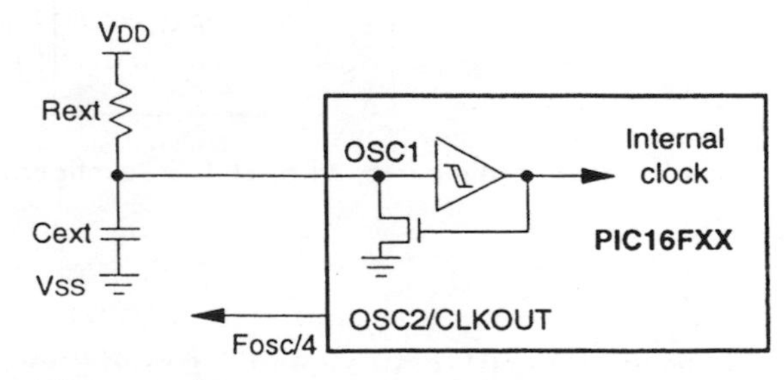

Figure A.22 RC oscillator mode.

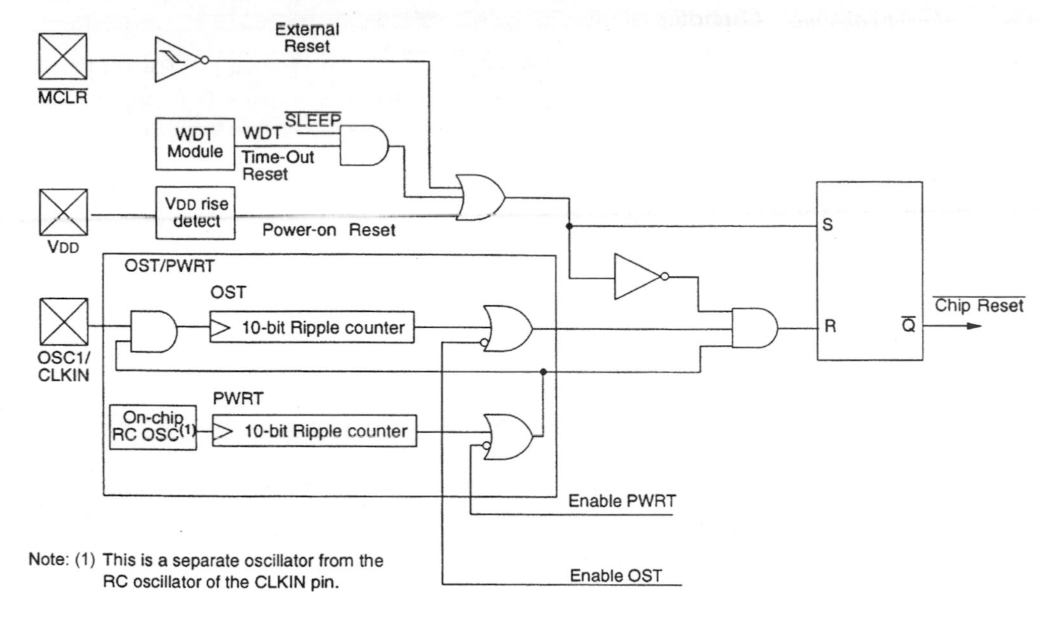

Figure A.23 Block diagram of on-chip reset circuit.

- $\overline{\text{MCLR}}$ reset during SLEEP
- WDT Reset (during normal operation)
- WDT Wake-up (during SLEEP).

Figure A.23 shows a simplified block diagram of the on-chip reset circuit. The $\overline{\text{MCLR}}$ reset path has a noise filter to ignore small pulses. The electrical specifications state the pulse-width requirements for the $\overline{\text{MCLR}}$ pin.

Some registers are not affected in any reset condition; their status is unknown on a POR reset and unchanged in any other reset. Most other registers are reset to a 'reset state' on POR, $\overline{\text{MCLR}}$ or WDT reset during normal operation and on $\overline{\text{MCLR}}$ reset during SLEEP. They are not affected by a WDT reset during SLEEP, since this reset is viewed as the resumption of normal operation.

A.8.4 Power-on Reset (POR)

A Power-on Reset pulse is generated on-chip when a VDD rise is detected (in the range of 1.2 V–1.7 V). To take advantage of the POR, just tie the $\overline{\text{MCLR}}$ pin directly (or through a resistor) to VDD. This will eliminate external RC components usually needed to create Power-on Reset. A minimum rise time for VDD must be met for this to operate properly.

The electrical specifications in the full data sheet state that when the device starts normal operation (exits the reset condition), device operating parameters (voltage, frequency, temperature, ...) must be met to ensure operation. If these conditions are not met, the device must be held in reset until the operating conditions are met.

The POR circuit does not produce an internal reset when VDD declines.

A.8.5 Power-up Timer (PWRT)

The Power-up Timer (PWRT) provides a fixed 72 ms nominal time-out (TPWRT) from POR. The Power-up Timer operates on an internal RC oscillator. The chip is kept in reset as long as the PWRT is active. The PWRT delay allows the VDD to rise to an acceptable level.

A configuration bit, PWRTE, can enable/disable the PWRT.

The power-up time delay TPWRT will vary from chip to chip due to VDD, temperature, and process variation. See DC parameters in the full data sheet for details.

A.8.6 Oscillator Start-up Timer (OST)

The Oscillator Start-up Timer (OST) provides a 1024 oscillator cycle delay (from OSC1 input) after the PWRT delay ends. This ensures the crystal oscillator or resonator has started and stabilized.

The OST time-out (TOST) is invoked only for XT, LP and HS modes and only on Power-on Reset or wake-up from SLEEP.

When VDD rises very slowly, it is possible that the TPWRT time-out and TOST time-out will expire before VDD has reached its final value. In this case, an external power-on reset circuit may be necessary.

A.8.7 Interrupts

The PIC 16F8X has four sources of interrupt:

1. external interrupt RB0/INT pin;
2. TMR0 overflow interrupt;
3. PORTB change interrupts (pins RB7:RB4);
4. data EEPROM write complete interrupt.

The interrupt control register (INTCON) records individual interrupt requests in flag bits. It also contains the individual and global interrupt enable bits.

The global interrupt enable bit, GIE (INTCON⟨7⟩) enables (if set) all un-masked interrupts or disables (if cleared) all interrupts. Individual interrupts can be disabled through their corresponding enable bits in INTCON register. Bit GIE is cleared on reset.

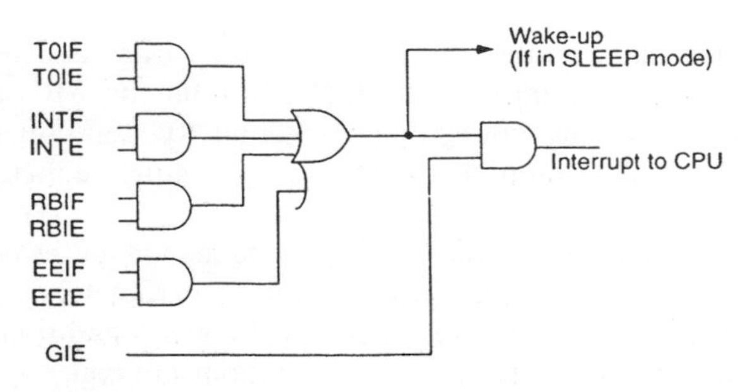

Figure A.24 Interrupt logic.

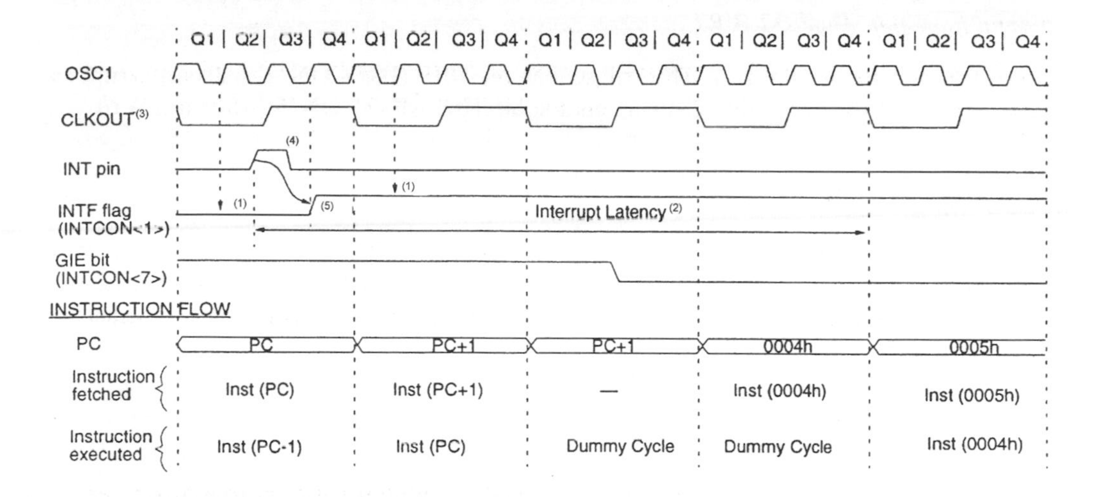

Notes: (1) INTF flag is sampled here (every Q1).
(2) Interrupt latency = 3–4Tcy where Tcy = instruction cycle time.
Latency is the same whether Inst (PC) is a single cycle or a two-cycle instruction.
(3) CLKOUT is available only in RC oscillator mode.
(4) For minimum width of INT pulse, refer to AC specs.
(5) INTF is enabled to be set anytime during the Q4–Q1 cycles.

Figure A.25 INT pin interrupt timing.

The 'return from interrupt' instruction, RETFIE, exits interrupt routine as well as sets the GIE bit, which re-enable interrupts.

The RBO/INT pin interrupt, the RB port change interrupt and the TMR0 overflow flags are contained in the INTCON register.

When an interrupt is responded to, the GIE bit is cleared to disable any further interrupt, the return address is pushed onto the stack, and the PC is loaded with 0004h. For external interrupt events, such as the RBO/INT pin or PORTB change interrupt, the interrupt latency will be three to four instruction cycles. The exact latency depends when the interrupt event occurs (Fig. A.25). The latency is the same for both one- and two-cycle instructions. Once in the interrupt service routine the source(s) of the interrupt can be determined by polling the interrupt flag bits. The interrupt flag bit(s) must be cleared in software before re-enabling interrupts to avoid infinite interrupt requests.

Note. Individual interrupt flag bits are set regardless of the status of their corresponding mask bit or the GIE bit.

INT interrupt

External interrupt on RB0/INT pin is edge triggered: either rising if INTEDG bit (OPTION⟨6⟩) is set, or falling, if INTEDG bit is clear. When a valid edge appears on the RB0/INT pin, the INTF bit (INTCON⟨1⟩) is set. This interrupt can be disabled by clearing control bit INTE (INTCON⟨4⟩). Flag bit INTF must be cleared in software via the interrupt service routine before re-enabling this interrupt. The INT interrupt can wake the processor from SLEEP (Section A.8.9) only if the INTE bit was set prior to going into SLEEP. The status of the GIE bit decides whether the processor branches to the interrupt vector following wake-up.

TMR0 interrupt

An overflow (FFh → 00h) in TMR0 will set flag bit T0IF (INTCON⟨2⟩). The interrupt can be enabled/disabled by setting/clearing enable bit T0IE (INTCON⟨5⟩) (Section A.6).

PORT RB interrupt

An input change on PORTB⟨7 : 4⟩ sets flag bit RBIF (INTCON⟨0⟩). The interrupt can be enabled/disabled by setting/clearing enable bit RBIE (INTCON⟨3⟩) (Section A.5.2).

Note. For a change on the I/O pin to be recognized, the pulse width must be at least TCY wide.

A.8.8 Watchdog Timer (WDT)

The Watchdog Timer is a free-running on-chip RC oscillator which does not require any external components. This RC oscillator is separate from the RC oscillator of the OSC1/CLKIN pin. That means that the WDT will run even if the clock on the OSC1/CLKIN and OSC2/CLKOUT pins of the device has been stopped, for example, by execution of a `SLEEP` instruction. During normal operation a WDT time-out generates a device RESET. If the device is in SLEEP mode, a WDT Wake-up causes the device to wake-up and continue with normal operation. The WDT can be permanently disabled by programming configuration bit WDTE as a '0' (Section A.8.1).

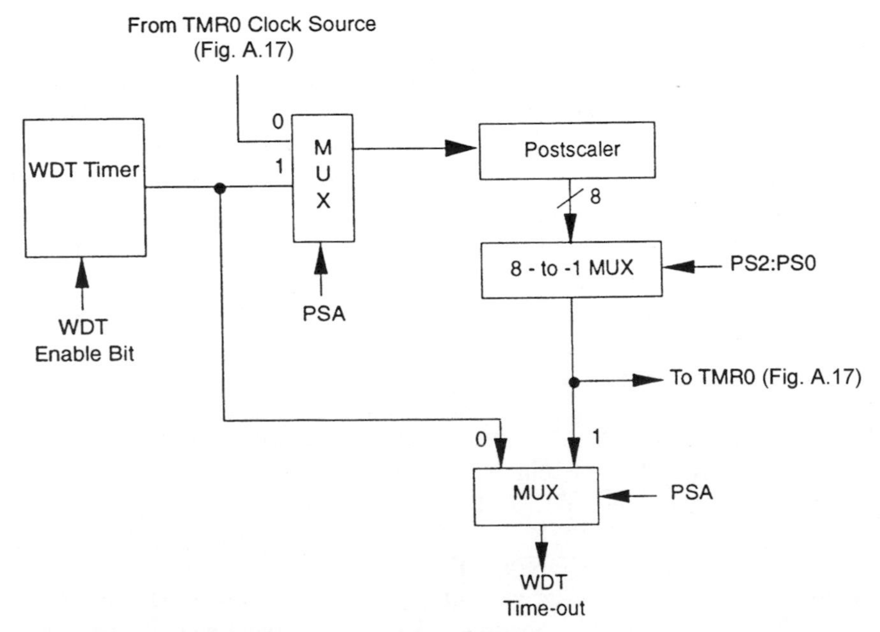

Figure A.26 Watchdog timer block diagram.

WDT period

The WDT has a nominal time-out period of 18 ms (with no prescaler). The time-out periods vary with temperature, VDD and process variations from part to part (see DC specifications). If longer time-out periods are desired, a prescaler with a division ratio of up to 1:128 can be assigned to the WDT under software control by writing to the OPTION register. Thus, time-out periods up to 2.3 seconds can be realized.

The `CLRWDT` and `SLEEP` instructions clear the WDT and the postscaler (if assigned to the WDT) and prevent it from timing out and generating a device RESET condition.

The $\overline{\text{TO}}$ bit in the STATUS register will be cleared upon a WDT time-out.

WDT programming considerations

It should also be taken into account that under worst case conditions (VDD = Min., Temperature = Max., max. WDT prescaler) it may take several seconds before a WDT time-out occurs.

A.8.9 Power-down Mode (SLEEP)

A device may be powered down (SLEEP) and later powered up (Wake-up from SLEEP).

SLEEP

The Power-down mode is entered by executing the `SLEEP` instruction.

If enabled, the Watchdog Timer is cleared (but keeps running), the $\overline{\text{PD}}$ bit (STATUS⟨3⟩) is cleared, the $\overline{\text{TO}}$ bit (STATUS⟨4⟩) is set, and the oscillator driver is turned off. The I/O ports maintain the status they had before the `SLEEP` instruction was executed (driving high, low, or hi-impedance).

For the lowest current consumption in SLEEP mode, place all I/O pins either at VDD or Vss, with no external circuitry drawing current from the I/O pins, and disable external clocks. I/O pins that are hi-impedance inputs should be pulled high or low externally to avoid switching currents caused by floating inputs. The T0CKI input should also be at VDD or Vss. The contribution from on-chip pull-ups on PORTB should be considered.

The $\overline{\text{MCLR}}$ pin must be at a logic high level (VIHMC).

It should be noted that a RESET generated by a WDT time-out does not drive the $\overline{\text{MCLR}}$ pin low.

Wake-up from SLEEP

The device can wake-up from SLEEP through one of the following events:

1. external reset input on $\overline{\text{MCLR}}$ pin;
2. WDT wake-up (if WDT was enabled);
3. interrupt from RB0/INT pin, RB port change, or data EEPROM write complete.

Peripherals cannot generate interrupts during SLEEP, since no on-chip Q clocks are present.

The first event ($\overline{\text{MCLR}}$ reset) will cause a device reset. The two latter events are considered as a continuation of program execution. The $\overline{\text{TO}}$ and $\overline{\text{PD}}$ bits can be used to

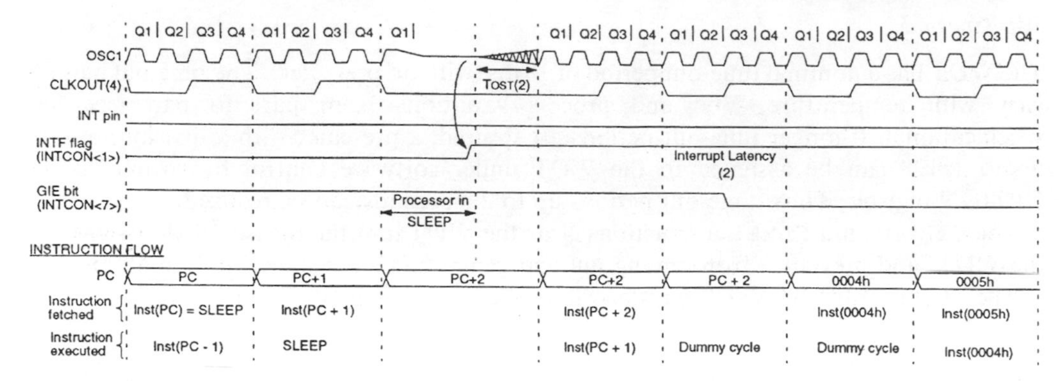

Notes: (1) XT, HS or LP oscillator mode assumed.
(2) TOST = 1024TOSC (drawing not to scale) This delay will not be there for RC osc mode.
(3) GIE = '1' assumed. In this case after wake- up, the processor jumps to the interrupt routine. If GIE = '0', execution will continue in-line.
(4) CLKOUT is not available in these osc modes, but shown here for timing reference.

Figure A.27 Wake-up from SLEEP through interrupt.

determine the cause of a device reset. The $\overline{PD}$ bit, which is set on power-up, is cleared when SLEEP is invoked. The $\overline{TO}$ bit is cleared if a WDT time-out occurred (and caused wake-up).

While the `SLEEP` instruction is being executed, the next instruction (PC + 1) is pre-fetched. For the device to wake-up through an interrupt event, the corresponding interrupt enable bit must be set (enabled). Wake-up occurs regardless of the state of the GIE bit. If the GIE bit is clear (disabled), the device continues execution at the instruction after the `SLEEP` instruction. If the GIE bit is set (enabled), the device executes the instruction after the `SLEEP` instruction and then branches to the interrupt address (0004h). In cases where the execution of the instruction following `SLEEP` is not desirable, the user should have a `NOP` after the `SLEEP` instruction.

Wake-up using interrupts

When global interrupts are disabled (GIE cleared) and any interrupt source has both its interrupt enable bit and interrupt flag bit set, one of the following will occur:

1. If the interrupt occurs *before* the execution of a `SLEEP` instruction, the `SLEEP` instruction will complete as a NOP. Therefore, the WDT and WDT postscaler will not be cleared, thc $\overline{TO}$ bit will not be set and $\overline{PD}$ bits will not be cleared.
2. If the interrupt occurs *during* or *after* the execution of a `SLEEP` instruction, the device will immediately wake from sleep. The `SLEEP` instruction will be completely executed before the wake-up. Therefore, the WDT and WDT postscaler will be cleared, the $\overline{TO}$ bit will be set and the $\overline{PD}$ bit will be cleared.

Even if the flag bits were checked before executing a `SLEEP` instruction, it may be possible for flag bits to become set before the `SLEEP` instruction completes. To determine whether a `SLEEP` instruction has been executed, test the $\overline{PD}$ bit. If the $\overline{PD}$ bit is set, the `SLEEP` instruction was executed as a NOP.

To ensure that the WDT is cleared, a `CLRWDT` instruction should be executed before a `SLEEP` instruction.

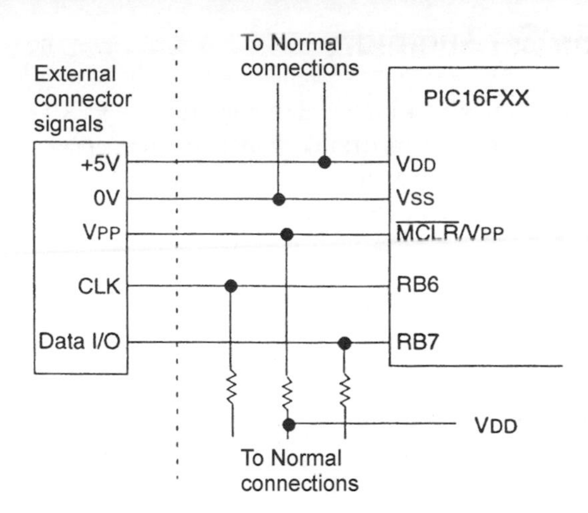

Figure A.28 Typical in-system serial programming connection.

A.8.10 Program Verification/Code Protection

If the code protection bit(s) have not been programmed, the on-chip program memory can be read out for verification purposes.

Note. Microchip does not recommend code protecting windowed devices.

A.8.11 ID Locations

Four memory locations (2000h–2003h) are designated as ID locations to store checksum or other code identification numbers. These locations are not accessible during normal execution but are readable and writable only during program/verify. Only the four least significant bits of ID locations are usable.

For ROM devices, these values are submitted along with the ROM code.

A.8.12 In-Circuit Serial Programming

PIC 16F8X microcontrollers can be serially programmed while in the end application circuit. This is simply done with two lines for clock and data, and three other lines for power, ground, and the programming voltage. Customers can manufacture boards with unprogrammed devices, and then program the microcontroller just before shipping the product, allowing the most recent firmware or custom firmware to be programmed.

The device is placed into a program/verify mode by holding the RB6 and RB7 pins low, while raising the $\overline{\text{MCLR}}$ pin from V_{IL} to V_{IHH} (see programming specification). RB6 becomes the programming clock and RB7 becomes the programming data. Both the RB6 and RB7 are Schmitt Trigger inputs in this mode.

After reset, to place the device into programming/verify mode, the program counter (PC) points to location 00h. A 6-bit command is then supplied to the device, 14-bits of program data is then supplied to or from the device, using load or read-type instructions. For complete details of serial programming, please refer to the PIC 16CXX Programming Specifications (Literature #DS30189).

For ROM devices, both the program memory and data EEPROM memory may be read, but only the data EEPROM memory may be programmed.

A.9 Instruction Set Summary

Each PIC 16FXX instruction is a 14-bit word divided into an OPCODE which specifies the instruction type and one or more operands which further specify the operation of the instruction. The PIC 16FXX instruction set summary in Table A.4 lists byte-oriented, bit-oriented, and literal and control operations.

Byte-oriented instructions

'f' represents a file register designator and 'd' represents a destination designator. The file register designator specifies which file register is to be used by the instruction.

The destination designator specifies where the result of the operation is to be placed. If 'd' is 0, the result is placed in the W register. If 'd' is 1, the result is placed in the file register specified by the instruction.

Bit-oriented instructions

'b' represents a bit field designator which selects the number of the bit affected by the operation, while 'f' represents the address of the file in which the bit is located.

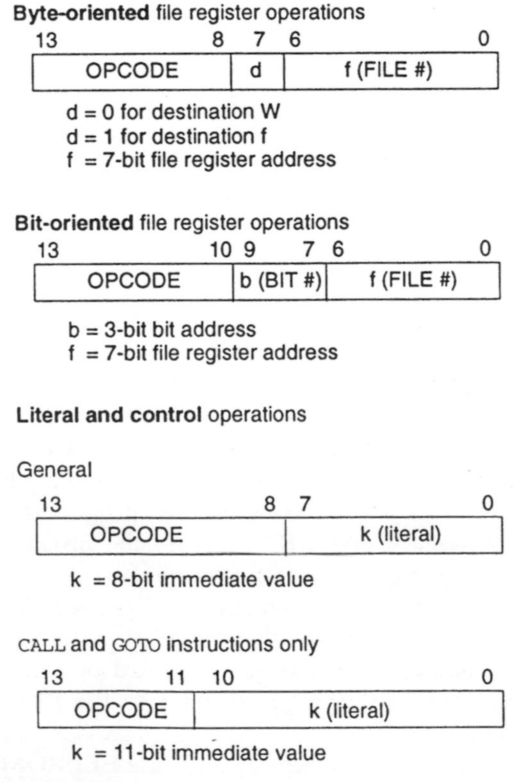

Figure A.29 General format for instructions.

Literal and control operations

'k' represents an 8- or 11-bit constant or literal value.

All instructions are executed within a single instruction cycle, unless a conditional test is true or the program counter is changed as a result of the instruction. The execution takes two instruction cycles with the second cycle executed as a NOP. Each cycle consists of four oscillator periods. Thus, for an oscillator frequency of 4 MHz, the normal instruction execution time is 1 μs. The instruction execution time is 2 μs for program branches.

Table A.4 lists the instructions recognized by Microchip's assembler (MPASM). Figure A.29 shows the three general formats of instructions.

Table A.4 Instruction set summary

Mnemonic, Operands		*Description*	*Cycles*	*14-Bit Opcode*				*Status Affected*	*Notes*
				MSb			*LSb*		
ADDWF	f, d	Add W and f	1	00	0111	dfff	ffff	C, DC, Z	1, 2
ANDWF	f, d	AND W with f	1	00	0101	dfff	ffff	Z	1, 2
CLRF	f	Clear f	1	00	0001	1ff	ffff	Z	2
CLRW	—	Clear W	1	00	0001	0000	0011	Z	
COMF	f, d	Complement f	1	00	1001	dfff	ffff	Z	1, 2
DECF	f, d	Decrement f	1	00	0011	dfff	ffff	Z	1, 2
DECFSZ	f, d	Decrement f, Skip if 0	1(2)	00	1011	dfff	ffff	None	1, 2, 3
INCF	f, d	Increment f	1	00	1010	dfff	ffff	Z	1, 2
INCFSZ	f, d	Increment f, Skip if 0	1(2)	00	1111	dfff	ffff	None	1, 2, 3
IORWF	f, d	Inclusive OR W with f	1	00	0100	dfff	ffff	Z	1, 2
MOVF	f, d	Move f	1	00	1000	dfff	ffff	Z	1, 2
MOVWF	f	Move W to f	1	00	0000	1fff	ffff	None	
NOP	—	No operation	1	00	0000	0xx0	0000	None	
RLF	f, d	Rotate left f through carry	1	00	1101	dfff	ffff	C	1, 2
RRF	f, d	Rotate f through carry	1	00	1100	dfff	ffff	C	1, 2
SUBWF	f, d	Subtract W from f	1	00	0010	dfff	ffff	C, DC, Z	1, 2
SWAPF	f, d	Swap nibbles in f	1	00	1110	dfff	ffff	None	1, 2
XORWF	f, d	Exclusive OR W with f	1	00	0110	ffff	ffff	Z	1, 2
Bit-oriented file register operations									
BCF	f, b	Bit Clear f	1	01	00bb	bfff	ffff	None	1, 2
BSF	f, b	Bit Set f	1	01	01bb	bfff	ffff	None	1, 2
BTFSC	f, b	Bit Test f, Skip if Clear	1(2)	01	10bb	bfff	ffff	None	3
BTFSS	f, b	Bit Test f, Skip if Set	1(2)	01	11bb	bfff	ffff	None	3
Literal and control operations									
ADDLW	k	Add literal and W	1	11	111x	kkkk	kkkk	C, DC, Z	
ANDLW	k	AND literal with W	1	11	1001	kkkk	kkkk	Z	
CALL	k	Call subroutine	2	10	0kkk	kkkk	kkkk		
CLRWDT	—	Clear Watchdog Timer	1	00	0000	0110	0100	$\overline{TO}$, $\overline{PD}$	
GOTO	k	Go to address	2	10	1kkk	kkkk	kkkk	None	
IORLW	k	Inclusive OR literal with W	1	11	1000	kkkk	kkkk	Z	
MOVLW	k	Move literal to W	1	11	00xx	kkkk	kkkk	None	
RETFIE	—	Return from interrupt	2	00	0000	0000	1001	None	
RETLW	k	Return from literal in W	2	11	01xx	kkkk	kkkk	None	
RETURN	—	Return from subroutine	2	00	0000	0000	1000	None	
SLEEP	—	Go into standby mode	1	00	0000	0110	0011	$\overline{TO}$, $\overline{PD}$	
SUBLW	k	Subtract W from literal	1	11	110x	kkkk	kkkk	C, DC, Z	
XORLW	k	Exclusive OR literal with W	1	11	1010	kkkk	kkkk	Z	

Notes: (1) When an I/O register is modified as a function of itself (i.e., `MOVF PORTB, 1`), the value used will be that value present on the pins themselves. For example, if the data latch is '1' for a pin configured as input and is driven low by an external device, the data will be written back with a '0'.

(2) If this instruction is executed on the TMRO register (and, where applicable, d = 1), the prescaler will be cleared if assigned to the TMRO.

(3) If Program Counter (PC) is modified or a conditional test is true, the instruction requires two cycles. The second cycle is executed as a NOP.

Note: To maintain upward compatibility with future PIC 16FXX products, do not use the `OPTION` and `TRIS` instructions.

All examples use the following format to represent a hexadecimal number, 0xhh, where h signifies a hexadecimal digit.

PIC16F8X

9.1 Instruction Descriptions

ADDLW **Add Literal and W**

Syntax:	[*label*] ADDLW k
Operands:	$0 \leq k \leq 255$
Operation:	(W) + k → (W)
Status Affected:	C, DC, Z
Encoding:	`11` `111x` `kkkk` `kkkk`
Description:	The contents of the W register are added to the eight bit literal 'k' and the result is placed in the W register.
Words:	1
Cycles:	1

Q Cycle Activity:

Q1	Q2	Q3	Q4
Decode	Read literal 'k'	Process data	Write to W

ANDLW **AND Literal with W**

Syntax:	[*label*] ANDLW k
Operands:	$0 \leq k \leq 255$
Operation:	(W) .AND. (k) → (W)
Status Affected:	Z
Encoding:	`11` `1001` `kkkk` `kkkk`
Description:	The contents of W register are AND'ed with the eight bit literal 'k'. The result is placed in the W register.
Words:	1
Cycles:	1

Q Cycle Activity:

Q1	Q2	Q3	Q4
Decode	Read literal "k"	Process data	Write to W

ADDWF **Add W and f**

Syntax:	[*label*] ADDWF f,d
Operands:	$0 \leq f \leq 127$ $d \in [0,1]$
Operation:	(W) + (f) → (destination)
Status Affected:	C, DC, Z
Encoding:	`00` `0111` `dfff` `ffff`
Description:	Add the contents of the W register with register 'f'. If 'd' is 0 the result is stored in the W register. If 'd' is 1 the result is stored back in register 'f'.
Words:	1
Cycles:	1

Q Cycle Activity:

Q1	Q2	Q3	Q4
Decode	Read register 'f'	Process data	Write to destination

ANDWF **AND W with f**

Syntax:	[*label*] ANDWF f,d
Operands:	$0 \leq f \leq 127$ $d \in [0,1]$
Operation:	(W) .AND. (f) → (destination)
Status Affected:	Z
Encoding:	`00` `0101` `dfff` `ffff`
Description:	AND the W register with register 'f'. If 'd' is 0 the result is stored in the W register. If 'd' is 1 the result is stored back in register 'f'.
Words:	1
Cycles:	1

Q Cycle Activity:

Q1	Q2	Q3	Q4
Decode	Read register 'f'	Process data	Write to destination

PIC16F8X

BCF — Bit Clear f

Syntax: [*label*] BCF f,b

Operands: 0 ≤ f ≤ 127
0 ≤ b ≤ 7

Operation: 0 → (f<b>)

Status Affected: None

Encoding:

01	00bb	bfff	ffff

Description: Bit 'b' in register 'f' is cleared.

Words: 1

Cycles: 1

Q Cycle Activity:

Q1	Q2	Q3	Q4
Decode	Read register 'f'	Process data	Write register 'f'

BSF — Bit Set f

Syntax: [*label*] BSF f,b

Operands: 0 ≤ f ≤ 127
0 ≤ b ≤ 7

Operation: 1 → (f<b>)

Status Affected: None

Encoding:

01	01bb	bfff	ffff

Description: Bit 'b' in register 'f' is set.

Words: 1

Cycles: 1

Q Cycle Activity:

Q1	Q2	Q3	Q4
Decode	Read register 'f'	Process data	Write register 'f'

BTFSC — Bit Test, Skip if Clear

Syntax: [*label*] BTFSC f,b

Operands: 0 ≤ f ≤ 127
0 ≤ b ≤ 7

Operation: skip if (f<b>) = 0

Status Affected: None

Encoding:

01	10bb	bfff	ffff

Description: If bit 'b' in register 'f' is '1' then the next instruction is executed.
If bit 'b', in register 'f', is '0' then the next instruction is discarded, and a NOP is executed instead, making this a 2TCY instruction.

Words: 1

Cycles: 1(2)

Q Cycle Activity:

Q1	Q2	Q3	Q4
Decode	Read register 'f'	Process data	No-Operat ion

If Skip: (2nd Cycle)

Q1	Q2	Q3	Q4
No-Operat ion	No-Operati on	No-Opera tion	No-Operat ion

BTFSS — Bit Test f, Skip if Set

Syntax: [*label*] BTFSS f,b

Operands: 0 ≤ f ≤ 127
0 ≤ b < 7

Operation: skip if (f<b>) = 1

Status Affected: None

Encoding:

01	11bb	bfff	ffff

Description: If bit 'b' in register 'f' is '0' then the next instruction is executed.
If bit 'b' is '1', then the next instruction is discarded and a NOP is executed instead, making this a 2TCY instruction.

Words: 1

Cycles: 1(2)

Q Cycle Activity:

Q1	Q2	Q3	Q4
Decode	Read register 'f'	Process data	No-Operat ion

If Skip: (2nd Cycle)

Q1	Q2	Q3	Q4
No-Operat ion	No-Operati on	No-Opera tion	No-Operat ion

PIC16F8X

CALL — Call Subroutine

Syntax:	[*label*] CALL k
Operands:	$0 \le k \le 2047$
Operation:	(PC)+ 1→ TOS, k → PC<10:0>, (PCLATH<4:3>) → PC<12:11>
Status Affected:	None
Encoding:	10 | 0kkk | kkkk | kkkk
Description:	Call Subroutine. First, return address (PC+1) is pushed onto the stack. The eleven bit immediate address is loaded into PC bits <10:0>. The upper bits of the PC are loaded from PCLATH. CALL is a two cycle instruction.
Words:	1
Cycles:	2

Q Cycle Activity:

	Q1	Q2	Q3	Q4
1st Cycle	Decode	Read literal 'k', Push PC to Stack	Process data	Write to PC
2nd Cycle	No-Operation	No-Operation	No-Operation	No-Operation

CLRF — Clear f

Syntax:	[*label*] CLRF f
Operands:	$0 \le f \le 127$
Operation:	00h → (f) 1 → Z
Status Affected:	Z
Encoding:	00 | 0001 | 1fff | ffff
Description:	The contents of register 'f' are cleared and the Z bit is set.
Words:	1
Cycles:	1

Q Cycle Activity:

Q1	Q2	Q3	Q4
Decode	Read register 'f'	Process data	Write register 'f'

CLRW — Clear W

Syntax:	[*label*] CLRW
Operands:	None
Operation:	00h → (W) 1 → Z
Status Affected:	Z
Encoding:	00 | 0001 | 0xxx | xxxx
Description:	W register is cleared. Zero bit (Z) is set.
Words:	1
Cycles:	1

Q Cycle Activity:

Q1	Q2	Q3	Q4
Decode	No-Operation	Process data	Write to W

CLRWDT — Clear Watchdog Timer

Syntax:	[*label*] CLRWDT
Operands:	None
Operation:	00h → WDT 0 → WDT prescaler, 1 → $\overline{TO}$ 1 → $\overline{PD}$
Status Affected:	$\overline{TO}$, $\overline{PD}$
Encoding:	00 | 0000 | 0110 | 0100
Description:	CLRWDT instruction resets the Watchdog Timer. It also resets the prescaler of the WDT. Status bits $\overline{TO}$ and $\overline{PD}$ are set.
Words:	1
Cycles:	1

Q Cycle Activity:

Q1	Q2	Q3	Q4
Decode	No-Operation	Process data	Clear WDT Counter

PIC16F8X

COMF **Complement f**

Syntax: [*label*] COMF f,d

Operands: $0 \le f \le 127$
$d \in [0,1]$

Operation: $(\bar{f}) \rightarrow$ (destination)

Status Affected: Z

Encoding:

00	1001	dfff	ffff

Description: The contents of register 'f' are complemented. If 'd' is 0 the result is stored in W. If 'd' is 1 the result is stored back in register 'f'.

Words: 1

Cycles: 1

Q Cycle Activity:

Q1	Q2	Q3	Q4
Decode	Read register 'f'	Process data	Write to destination

DECF **Decrement f**

Syntax: [*label*] DECF f,d

Operands: $0 \le f \le 127$
$d \in [0,1]$

Operation: (f) - 1 $\rightarrow$ (destination)

Status Affected: Z

Encoding:

00	0011	dfff	ffff

Description: Decrement register 'f'. If 'd' is 0 the result is stored in the W register. If 'd' is 1 the result is stored back in register 'f'.

Words: 1

Cycles: 1

Q Cycle Activity:

Q1	Q2	Q3	Q4
Decode	Read register 'f'	Process data	Write to destination

DECFSZ **Decrement f, Skip if 0**

Syntax: [*label*] DECFSZ f,d

Operands: $0 \le f \le 127$
$d \in [0,1]$

Operation: (f) - 1 $\rightarrow$ (destination);
skip if result = 0

Status Affected: None

Encoding:

00	1011	dfff	ffff

Description: The contents of register 'f' are decremented. If 'd' is 0 the result is placed in the W register. If 'd' is 1 the result is placed back in register 'f'.
If the result is 1, the next instruction, is executed. If the result is 0, then a NOP is executed instead making it a 2TCY instruction.

Words: 1

Cycles: 1(2)

Q Cycle Activity:

Q1	Q2	Q3	Q4
Decode	Read register 'f'	Process data	Write to destination

If Skip: (2nd Cycle)

Q1	Q2	Q3	Q4
No-Operation	No-Operation	No-Operation	No-Operation

GOTO **Unconditional Branch**

Syntax: [*label*] GOTO k

Operands: $0 \le k \le 2047$

Operation: k $\rightarrow$ PC<10:0>
PCLATH<4:3> $\rightarrow$ PC<12:11>

Status Affected: None

Encoding:

10	1kkk	kkkk	kkkk

Description: GOTO is an unconditional branch. The eleven bit immediate value is loaded into PC bits <10:0>. The upper bits of PC are loaded from PCLATH<4:3>. GOTO is a two cycle instruction.

Words: 1

Cycles: 2

Q Cycle Activity:

	Q1	Q2	Q3	Q4
1st Cycle	Decode	Read literal 'k'	Process data	Write to PC
2nd Cycle	No-Operation	No-Operation	No-Operation	No-Operation

PIC16F8X

INCF — Increment f

Syntax: [*label*] INCF f,d

Operands: $0 \le f \le 127$; $d \in [0,1]$

Operation: (f) + 1 → (destination)

Status Affected: Z

Encoding:

00	1010	dfff	ffff

Description: The contents of register 'f' are incremented. If 'd' is 0 the result is placed in the W register. If 'd' is 1 the result is placed back in register 'f'.

Words: 1

Cycles: 1

Q Cycle Activity:

Q1	Q2	Q3	Q4
Decode	Read register 'f'	Process data	Write to destination

INCFSZ — Increment f, Skip if 0

Syntax: [*label*] INCFSZ f,d

Operands: $0 \le f \le 127$; $d \in [0,1]$

Operation: (f) + 1 → (destination), skip if result = 0

Status Affected: None

Encoding:

00	1111	dfff	ffff

Description: The contents of register 'f' are incremented. If 'd' is 0 the result is placed in the W register. If 'd' is 1 the result is placed back in register 'f'.
If the result is 1, the next instruction is executed. If the result is 0, a NOP is executed instead making it a 2TCY instruction.

Words: 1

Cycles: 1(2)

Q Cycle Activity:

Q1	Q2	Q3	Q4
Decode	Read register 'f'	Process data	Write to destination

If Skip: (2nd Cycle)

Q1	Q2	Q3	Q4
No-Operation	No-Operation	No-Operation	No-Operation

IORLW — Inclusive OR Literal with W

Syntax: [*label*] IORLW k

Operands: $0 \le k \le 255$

Operation: (W) .OR. k → (W)

Status Affected: Z

Encoding:

11	1000	kkkk	kkkk

Description: The contents of the W register is OR'ed with the eight bit literal 'k'. The result is placed in the W register.

Words: 1

Cycles: 1

Q Cycle Activity:

Q1	Q2	Q3	Q4
Decode	Read literal 'k'	Process data	Write to W

IORWF — Inclusive OR W with f

Syntax: [*label*] IORWF f,d

Operands: $0 \le f \le 127$; $d \in [0,1]$

Operation: (W) .OR. (f) → (destination)

Status Affected: Z

Encoding:

00	0100	dfff	ffff

Description: Inclusive OR the W register with register 'f'. If 'd' is 0 the result is placed in the W register. If 'd' is 1 the result is placed back in register 'f'.

Words: 1

Cycles: 1

Q Cycle Activity:

Q1	Q2	Q3	Q4
Decode	Read register 'f'	Process data	Write to destination

PIC16F8X

MOVF **Move f**

Syntax: [*label*] MOVF f,d

Operands: $0 \le f \le 127$
$d \in [0,1]$

Operation: (f) → (destination)

Status Affected: Z

Encoding:

00	1000	dfff	ffff

Description: The contents of register f is moved to a destination dependant upon the status of d. If d = 0, destination is W register. If d = 1, the destination is file register f itself. d = 1 is useful to test a file register since status flag Z is affected.

Words: 1

Cycles: 1

Q Cycle Activity:

Q1	Q2	Q3	Q4
Decode	Read register 'f'	Process data	Write to destination

MOVLW **Move Literal to W**

Syntax: [*label*] MOVLW k

Operands: $0 \le k \le 255$

Operation: k → (W)

Status Affected: None

Encoding:

11	00xx	kkkk	kkkk

Description: The eight bit literal 'k' is loaded into W register. The don't cares will assemble as 0's.

Words: 1

Cycles: 1

Q Cycle Activity:

Q1	Q2	Q3	Q4
Decode	Read literal 'k'	Process data	Write to W

MOVWF **Move W to f**

Syntax: [*label*] MOVWF f

Operands: $0 \le f \le 127$

Operation: (W) → (f)

Status Affected: None

Encoding:

00	0000	1fff	ffff

Description: Move data from W register to register 'f'.

Words: 1

Cycles: 1

Q Cycle Activity:

Q1	Q2	Q3	Q4
Decode	Read register 'f'	Process data	Write register 'f'

NOP **No Operation**

Syntax: [*label*] NOP

Operands: None

Operation: No operation

Status Affected: None

Encoding:

00	0000	0xx0	0000

Description: No operation.

Words: 1

Cycles: 1

Q Cycle Activity:

Q1	Q2	Q3	Q4
Decode	No-Operation	No-Operation	No-Operation

PIC16F8X

OPTION	Load Option Register
Syntax:	[*label*] OPTION
Operands:	None
Operation:	(W) → OPTION
Status Affected:	None
Encoding:	00 | 0000 | 0110 | 0010
Description:	The contents of the W register are loaded in the OPTION register. This instruction is supported for code compatibility with PIC16C5X products. Since OPTION is a readable/writable register, the user can directly address it.
Words:	1
Cycles:	1
Example	

To maintain upward compatibility with future PIC16CXX products, do not use this instruction.

RETLW	Return with Literal in W
Syntax:	[*label*] RETLW k
Operands:	$0 \leq k \leq 255$
Operation:	k → (W); TOS → PC
Status Affected:	None
Encoding:	11 | 01xx | kkkk | kkkk
Description:	The W register is loaded with the eight bit literal 'k'. The program counter is loaded from the top of the stack (the return address). This is a two cycle instruction.
Words:	1
Cycles:	2

Q Cycle Activity:	Q1	Q2	Q3	Q4
1st Cycle	Decode	Read literal 'k'	No-Operation	Write to W, Pop from the Stack
2nd Cycle	No-Operation	No-Operation	No-Operation	No-Operation

RETFIE	Return from Interrupt
Syntax:	[*label*] RETFIE
Operands:	None
Operation:	TOS → PC, 1 → GIE
Status Affected:	None
Encoding:	00 | 0000 | 0000 | 1001
Description:	Return from Interrupt. Stack is POPed and Top of Stack (TOS) is loaded in the PC. Interrupts are enabled by setting Global Interrupt Enable bit, GIE (INTCON<7>). This is a two cycle instruction.
Words:	1
Cycles:	2

Q Cycle Activity:	Q1	Q2	Q3	Q4
1st Cycle	Decode	No-Operation	Set the GIE bit	Pop from the Stack
2nd Cycle	No-Operation	No-Operation	No-Operation	No-Operation

RETURN	Return from Subroutine
Syntax:	[*label*] RETURN
Operands:	None
Operation:	TOS → PC
Status Affected:	None
Encoding:	00 | 0000 | 0000 | 1000
Description:	Return from subroutine. The stack is POPed and the top of the stack (TOS) is loaded into the program counter. This is a two cycle instruction.
Words:	1
Cycles:	2

Q Cycle Activity:	Q1	Q2	Q3	Q4
1st Cycle	Decode	No-Operation	No-Operation	Pop from the Stack
2nd Cycle	No-Operation	No-Operation	No-Operation	No-Operation

PIC16F8X

RLF — Rotate Left f through Carry

Syntax: [*label*] RLF f,d

Operands: $0 \le f \le 127$
$d \in [0,1]$

Operation: See description below

Status Affected: C

Encoding:

00	1101	dfff	ffff

Description: The contents of register 'f' are rotated one bit to the left through the Carry Flag. If 'd' is 0 the result is placed in the W register. If 'd' is 1 the result is stored back in register 'f'.

C ← Register f

Words: 1

Cycles: 1

Q Cycle Activity:

Q1	Q2	Q3	Q4
Decode	Read register 'f'	Process data	Write to destination

SLEEP

Syntax: [*label*] SLEEP

Operands: None

Operation: 00h → WDT,
0 → WDT prescaler,
1 → $\overline{TO}$,
0 → $\overline{PD}$

Status Affected: $\overline{TO}$, $\overline{PD}$

Encoding:

00	0000	0110	0011

Description: The power-down status bit, $\overline{PD}$ is cleared. Time-out status bit, $\overline{TO}$ is set. Watchdog Timer and its prescaler are cleared.
The processor is put into SLEEP mode with the oscillator stopped. See Section 14.8 for more details.

Words: 1

Cycles: 1

Q Cycle Activity:

Q1	Q2	Q3	Q4
Decode	No-Operation	No-Operation	Go to Sleep

RRF — Rotate Right f through Carry

Syntax: [*label*] RRF f,d

Operands: $0 \le f \le 127$
$d \in [0,1]$

Operation: See description below

Status Affected: C

Encoding:

00	1100	dfff	ffff

Description: The contents of register 'f' are rotated one bit to the right through the Carry Flag. If 'd' is 0 the result is placed in the W register. If 'd' is 1 the result is placed back in register 'f'.

C → Register f

Words: 1

Cycles: 1

Q Cycle Activity:

Q1	Q2	Q3	Q4
Decode	Read register 'f'	Process data	Write to destination

SUBLW — Subtract W from Literal

Syntax: [*label*] SUBLW k

Operands: $0 \le k \le 255$

Operation: k - (W) → (W)

Status Affected: C, DC, Z

Encoding:

11	110x	kkkk	kkkk

Description: The W register is subtracted (2's complement method) from the eight bit literal 'k'. The result is placed in the W register.

Words: 1

Cycles: 1

Q Cycle Activity:

Q1	Q2	Q3	Q4
Decode	Read literal 'k'	Process data	Write to W

PIC16F8X

SUBWF	Subtract W from f			
Syntax:	[*label*] SUBWF f,d			
Operands:	0 ≤ f ≤ 127 d ∈ [0,1]			
Operation:	(f) - (W) → (destination)			
Status Affected:	C, DC, Z			
Encoding:	00	0010	dfff	ffff
Description:	Subtract (2's complement method) W register from register 'f'. If 'd' is 0 the result is stored in the W register. If 'd' is 1 the result is stored back in register 'f'.			
Words:	1			
Cycles:	1			
Q Cycle Activity:	Q1	Q2	Q3	Q4
	Decode	Read register 'f'	Process data	Write to destination

SWAPF	Swap Nibbles in f			
Syntax:	[*label*] SWAPF f,d			
Operands:	0 ≤ f ≤ 127 d ∈ [0,1]			
Operation:	(f<3:0>) → (destination<7:4>), (f<7:4>) → (destination<3:0>)			
Status Affected:	None			
Encoding:	00	1110	dfff	ffff
Description:	The upper and lower nibbles of register 'f' are exchanged. If 'd' is 0 the result is placed in W register. If 'd' is 1 the result is placed in register 'f'.			
Words:	1			
Cycles:	1			
Q Cycle Activity:	Q1	Q2	Q3	Q4
	Decode	Read register 'f'	Process data	Write to destination

TRIS	Load TRIS Register			
Syntax:	[*label*] TRIS f			
Operands:	5 ≤ f ≤ 7			
Operation:	(W) → TRIS register f;			
Status Affected:	None			
Encoding:	00	0000	0110	0fff
Description:	The instruction is supported for code compatibility with the PIC16C5X products. Since TRIS registers are readable and writable, the user can directly address them.			
Words:	1			
Cycles:	1			
Example				

To maintain upward compatibility with future PIC16CXX products, do not use this instruction.

XORLW	Exclusive OR Literal with W			
Syntax:	[*label*] XORLW k			
Operands:	0 ≤ k ≤ 255			
Operation:	(W) .XOR. k → (W)			
Status Affected:	Z			
Encoding:	11	1010	kkkk	kkkk
Description:	The contents of the W register are XOR'ed with the eight bit literal 'k'. The result is placed in the W register.			
Words:	1			
Cycles:	1			
Q Cycle Activity:	Q1	Q2	Q3	Q4
	Decode	Read literal 'k'	Process data	Write to W

PIC16F8X

XORWF	**Exclusive OR W with f**
Syntax:	[*label*] XORWF f,d
Operands:	$0 \le f \le 127$ $d \in [0,1]$
Operation:	(W) .XOR. (f) → (destination)
Status Affected:	Z
Encoding:	00 | 0110 | dfff | ffff
Description:	Exclusive OR the contents of the W register with register 'f'. If 'd' is 0 the result is stored in the W register. If 'd' is 1 the result is stored back in register 'f'.
Words:	1
Cycles:	1
Q Cycle Activity:	

Q1	Q2	Q3	Q4
Decode	Read register 'f'	Process data	Write to destination

Appendix B
DIZI-2 Demonstration Board

B.1 Design, Construction and Testing
B.2 CR-ADC Input and EEPROM Storage
B.2 LOCK Application, Design and List File

A circuit was required to demonstrate a range of basic microcontroller programming techniques via a set of simple applications for the PIC 16F84. The DIZI (display, buzzer and interrupt) board was designed to allow the special hardware features of the PIC chip to be exercised, including interrupts, timer and EEPROM memory. In-circuit programming was not incorporated, in order to emphasize the stand-alone operation of the microcontroller. The chip would be programmed separately and then physically transferred to the target system. The enhanced DIZI-2 board described in this appendix incorporates an on-board battery supply, a finger pot to provide an analogue input and hardware switch debouncing to improve the reliability of the push button operation.

The circuit is built on a 100 × 100 mm piece of stripboard which has copper tracks for making the component connections on a standard 0.1-inch grid. The design includes a 2 × 1.5 V battery pack on the board. The power is switched on via a non-latching push button so that it cannot be left on accidentally, and thereby exhaust the batteries; it must be held on manually while the circuit is in operation.

The basic demonstration programs in Part B of this book can be run on the DIZI-2 board, while the motor programs must be run on the MOTA demo board. The simple introductory circuits in Part B and the motor board can be constructed using the same techniques as will be described for the DIZI board. The reader who is inexperienced in prototype construction is encouraged to attempt these tasks. The binary output counts from the BINx programs will be seen on the corresponding LED segments of the DIZI display, although the binary number is not displayed so clearly as it would be on a set of eight discrete LEDs, or a LED bar graph module.

B.1 Design, Construction and Testing

B.1.1 DIZI-2 Circuit Design (Fig. B.1)

A seven-segment display allows decimal and hexadecimal digits to be shown. A range of applications with a numerical output can thus be demonstrated, for example, the electronic

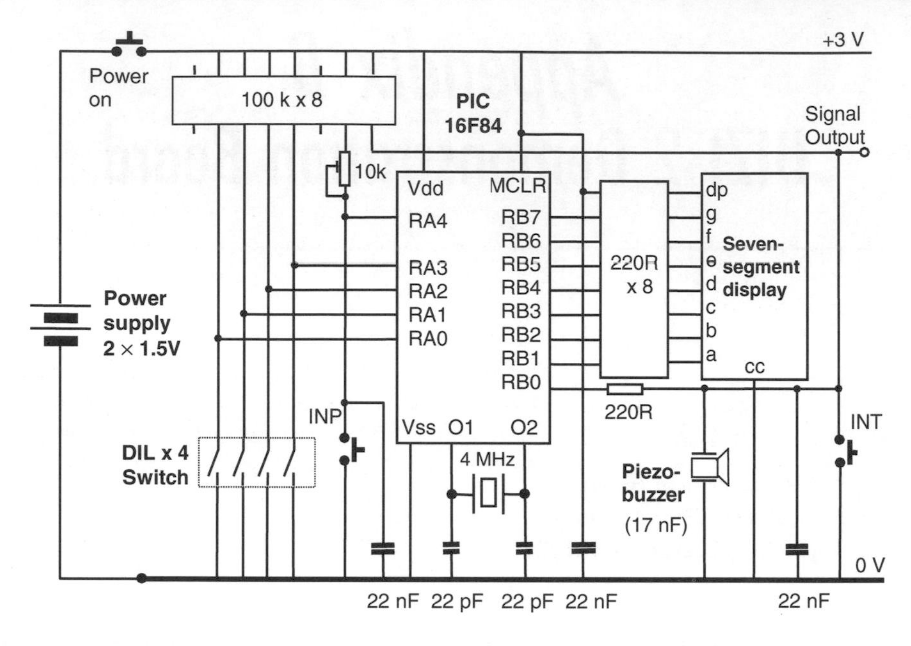

Figure B.1 DIZI-2 board circuit diagram.

DICE in Chapter 12 and the LOCK application detailed below. Port B has eight I/O bits; seven are used for the LED display, leaving RB0 free for use as both an audio output and push button (interrupt) input. A small audio transducer, a piezo electric buzzer, provides a simple and effective way of monitoring audio output frequencies or generating status signals.

Port A has five pins. RA4 can be used as an input to the TMR0 counter, so this was allocated as another push button input. A 4-bit switch bank is useful for setting coded inputs (for example, BCD inputs for the LOCK application), so RA0 to RA3 were allocated for this purpose. The switch and push button inputs have 100 k pull-up resistors, and the push buttons have 22 μF debouncing capacitors. These are often fitted to such inputs because, when a switch closes, the metal contacts may bounce open again several times before finally closing. The CR network prevents this causing multiple transitions on the logic signal input to which the switch is connected, because after the capacitor has quickly discharged on the first contact, it must recharge via the 100 k. This takes a relatively long time, preventing the voltage from jumping back to the high level when the contacts re-open. By the same process, the CR network also ensures a smooth transition from low to high logic levels when the push button is released.

In addition, the CR network on RA4 was modified with a potentiometer (pot) connected as a variable resistance in series with the 100 k, so that it could also be used to demonstrate the analogue input process in the LOCK program.

Developing circuits for microcontrollers is not always too difficult, because the same circuit elements can often be re-used in different designs. Thus, in the DIZI circuit, the switch inputs, display outputs and the clock circuit are standard arrangements of components. Remember, however, that, unlike the PIC, many microcontroller outputs cannot drive LEDs directly, but need a current driver stage inserted between each chip output and LED (Fig. B.1).

B.1.2 DIZI-2 Board Layout (Fig. B.2)

The layout of a PCB or prototype circuit is derived from the circuit diagram. The pins on DIL (dual in-line) chips are spaced 0.1 inch apart, so the circuits must be laid out on a 0.1-inch grid. This can be a bit inconvenient in Europe, because component dimensions are quoted in millimetres (0.1 inch = 2.54 mm). When the pin-out of each component has been established by reference to the data sheet or catalogue information, the connections can be mapped out on a square grid on paper. Alternatively, it is not too difficult to use the basic drawing tools in a wordprocessor to do the same job. Examples of such connection diagrams are given in the text.

The board is viewed from the front (component) side, with the tracks on the back shown vertically. The components are numbered for reference to the parts list below. The chips are all placed in the same orientation, so that pin 1 is bottom left. The seven-segment display has the decimal point bottom right. The ICs must obviously be fitted across the tracks, so that their pin connections are separated. The PIC chip should be fitted in a socket so that it can be removed for programming.

Horizontal links of tinned copper wire (TCW) complete the connections required. A solder joint is shown as a solid black dot. The broader solid lines indicate a continuous link across the tracks on the back of the board, where a set of adjacent tracks must be connected. Where required, the tracks are cut with a hand drill; these positions are shown as small white circles. The tracks must also be cut between the opposite pairs of pins on each DIL component, and, in this case, under the clock circuit capacitors (11).

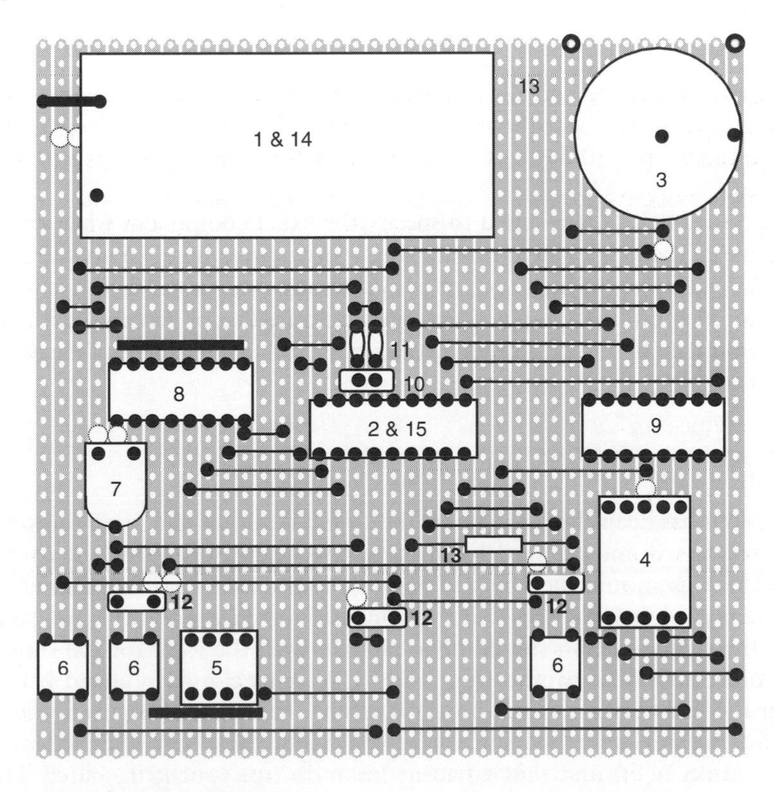

Figure B.2 DIZI stripboard layout.

Table B.1 DIZI board parts list

Layout Ref.	*Description*	*Maplin Order code*
1	Battery Box, 2 × AA cells, PCB mounting	CL17T
2	Microcontroller, PIC 16F84-04/P (Arizona Microchip)	NT48C
3	Piezo Electric Sounder, PCB mounting	JH24B
4	Seven Segment LED Display, 0.5″, Common Cathode	FR41U
5	Piano DIL Switch, 4-way	PM75S
6	Tactile Switch, PCB mounting (3 off)	PM46A
	Caps for above: Red	PM49D
	Blue	PM50E
	Yellow	PM51F
7	Preset Potentiometer, 10 k, H-mount	DT44X
	Short Shaft for above	DT48C
8	DIL Isolated Resistor Network, 100 k × 8	DL93B
9	DIL Isolated Resistor Network, 220 R × 8	DL84F
10	Quartz Crystal, General Purpose, 4 MHz	FY82D
11	Capacitor, 22 pF, Ceramic (2 off)	WX48C
12	Capacitor, 22 nF, Polyester (3 off)	DT96E
13	Stripboard, SRBP 3939 100 mm × 100 mm	JP49D
14	Batteries, 1.5 V, size AA, Duracell (2 off)	JY48C
15	18-pin DIL IC Socket	HQ76H

A computer drawing method allows component positioning to be easily adjusted so that the minimum area of stripboard is used. However, with experience, the circuit may be built directly onto the board without necessarily drawing the layout, perhaps with some wastage of board area.

A parts list (Table B.1) is required to specify the exact component when ordering from a suitable supplier. The availability of components varies over time, so updating is sometimes necessary. For example, the finger preset pot originally used in the design is no longer listed by the UK supplier quoted, Maplin Electronics, and had to be replaced with an equivalent part. The layout then had to be amended because the pin arrangement of the new component was different.

B.1.3 Construction

When the layout has been checked against the circuit diagram, the main components can be inserted in the board and retained by, if necessary, slightly bending the corner pins outwards. All the pins should then be soldered to the tracks using the minimum amount of solder necessary, whilst ensuring that the joint is covered evenly with no cavities. At the same time, the soldering iron should be in contact with the joint for the shortest possible time, to avoid component overheating. The TCW links can be retained before soldering by bending the ends towards each other. If a very neat job is required, one end can be soldered and the link stretched slightly before fixing the other end, to ensure that the link has no kinks in it, and that adjacent links do not touch; insulated TCW may be used on longer links if necessary. The tracks should then be cut where necessary, and the track side brushed with a small stiff brush to clear any debris. Rake between the tracks with a small screwdriver or knife to ensure that there are no short circuits left between adjacent tracks and solder joints.

B.1.4 Static Testing

Thoroughly re-inspect the board for correct connections, and to ensure that there are no debris, solder splashes or whiskers or dry joints. With the batteries not yet fitted, check with a multimeter that there is no short circuit between the power supplies. Fit the batteries, but not the PIC chip, and hold down the power button. The display decimal point should light. Check the supply voltages on the supply tracks and PIC socket pins: pin 5 = 0 V and pin 14 = +3 V. Check that the voltages at the PIC inputs change correctly as the switches are toggled. A DMM (digital multimeter) or oscilloscope is required for this test, because of the high impedance of the pull-up resistors. Connect a temporary link between pin 14 (+3 V) on the PIC IC socket and each PIC output in turn, RB0–RB7. The piezo buzzer should produce an audible 'tick' and the LED segments should light.

B.1.5 Test Program

To complete the testing of the DIZI-2 board, a program should be blown into the PIC which exercises all the hardware, while remaining as simple as possible so that there is no question of the software being faulty. A suitable program is listed on page 288 (Program B.1); it does not test the analogue input operation, which will be covered later. When this program has been loaded, the following test procedure will confirm correct hardware operation.

Step	*Test*	*Result*
1	Power button on	Decimal point on
2	Button B pressed and released	All display segments on
3	Button A pressed and released	Buzzer sounds
4	Operate DIL switches	Segments a, b, c, d change

If faults are found, it is quite possible that there are still short circuits on the board. Check also that all the tracks have been cut as required, and that all connections are correct.

B.2 CR-ADC Input and EEPROM Storage

B.2.1 Analogue to Digital Conversion

One feature of the DIZI board not described in the main text is the analogue input. A similar input is available on the MOTA board, so the method for using it will now be explained. Some PIC chips, and other microcontrollers, have built in analogue to digital converters (ADCs), which allow analogue voltages to be converted to binary form for input to the processor. An ADC would be needed, for example, if a temperature is to be measured by the controller in a process system. The general block diagram for a counting ADC is shown in Fig. B.3.

Program B.1 DIZI board test program

```
; diz1.asm

; Test DIZI hardware ..................................

            GOTO      inter     ; jump over delay

; Delay Subroutine ....................................

delay       MOVLW     0FF       ; Load FF
            MOVWF     0C        ; into counter
down        DECFSZ    0C        ; and decrement
            GOTO      down      ; until zero
            RETURN

; Check Interrupt Button ..............................

inter       BTFSC     06,0      ; Test Button RB0
            GOTO      inter     ; until pressed

; Check Display .......................................

            MOVLW     00        ; Set PortB bits
            TRIS      06        ; as outputs
            MOVLW     0FF       ; Switch on all
            MOVWF     06        ; display segments

; Check Input Button ..................................

input       BTFSC     05,4      ; Test Button RA4
            GOTO      input     ; until pressed

; Check DIP Switches and Buzzer .......................

again       MOVF      05,W      ; get DIL input &
            MOVWF     06        ; send to display
            RLF       06        ; rotate bits left

            BSF       06,0      ; set buzzer high
            CALL      delay     ; delay about 1ms
            BCF       06,0      ; reset buzzer low
            CALL      delay     ; delay about 1ms

            GOTO      again     ; and keep going..
            END                 ; End of code
```

The 8-bit ADC shown converts the analogue voltage present at its input to an 8-bit binary number, which means it can detect 256 different voltage levels. If the input range is set to, say, 0–2.55 V, then 255 steps of 10 mV can be detected. This type of ADC uses an 8-bit binary up-counter and digital to analogue converter (DAC). When the 'Start conversion' signal goes active, the counter starts counting up. The DAC converts the binary number in the counter to a corresponding voltage. This voltage is compared with the input voltage to be measured. The output of the voltage comparator changes when the DAC output exceeds the input voltage. The counter is then stopped, because it now

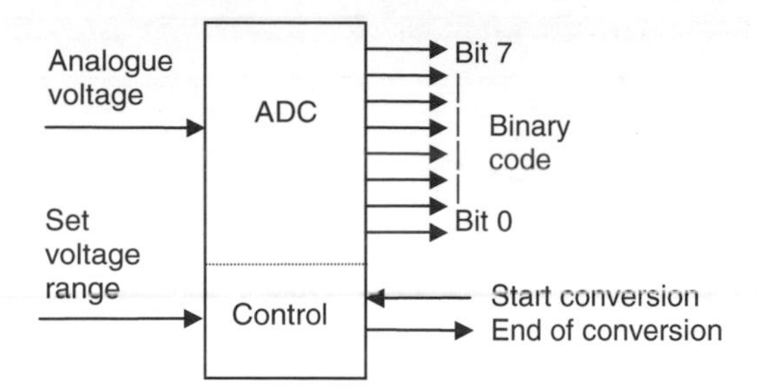

Figure B.3 General analogue to digital converter.

contains the binary equivalent of the input voltage. The comparator output change also provides the 'End of conversion' signal to the processor to indicate completion of the conversion count.

B.2.2 CR-ADC

The hardware for the standard ADC above is fairly complex, while the control algorithm is relatively simple. An alternative approach to acquiring analogue inputs is to use simpler hardware with a more complex software conversion procedure. The CR-ADC is based on the measurement, using a counter register, of the rise time in a CR network connected to the processor system input. The CR converter will generally be slower and less accurate than a hardware-based ADC, but may be quite adequate in simple applications, where a dedicated ADC port is not available.

The components connected to RA4 in the DIZI circuit are shown in Fig. B.4. The capacitor charging curve (Fig. B.5) shows the time constant of the circuit as 2.3 ms, assuming that the pot is set midway. This is the time taken to reach 63% of the final value (3 V) as the capacitor C charges via R. The PIC chip is a CMOS device, so the voltage level at which an input changes from logic 0 to 1, the threshold voltage, is around half the supply voltage, 1.5 V. Therefore the time taken to reach this level, here called the charging

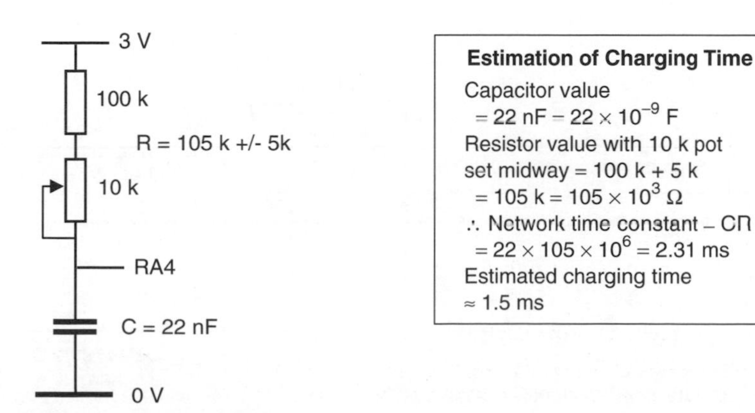

Figure B.4 CR conversion network.

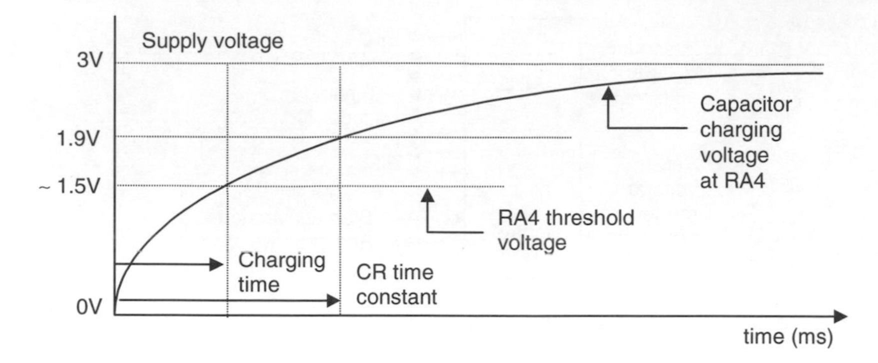

Figure B.5 CR network characteristic.

time, is estimated at 1.5 ms. This could be calculated more accurately from the formula for the charging of a capacitor, but as long as the circuit operation is consistent, it is not necessary for this application.

The resistance, R, varies between 100 k and 110 k, depending on the position of the pot. The variation in the pot value will produce a corresponding variation in the rise time of the circuit. The rise time can be measured by discharging the capacitor and then counting while the voltage rises back towards the threshold. The capacitor is discharged by setting RA4 as an output and then setting the port data bit to zero. RA4 is then reconfigured as an input and checked at fixed time intervals while a register is incremented. The count is stopped when RA4 goes high.

The waveform which will be seen at RA4 is illustrated in Fig. B.6 (not to scale). A register labelled PotVal is incremented, and RA4 checked, within a loop taking, in the LOCK program, 20 μs to execute. An adjustable delay routine allows the timing to be modified to suit the application and CR component values.

The result of the process is that a count is obtained which represents the setting of the pot. This could be converted to a resistance value if required, but in the LOCK program all we need is a variation in the displayed digit between 0 and 9, to allow the user to input a decimal combination. Therefore, the delay associated with the count was simply adjusted to give one decade on the display with one turn of the pot. Only the low four

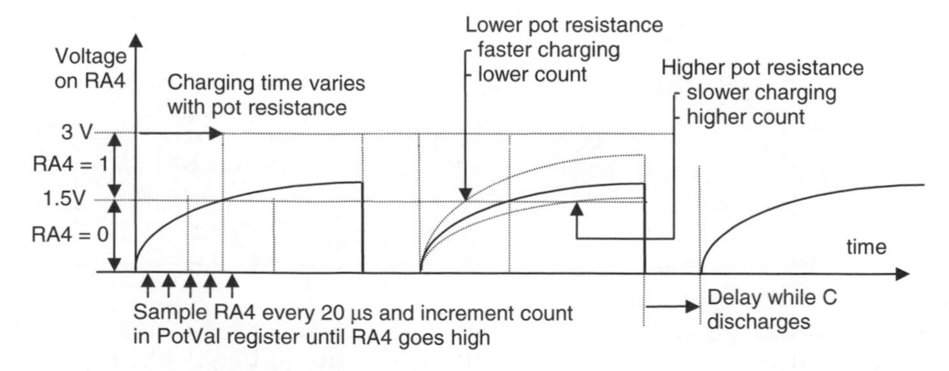

Figure B.6 Conversion waveform at RA4.

Table B.2 CR-ADC algorithm

```
Set RA4 as output
Clear RA4 to 0V to discharge C
Clear Counter Register
Set RA4 as input
Test RA4 while C charges through R:
     Increment Counter Register
     Delay 20µs
     Until RA4 = 1
Convert Count to Resistance or Pot Position
```

bits of the count were required, so any decade of values could be used. The upper end of the 4-bit range, hex numbers A to F, are displayed as '-'. These can be used as 'hidden' digits for extra security, if required.

A similar method can be used to operate the analogue input on the MOTA board (Fig. 13.2). In this case, RA2 and RA3 are connected to the CR network. Again, a pot is available to provide a manual analogue input, or an external voltage can be measured. To convert the pot value to a digital equivalent, C5 would be discharged via RA3, RA2 set high, and the count made until RA3 goes high. To measure a voltage at the analogue input, C5 would be discharged by RA3, and then RA3 monitored. The time taken for RA3 to go high will be inversely proportional to the voltage applied at the input. An external amplifier or attenuator might be needed to bring the voltage into the correct range, but care must be taken not to exceed the maximum rated voltages at RA3. The motor speed could then be controlled by a voltage source or sensor, or analogue feedback employed to implement closed loop control.

B.2.3 EEPROM Memory

Non-volatile read-and-write memory is very useful because data input by the user or acquired by the processor during its operation can be retained while the power is off. One important application area is data security and encryption. PIC devices are used, for example, in smart cards for controlling access to satellite television broadcast channels. The LOCK application illustrates this feature of the PIC 16F84 by using the EEPROM memory to store a 4-digit security code.

The PIC 16F84 has 64 bytes of EEPROM, with addresses 00–3F. The memory is accessed via EEDATA and EEADR in the special function register (SFR) set. The EEPROM address is loaded into EEADR, and the data byte to be stored in EEDATA. An artificially complex write initialization sequence is then executed to actually write to the EEPROM memory, using EECON1 and EECON2 page 1 SFRs. The sequence is designed to reduce the possibility of an accidental write to the EEPROM, because a high level of reliability is required for security applications. This code sequence is given in the data sheet and LOCK program listing.

The read sequence, for retrieving the data, is more straightforward. Using EECON1, the data in the address pointed to by EEADR is returned in EEDATA. For accessing sequences of locations, EEADR can be incremented directly.

B.3 LOCK Application, Design and List File

B.3.1 LOCK Application

In this demonstration application, a sequence of four decimal digits is stored in the PIC EEPROM memory from the DIL switch inputs. This sets the combination for the lock. To 'open' the lock, the pot is rotated, and the input decimal digits are displayed and entered. This simulates the rotary action of mechanical combination locks. If the sequence of four input digits matches those previously stored in EEPROM, a siren sound is made to indicate the opening of the lock.

In the actual application, a solenoid operated lock mechanism would be activated from this output, by replacing the siren sequence with an instruction to set bit RB0. A suitable current driver interface for the solenoid would be required. Only the Power button, Enter button, Digit Select pot and display would be accessible to the user in the final design. The hardware would need to be reconfigured so that the unit would appear as shown in Figure B.7 to the user. The DIL switch bank and its button for setting the entry code would be concealed.

B.3.2 Program Structure

The program contains the following blocks:

1. Declaration of register and bit labels
2. Initialization of registers
3. Sequence 1 – Store combination
4. Sequence 2 – Check combination
5. End 1 – Continuous siren output
6. End 2 – Sleep
7. Subroutine 1 – Display code table
8. Subroutine 2 – Variable delay
9. Subroutine 3 – Output one tone cycle
10. Subroutine 4 – Get digit from pot

The program blocks should be ordered in such a way that labels referenced have already been defined when the program is assembled. Thus, the subroutines should be placed *before* the main sequences in the source code. However, when actually developing the code, if you are working 'top down', the subroutines may actually be written after the main sequences. To place them correctly in the source code, they can then be inserted when editing, or cut and pasted later.

The program has two main sequences, for inputting and checking a combination, and two alternate endings. The processor goes to sleep after completion of the input sequence,

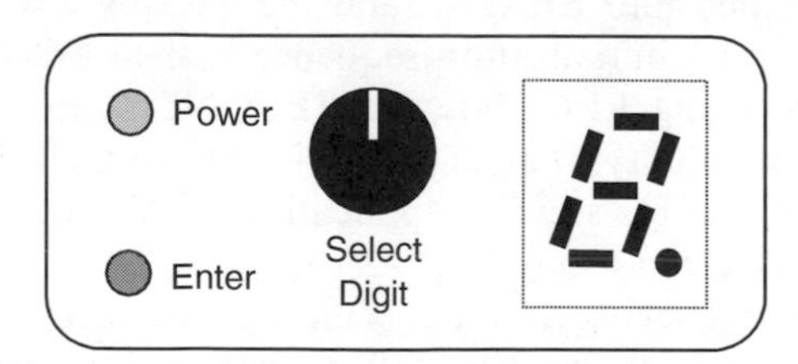

Figure B.7 Lock user interface.

or an incorrect digit match. The DIZI board must be re-powered to try again, as there are no other interrupts enabled to restart it. If the combination checks out correctly, the siren ending is used, which continues until the power goes off.

B.3.3 Pseudocode (*Table B.3*)

The program is outlined below using 'pseudocode', which is a text method of designing the program, which may be used instead of a flowchart. The pseudocode is developed in a word-processor or the program source code editor until the statements are detailed enough to be converted into assembly code statements. In this case, it must be written in a form which allows it to be easily converted to PIC assembly language. The program structure and logic can thus be worked out before attempting to write the source code itself. To use pseudocode effectively, the programmer must be reasonably expert in using the language syntax.

The conventions used in the pseudocode are as follows:

- Block structure applied
- Target hardware specified
- Register and bit labels defined
- User inputs included in the sequence
- Jump destination labels defined
- CALL [subname]
  – Call subroutine at address label 'subname'
- GOTO [addresslabel] UNTIL (condition)
  – implemented using Bit Test, Skip and GoTo operation
- (regname)
  = contents of register labelled 'regname'
- Program block type defined:
  INIT = Initialize
  MAIN = Main program
  SEQn = Sequence ending with GOTO
  ENDn = End operation
  SUBn = Subroutine, operationally receiving and/or returning values

B.3.4 LOCK List File (*Program B.2*)

The list file for the LOCK program contains the source code and machine code. If the reader wishes to test the program, the machine code (column 2 of the list file) can be entered directly into the program memory buffer in the programming software. This avoids the need to type in the source code, if the hex file itself is not available.

The source code file uses the following conventions:

- Full details of hardware and operation of application in source code
- SFR, user and bit labels defined in separate blocks for clarity
- Block and line comments in source code
- Lower case for address labels
- Upper case for instruction mnemonics and SFR labels
- Capitalization of user register labels
- Identification and separation of block types

Table B.3 Pseudocode for LOCK program

```
LOCK PROGRAM PSEUDOCODE                        MPB         29/8/99
**********************************************************
Hardware: DIZI PIC 16F84 Demo Board
************************************************
General Purpose Register Labels:
      0C = Period = Delay Period Preload Value
      0D = Count = Delay Counter
      0E = PotVal = Count from ADC conversion
      0F = DigVal = Low 4 bits of PotVal
User Bit Labels:
      butA (RA4 input) - Normally 1
      butB (RB0 input) - Normally 1
      buzO (RB0 output)
See Data Sheet for SFR Labels and addresses

{Power Button On}
INIT: Initialize Port B *********************************

      Port A defaults to inputs
      RA0 - RA3 = DIL Switches = 4-bit input
      RA4 = Input = butA = INP Button
      RB0 = Input = butB = INT Button
      RB1 - RB7 = Output = 7 Seg Display

MAIN: Select Set or Check Combination ********************

select          {Press Button A or B}
                If (butA)=0, GOTO [stocom]
                If (butB)=0, GOTO [checom]
                GOTO [select]

SEQ1: Store 4 digits in EEPROM, beep after each ***********

stocom          {Release Button A}
                CALL [delay] with (W)=FF
                GOTO [stocom] UNTIL (butA)=1

                Clear (EEADR)
getdil          {Set DIL Switches or Press A}
                Read (PORTA) into (W)
                Calc (W) AND 0F
                Store (W) in (EEDATA)
                CALL [codtab] with (W)=00-0F
                {Returns with '7SegCode' in (W)}
                Output (W) to (PORTB)
                GOTO [getdil] UNTIL (butA)=0

waita {Release Button A}
                GOTO [waita] UNTIL (butA)=1
                Store (EEDATA) in (EEADR)
                CALL [beep]
                Increment (EEADR) from 00 to 04
                GOTO [getdil] UNTIL (EEADR)=4
                CALL [beep]
                CALL [beep]
                GOTO [done]

                                                  continued...
```

```
SEQ2: Check 4 digits from pot for match *******************

checom          {Release Button B}
                CALL [delay] with (W)=FF

                GOTO [checom] UNTIL (butB)=1

                Clear (EEADR)
potin {Adjust Pot or Press Button B}
                CALL [getpot] for (DigVal)
                {Returns with (DigVal)=00-0F}
                GOTO [potin] UNTIL (butB)=0

                Read (EEDATA) at (EEADR)
                Compare (EEDATA) with (DigVal)
                If (Z)=0 GOTO [done]

waitb {Release Button B}
                GOTO [waitb] UNTIL (butB)=1
                CALL [beep]
                Increment (EEADR)
                GOTO [potin] UNTIL EEADR=4
                GOTO [siren]


END1: Sequences matches, sound siren **********************

      siren     CALL [beep]
                GOTO [siren]


END2: Digit compare failed, finish ************************

      done      Clear (PORTB)
                Sleep


SUBROUTINES ***********************************************

SUB1: Get Display Code
      Receives: Table Offset in W
      Returns: 7-Segment Display Code in W

codtab          Add (W) to (PCL)
                RETURN with '7SegCode' in (W)

SUB2: Variable Delay
      Receives: (Count) in W

delay           Load (Count) from (W)
                Decrement (Count) UNTIL (Count)=0
                RETURN

SUB3: Outputs one cycle of sound output
      Receives: (Period)

beep            Load (Period) with FF
                Set RB0 as Output
```

continued...

```
cycle           Set (BuzO)=1
                CALL [delay] with (Period) in W
                Set BuzO=0
                CALL [delay] with (Period) in W
                Decrement (Period) from FF to 00
                GOTO [cycle] UNTIL (Period)=0

                Reset RB0 as Input
                RETURN

SUB4: Get Pot Value using CR ADC method
      Returns: (DigVal)=00-0F

getpot          Set RA4 as Output
                Clear (RA4)
                CALL [delay] with (W)=FF
                Reset RA4 as Input

                Clear (PotVal)
check           Increment (PotVal) from 00 to XX
                CALL [delay] with (W)=3
                GOTO [check] UNTIL (RA4)=1

                (DigVal) = (PotVal) AND 0F
                CALL [codtab] with (DigVal)=00-0F
                RETURN

END OF LOCK PROGRAM ***************************************
```

Program B.2 LOCK program list file

```
MPASM 01.21 Released          LOCK.ASM          8-17-1999          20:27:58          PAGE 1

LOC OBJECT CODE     LINE SOURCE TEXT
  VALUE
                    00001 ; ***************************************************************
                    00002 ; LOCK.ASM                                                MPB 17/8/99
                    00003 ; ***************************************************************
                    00004 ;
                    00005 ; Four digit combination lock simulation demonstrates the hardware
                    00006 ; features of the DIZI demo board and the PIC 16F84.
                    00007 ;
                    00008 ; Hardware: DIZI Demo Board with PIC 16F84 (4MHz)
                    00009 ; Setup: RA0-RA3 DIL Switch Inputs
                    00010 ; RA4 Push Button Input / Analogue Input
                    00011 ; RB0 Push Button Input / Audio Output
                    00012 ; RB1-RB7 7-Segment Display Output
                    00013 ; Fuses: WDT off, PuT on, CP off
                    00014 ;
                    00015 ; Operation - - - - - - - - - - - - - - - - - - - - - - - - - - - - - -
                    00016 ;
                    00017 ; To set the combination, a sequence of 4 digits is input on the DIL
                    00018 ; piano switches; this is retained in the EEPROM when the power is off.
                    00019 ; To 'open' the lock, a sequence of 4 digits is input via
                    00020 ; the potentiometer. These are compared with the stored data, and
                    00021 ; an audio output generated to indicate the correct sequence.
                    00022 ; The processor halts if any digit fails to match, and the program
                    00023 ; must be restarted.
                    00024 ;
                    00025 ; To set a combination:
                    00026 ;        1.        Hold Power On Button
                    00027 ;        2.        Press Button A
                    00028 ;        3.        Set a digit on DIL switches and Press A - beeps
                    00029 ;        4.        Repeat step 3 for 3 more digits
                    00030 ;        5.        Release Power Button
                    00031 ;
                    00032 ; To check a combination:
                    00033 ;        1.        Hold Power On Button
                    00034 ;        2.        Press Button B
                    00035 ;        3.        Set a digit on pot and Press B - beeps if matched
                    00036 ;        4.        Repeat step 3 for 3 more digits
                    00037 ;                    - if digits all match, siren is sounded
                    00038 ;                    - if any digit fails to match, the processor halts
                    00039 ;        5.        Release Power Button
                    00040 ;
                    00041 ; ******************************************************************
                    00042        PROCESSOR 16F84        ; Processor Type Directive
                    00043 ; ******************************************************************
                    00044
                    00045 ; EQU: Special Function Register Equates...........................
                    00046
  0002              00047 PCL      EQU      02         ; Program Counter Low
  0005              00048 PORTA    EQU      05         ; Port A Data
  0006              00049 PORTB    EQU      06         ; Port B Data
  0003              00050 STATUS   EQU      03         ; Flags
  0008              00051 EEDATA   EQU      08         ; EEPROM Memory Data
  0009              00052 EEADR    EQU      09         ; EEPROM Memory Address
  0008              00053 EECON1   EQU      08         ; EEPROM Control Register 1
  0009              00054 EECON2   EQU      09         ; EEPROM Control Register 2
                    00055
                    00056 ; EQU: User Register Equates.......................................
                    00057
  000C              00058 Period   EQU      0C         ; Period of Output Sound
  000D              00059 Count    EQU      0D         ; Delay Down Counter
  000E              00060 PotVal   EQU      0E         ; Analogue Input Value
```

continued...

```
  000F          00061 DigVal  EQU       0F          ; Current Digit Value 00 to 09
                00062
                00063 ; EQU: SFR Bit Equates.............................................
                00064
  0005          00065 RP0     EQU       5           ; STATUS - Register Page Select
  0000          00066 RD      EQU       0           ; EECON1 - EEPROM Memory Read Byte
                                                      Initiate
  0001          00067 WR      EQU       1           ; EECON1 - EEPROM Memory Write Byte
                                                      Initiate
  0002          00068 WREN    EQU       2           ; EECON1 - EEPROM Memory Write Enable
  0002          00069 Z       EQU       2           ; STATUS - Zero Flag
                00070
                00071 ; EQU: User Bit Equates............................................
                00072
  0004          00073 butA    EQU       4           ; PORTA - RA4 Input Button
  0000          00074 butB    EQU       0           ; PORTB - RB0 Input Button
  0000          00075 buzO    EQU       0           ; PORTB - RB0 Output Buzzer
                00076
                00077 ; ****************************************************************
                00078
                00079 ; INIT: Initialize Port B (Port A defaults to inputs)
                00080
0000 3001       00081 start   MOVLW     001         ; RB0 = Input, RB1-RB7 = Outputs
0001 0066       00082         TRIS      PORTB       ; Set Data Direction
0002 0086       00083         MOVWF     PORTB       ; Clear Data
0003 286D       00084         GOTO      select      ; Select Combination Read or Write
                00085
                00086 ; SUBROUTINES ****************************************************
                00087
                00088 ; SUB1:   Seven-Segment Code Table using PCL + offset in W
                00089 ;         Returns digit display codes, with '-' for numbers A to F
                00090
0004 0782       00091 codtab  ADDWF PCL ; Add offset to Program Counter
0005 347E       00092         RETLW     B'01111110' ; Return with display code for '0'
0006 340C       00093         RETLW     B'00001100' ; Return with display code for '1'
0007 34B6       00094         RETLW     B'10110110' ; Return with display code for '2'
0008 349E       00095         RETLW     B'10011110' ; Return with display code for '3'
0009 34CC       00096         RETLW     B'11001100' ; Return with display code for '4'
000A 34DA       00097         RETLW     B'11011010' ; Return with display code for '5'
000B 34FA       00098         RETLW     B'11111010' ; Return with display code for '6'
000C 340E       00099         RETLW     B'00001110' ; Return with display code for '7'
000D 34FE       00100         RETLW     B'11111110' ; Return with display code for '8'
000E 34DE       00101         RETLW     B'11011110' ; Return with display code for '9'
000F 3480       00102         RETLW     B'10000000' ; Return with display code for '-'
0010 3480       00103         RETLW     B'10000000' ; Return with display code for '-'
0011 3480       00104         RETLW     B'10000000' ; Return with display code for '-'
0012 3480       00105         RETLW     B'10000000' ; Return with display code for '-'
0013 3480       00106         RETLW     B'10000000' ; Return with display code for '-'
0014 3480       00107         RETLW     B'10000000' ; Return with display code for '-'
                00108
                00109 ; ----------------------------------------------------------------
                00110 ; SUB2:   Delay routine
                00111 ;         Receives delay count in W
                00112
0015 008D       00113 delay   MOVWF     Count       ; Load counter from W
0016 0B8D       00114 loop    DECFSZ    Count       ; and decrement
0017 2816       00115         GOTO      loop        ; until zero
0018 0008       00116         RETURN    ;           and return
                00117
                00118 ; - - - - - - - - - - - - - - - - - - - - - - - - - - - - - - - -
                00119 ; SUB3: Output One Beep Cycle to BuzO
                00120
0019 30FF       00121 beep    MOVLW     0FF         ; Load FF into
001A 008C       00122         MOVWF     Period      ; Period counter
                00123
001B 3000       00124         MOVLW     B'00000000' ; Set RB0
```

continued...

```
001C 0066           00125           TRIS      PORTB          ; as output
                    00126
                    00127 ; Do one cycle of rising tone....
                    00128
001D 1406           00129 cycle     BSF       PORTB,buzO     ; Output High
001E 080C           00130           MOVF      Period,W       ; Load W with Period value
001F 2015           00131           CALL      delay          ; and delay for Period
                    00132
0020 1006           00133           BCF       PORTB,buzO     ; Output Low
0021 2015           00134           CALL      delay          ; and delay for same Period
0022 0B8C           00135           DECFSZ    Period         ; Decrement Period
0023 281D           00136           GOTO      cycle          ; and do next cycle until 0
                    00137
                    00138 ; Set RB0 to input again.........
                    00139
0024 3001           00140           MOVLW     B'00000001'    ; Reset RB0
0025 0066           00141           TRIS      PORTB          ; as input
0026 0008           00142           RETURN                   ; from tone cycle
                    00143
                    00144 ; - - - - - - - - - - - - - - - - - - - - - - - - - - - - - - - - - -
                    00145 ; SUB4:   Get pot value (Rv) using rise time due to C and R on RA4
                    00146 ;         Returns with digit value (0-F) in DigVal
                    00147
                    00148 ; Discharge external capacitor on RA4
                    00149
0027 300F           00150 getpot    MOVLW     B'00001111'    ; Set RA4
0028 0065           00151           TRIS      PORTA          ; as output
0029 1205           00152           BCF       PORTA,4        ; and discharge C setting output low
002A 30FF           00153           MOVLW     0FF            ; Delay for about 1ms
002B 2015           00154           CALL      delay          ; to ensure C is discharged
002C 301F           00155           MOVLW     B'00011111'    ; Reset RA4
002D 0065           00156           TRIS      PORTA          ; as input
                    00157
                    00158 ; Increment a counter until RA4 goes high due to charging of C
                    00159
002E 018E           00160           CLRF      PotVal         ; Clear input value counter
002F 0A8E           00161 check     INCF      PotVal         ; increment counter
0030 3003           00162           MOVLW     03             ; Set delay count to 3
0031 2015           00163           CALL      delay          ; and delay between input checks
0032 1E05           00164           BTFSS     PORTA,4        ; Check input bit RA4
0033 282F           00165           GOTO      check          ; and repeat if not yet high
                    00166
                    00167 ; Mask out high bits of count value, and store & display
                    00168 ; 4-bit digit value, 0-F
                    00169
0034 080E           00170           MOVF      PotVal,W       ; Put count value in W
0035 390F           00171           ANDLW     00F            ; and set high 4 bits to 0
0036 008F           00172           MOVWF     DigVal         ; Store 4-bit value
0037 2004           00173           CALL      codtab         ; Get 7-segment code, 0-9
0038 0086           00174           MOVWF     PORTB          ; and display
                    00175
0039 0008           00176           RETURN                   ; with DigVal from setting of pot
                    00177
                    00178 ; MAIN SEQUENCES ****************************************************
                    00179
                    00180 ; SEQ1:   Store 4 digits in non volatile EEPROM
                    00181 ;         Beep after each digit, and twice when 4 done
                    00182
                    00183 ; Complete Button A input operation
                    00184
003A 30FF           00185 stocom    MOVLW     0FF            ; Delay for about 1ms
003B 2015           00186           CALL      delay          ; to avoid Button A switch bounce
003C 1E05           00187           BTFSS     PORTA,butA     ; Wait for Button A
003D 283A           00188           GOTO      stocom         ; to be released
                    00189
                    00190 ; Read 4-bit binary number from DIL switches into EEDATA and display
```

continued...

```
                  00191
003E 0189         00192           CLRF      EEADR           ; Zero EEPROM address register
003F 0805         00193 getdil    MOVF      PORTA,W         ; Read DIL switches
0040 390F         00194           ANDLW     0F              ; and set high 4 bits to 0
0041 0088         00195           MOVWF     EEDATA          ; Put DIL value in EEPROM data
                  00196
0042 2004         00197           CALL      codtab          ; Display DIL input as decimal
0043 0086         00198           MOVWF     PORTB           ;
                  00199
0044 1A05         00200           BTFSC     PORTA,butA      ; Check if Button A pressed
0045 283F         00201           GOTO      getdil          ; If not, keep reading DIL input
                  00202
                  00203 ; Store the current DIL input in EEPROM at current address
                  00204
0046 1683         00205 store     BSF       STATUS,RP0      ; Select Register Bank 1
0047 1508         00206           BSF       EECON1,WREN     ; Enable EEPROM write
0048 3055         00207           MOVLW     055             ; Write initialization sequence
0049 0089         00208           MOVWF     EECON2          ;
004A 30AA         00209           MOVLW     0AA             ;
004B 0089         00210           MOVWF     EECON2          ;
004C 1488         00211           BSF       EECON1,WR       ; Write data into current address
004D 1283         00212           BCF       STATUS,RP0      ; Re-select Register Bank 0
                  00213
004E 1E05         00214 waita     BTFSS     PORTA,butA      ; Wait for Button A to be released
004F 284E         00215           GOTO      waita           ;
0050 2019         00216           CALL      beep            ; Beep to indicate digit write done
                  00217
                  00218 ; Check if 4 digits have been stored yet, if not, get next
0051 0A89         00220           INCF      EEADR           ; Select next EEPROM address
0052 1D09         00221           BTFSS     EEADR,2         ; Is the address now = 4?
0053 283F         00222           GOTO      getdil          ; If not, get next digit
0054 2019         00224           CALL      beep            ; Beep twice when 4 digits stored
0055 2019         00225           CALL      beep            ;
0056 2874         00226           GOTO      done            ; Go to sleep when done
                  00228 ; - - - - - - - - - - - - - - - - - - - - - - - - - - - - - - - - - - - - -
                  00230 ; SEQ2: Check PotVal v EEPROM
                  00231
0057 30FF         00232 checom    MOVLW     0FF             ; Delay for about 1ms
0058 2015         00233           CALL      delay           ; to avoid Button B switch bounce
0059 1C06         00234           BTFSS     PORTB,butB      ; Wait for Button B to be released
005A 2857         00235           GOTO      checom          ;
                  00236
                  00237 ; Read the value set on the input pot
                  00238
005B 0189         00239           CLRF      EEADR           ; Zero EEPROM address
005C 2027         00240 potin     CALL      getpot          ; Get a digit value set on pot (Rv)
005D 1806         00241           BTFSC     PORTB,butB      ; Check in Button pressed again
005E 285C         00242           GOTO      potin           ; If not, keep reading the pot
                  00243
                  00244 ; Get a digit value from EEPROM and compare with the pot input
                  00245
005F 1683         00246           BSF       STATUS,RP0      ; Select Register Bank 1
0060 1408         00247           BSF       EECON1,RD       ; Read selected EEPROM location
0061 1283         00248           BCF       STATUS,RP0      ; Re-select Register Bank 0
0062 0808         00249           MOVF      EEDATA,W        ; Copy EEPROM data to W
                  00250
0063 068F         00251           XORWF     DigVal          ; Compare the input with EEPROM data
0064 1D03         00252           BTFSS     STATUS,Z        ; If it does not match, go to sleep
0065 2874         00253           GOTO      done            ;
                  00254
                  00255 ; If digit match obtained, check if 4 done and do next if not
                  00256
0066 1C06         00257 waitb     BTFSS     PORTB,butB      ; Wait for Button B to be released
0067 2866         00258           GOTO      waitb           ;
0068 2019         00259           CALL      beep            ; Beep to confirm successful match
                  00260
```

continued...

```
0069 0A89      00261        INCF     EEADR       ; Select next EEPROM location
006A 1D09      00262        BTFSS    EEADR,2     ; 4 digits checked yet?
006B 285C      00263        GOTO     potin       ; If not, do the next
006C 2872      00264        GOTO     siren       ; When 4 digits done, run siren
               00265
               00266 ; *************************************************************
               00267
               00268 ; MAIN: Select Set or Check Combination
               00269
006D 1E05      00270 select BTFSS    PORTA,butA  ; Button A pressed?
006E 283A      00271        GOTO     stocom      ; If so, store a combination
006F 1C06      00272        BTFSS    PORTB,butB  ; Button B pressed?
0070 2857      00273        GOTO     checom      ; If so, check a combination
0071 286D      00274        GOTO     select      ; repeat endlessly
               00275
               00276 ; *************************************************************
               00277
               00278 ; END1: When combination successfully matched, make siren sound
               00279
0072 2019      00280 siren  CALL     beep        ; Do a tone cycle
0073 2872      00281        GOTO     siren       ; and repeat endlessly
               00282
               00283 ; - - - - - - - - - - - - - - - - - - - - - - - - - - - - - - -
               00284
               00285 ; END2: When a digit check fails, go to sleep, and try again
               00286
0074 0186      00287 done   CLRF     PORTB       ; Switch off display
0075 0063      00288        SLEEP                ; Processor halts
               00289
               00290 ; *************************************************************
               00291        END                  ; of program source code
```

```
SYMBOL TABLE
LABEL             VALUE

Count             0000000D
DigVal            0000000F
EEADR             00000009
EECON1            00000008
EECON2            00000009
EEDATA            00000008
PCL               00000002
PORTA             00000005
PORTB             00000006
Period            0000000C
PotVal            0000000E
RD                00000000
RP0               00000005
STATUS            00000003
WR                00000001
WREN              00000002
Z                 00000002
__16C84           00000001
beep              00000019
butA              00000004
butB              00000000
buzO              00000000
check             0000002F
checom            00000057
codtab            00000004
cycle             0000001D
delay             00000015
done              00000074
getdil            0000003F
getpot            00000027
loop              00000016
potin             0000005C
select            0000006D
siren             00000072
start             00000000
stocom            0000003A
store             00000046
waita             0000004E
waitb             00000066

MEMORY USAGE MAP ('X' = Used, '-' = Unused)
0000 : XXXXXXXXXXXXXXXX XXXXXXXXXXXXXXXX XXXXXXXXXXXXXXXX XXXXXXXXXXXXXXXX
0040 : XXXXXXXXXXXXXXXX XXXXXXXXXXXXXXXX XXXXXXXXXXXXXXXX XXXXXX----------

All other memory blocks unused.
```

Index